The Hand, an Organ of the Mind

THE HAND, AN ORGAN OF THE MIND

What the Manual Tells the Mental

edited by Zdravko Radman

The MIT Press
Cambridge, Massachusetts
London, England

MIT Press books may be purchased at special quantity discounts for business or sales promotional use. For information, please email special_sales@mitpress.mit.edu or write to Special Sales Department, The MIT Press, 55 Hayward Street, Cambridge, MA 02142.

This book was set in Stone Sans and Stone Serif by Toppan Best-set Premedia Limited, Hong Kong. Printed and bound in the United States of America.

Library of Congress Cataloging-in-Publication Data

The hand, an organ of the mind : what the manual tells the mental / edited by Zdravko Radman.
    p. cm.
Includes bibliographical references and index.
ISBN 978-0-262-01884-5 (alk. paper)
1. Human engineering. 2. Gesture. 3. Handicraft. I. Radman, Zdravko, 1951–.
T59.7.H337   2013
121'.35—dc23
2012034870

10 9 8 7 6 5 4 3 2 1

This volume is dedicated to the memory of Marc Jeannerod, a friend and outstanding researcher who had a firm grasp on the nature of mind. His insights continue to inspire and invite further investigation.

Contents

Foreword: Hand Manifesto

Jesse J. Prinz

We live an age of ocular domination. Ours is said to be a visual culture, and researchers who study perception focus overwhelmingly on vision. But this ocularcentric view casts a shadow on an equally important organ: the hand. This handbook (or hand book) serves as an overdue corrective, and Zdravko Radman's efforts will help us grasp the magnitude of this neglect. He has assembled a compendium of cutting-edge research on our most malleable appendage, and no reader who thumbs through these pages will emerge unchanged. Radman has entrusted me to introduce his groundbreaking collection with a manifesto.

Let me begin with a simple observation. In societies that wear clothing to cover the body, the hand is second only to the face in being characteristically unclothed. When only limited flesh is visible, we see features and fingers. This is not because hands are more respectable than all other body parts or less vulnerable to the elements. It is because they are extraordinarily useful in human life. In this manifesto, I will point out six ways in which our hands serve us. In the chapters that follow, this list will be illustrated and extended.

Feeling: The Hand and Perception

First, we must never forget that hands are perceptual organs. They play a crucial role in gathering information about the world. Indeed, the hand is several sense organs in one. It is layered with tactile transducers that tell us about touch, pressure, heat, and cold. These are especially concentrated in our fingertips, each of which has 3,000 receptors, comparable to the number of receptors in the trunk of the body. Fingers can tell us whether a surface is smooth or rough, wet or dry, hot or cold, soft or hard, and so on. They also contain nociceptors, which are implicated in pain responses. Thus, the hand is not a neutral observer, but an instrument of value.

These aspects do not exhaust the sensory information conveyed by the hand. We also receive proprioceptive information—the position of our fingers and wrists—and kinesthetic information—the tension in our muscles. These contribute crucially to our capacity to derive information about objects. For example, when the hand can extend without bending on a surface, we know through a combination of touch and proprioception that the surface is flat. When the fingers bend against a surface, we infer curvature. When muscles tense while hefting an object, we can discern its weight, and even its approximate shape from kinesthesia: imagine sitting in the dark and trying to tell whether you are holding the handle of a ladle or a carving fork: by touch they are identical, but kinesthetically they differ.

The various senses of the hand can work together to tell us precise information about object identity. Within a single moment of contact, we can often discern enough information about a contour to know what we are feeling; imagine putting your hand on a face or a tennis ball in the dark. A single sampling can also obtain information about texture that reveals object identity: a towel, water, sand, or stone. And information about pain and temperature can be revelatory as well: a flame, a thorn, or an ice cube. Acuity increases when we combine the various haptic senses with multiple samplings. As fingers explore a finite surface, we can piece together complex forms. In the Japanese horror film *Blind Beast* (dir. Yasuzo Masumura), a blind artist kidnaps a woman and sculpts a realistic portrait of her body by touch alone. The world takes form as we grope, stroke, and wander with our hands.

Our haptic senses also interact with vision and other modalities. We see an object and spontaneously imagine what it would be like to touch it. A knife may look sharp, an icicle may look cold, and a hammer may look graspable. Plausibly, these cases involve the formation of tactile imagery. Vision is especially prone to generate haptic imagery, since both senses are, more than any others, capable of deriving information about shape. The associations can move in the opposite direction as well, from touch to visual imagery. Upon feeling the body of a loved one at night, one might visualize it. Upon reading Braille with one's fingertips, one might imagine what the dots look like visually arrayed. This may be why the visual cortex is active during Braille reading. These associations may even be innate, given that this visual cortical response is also demonstrated in the congenitally blind.

Associations with the haptic senses are not limited to vision. On hearing the sound of a crinkly plastic wrapper, we might imagine how it feels. The sound of sandpaper, the smell of sage, and the taste of lemon rind may all trigger tactile associations.

In these cases, we can see how touch can serve to fill in information that is missing from another sense. When we hear sandpaper, we fill in the feeling, and that knowledge of texture, which is appended to the sound, may play a role in our inferring the effects of the sanding that we are hearing. Or, to take an example from phenomenology and enactive perception, upon seeing one surface of a ball, we may use tactile imagery to image its sphericality. Thus, touch can turn an emitted sound or partial surface into a whole object.

In this capacity, the haptic senses may also play a role in giving the world a sense of objectivity. Sight and sound are spatial senses in that they can be used to locate objects along spatial dimensions, but it is not clear phenomenology that these senses alone give us the impression of objects existing in a space that is external to the mind. Haptic senses, because they often exploit bodily movement, may present space in a way that is more decidedly external. When we see an object and imagine reaching for it or exploring it with our hands, we convert the visual information into an action that is quite literally extended in the world. This may contribute to our impression that things are out there: They are there to be touched.

Manipulating: The Hand and Action

In exploring the sensory aspects of the hand, we have already seen that hands are not passive. The word "feeling" itself can refer to a sensation or an activity, and these are usually linked. We feel something in order to have a feeling of it. In cases of sensory association, the active nature of the hand is equally apparent. When we see something as graspable, we imagine what it will feel like when we take hold of it. This is what J. J. Gibson called an *affordance*: an immediate registration of an object's potential for interaction. This takes us beyond the merely sensory aspects of the hand to a second interesting feature: Hands are things with which we act.

Manual actions are extraordinarily important for our species. We radically alter our environments by building and using tools, making shelters, crafting clothing, hunting, gathering, cooking, and consuming. All these things we do with our hands. The evolution of the prehensile thumb and the bipedal gate has made us especially skilled at manipulating the world. Arguably, these two advances are the physical underpinnings of human uniqueness. We are, more than any other creature, a handed species. Being handed even distinguishes us from the great apes, whose hands double as feet, limiting their role considerably. Apes use tools, but only while sitting or being relatively stationary. We can track an animal while holding a spear poised for throwing or plant seeds while walking across a field. With easy access to our hands, we also achieve unprecedented manual dexterity: We can tie knots, play instruments, and paint pictures. Hands are a *sine qua non* for the manufacture of artifacts that are distinctively human. Hands also allow us to do many things that other creatures do with their mouths, such as fighting, foraging, tearing, and grasping food. This new division of labor may serve to make mouths more available for communicating.

These considerations point to an important evolutionary role for the hand, but it is important to see that manual manipulation is also central to life in the present. Almost every form of human labor, whether it involves working the fields or writing poems, uses the hand as a primary instrument. Those who lose their hands know this only too well. For those of us fortunate enough to have hands, it is almost impossible to imagine going through a day without using them. The word "handicap" derives from an old game in which betters placed their hands in cap, putting themselves as a serious disadvantage.

Grasping: The Hand and Thought

Perceiving and acting are the most obvious functions of the hand, but not the only ones. Hands and haptic imagery are also important for thinking and problem solving. The most obvious case of this is calculation. We can count on our fingers. We can add and subtract. And, with the advent of writing, we can carry out symbolic calculations such as long division. Hands also allow us to measure spatial extent and to pinpoint targets (think of the use of hand positions when playing billiards). These are examples of extended cognition: We can perform operations with our hands that are akin to those we perform in our heads.

Moving into our heads, we find that the hands continue to be important for thought. The hands are not inside our heads, or course, but haptic imagery, including imagined touch and hand movements, is important for some kinds of thinking. Arthur Glenberg and Michael Kashak describe what they call the *action-language compatibility effect*. They asked participants in an experiment to press a button when they each item on a list of sentences. The buttons were arranged either close to the body or far away from it, and Glenberg and Kashak found that comprehension was stalled when sentences described actions that were in tension with the button location. For example, participants were slower to comprehend "Andy delivered the pizza to you" when they had to press a button that was located farther away from their bodies. The effect even worked for more abstract sentences, such as "Liz told you the story." This suggests that people understand sentences by forming simulations of them, and these simulations often involve imagined hand gestures. It is especially striking that this effect works for the abstract cases, suggesting that we grasp (literally) abstract ideas by grounding them in bodily movements.

Mark Chen and John Bargh obtained similar results when they presented emotionally charged words. Participants were asked to either push a lever away or pull it toward them when reading words that were either negative or positive. When the direction of motion failed to match the valence of the word (e.g., a negative word paired with pulling the lever toward the body), response times were slower. This suggests that hearing negative things disposes us to push and hearing positive things disposes us to pull. Comprehension of valence may be facilitated by our ability to feel or imagine these behavioral dispositions. Once again, the hands are instruments of evaluation, not just perception.

Presumably, imagined hand movements are not necessary for comprehension of valence, but they arise spontaneously and contribute to understanding. This fits into a broader Jamesian framework, on which we use bodily perception to determine how we feel emotionally. Perceiving our fists clench may help us realize that we are angry. We may even make metaphorical use of tactile information in the emotional domain. In a study with Lawrence Williams, John Bargh found that grasping a warm or cold cup of coffee can increase the likelihood that people will attribute warm or cold personality traits to others. This suggests that we may grasp some personality traits by mapping them onto bodily sensations, and we attribute such traits by imagining how people make us feel. In this way, hands can become windows to the soul.

Touching: The Hand and Sociality

The last example of how we use hands in thinking suggests that hands play an important role in social cognition. A more widely discussed example involves so called mirror neurons. These are cells in the primate brain that fire both when planning manual actions and when observing them. Theorists such as Alvin Goldman and Vittorio Gallese have surmised that we understand another person's intentions by use of such mirror neurons. If we see someone grasping something, we spontaneously imagine what it is like to plan a grasping action, and that helps us simulate their psychological state as they perform the action. Even if one doubts that simulation is the primary means by which we understand other minds, this particular kind of example is highly plausible. It is difficult to observe someone doing something with his or her hands without feeling an impulse to do it ourselves, and this impulse can clue us in to what someone is trying to achieve even if he or she has not yet managed to succeed. Imagine what it is like to see someone unsuccessfully thread a needle; the compulsion to help is almost irresistible.

The role of hands in human social life extends far beyond mental-state attribution. We use hands to greet, to console, to caress, and to fight. A hand on the shoulder can bring comfort, a handshake can promote trust, and a raised fist can be seen as an act of aggression. Hands bind us together and set us against each other. Their ability to do so is not merely instrumental. Touching another person and being touched have strong psychological effects. The manner of touch can cause anything from arousal to alarm. There are even feelings that are produced almost exclusively by manual contact. Tickling is the prime example. Hands can easily tickle, but things that are not handlike or controlled by hands cannot. This may stem from the fact that the tickling response is a kind of submission signal that we issue during rough-and-tumble play. If a friend or loved one touches your vulnerable parts, you will laugh. Had a stranger touched those parts, you would scream. The tickle laugh says: You have beaten me physically, but it's okay because we are only playing.

Gesturing: The Hand and Communication

Consoling and tickling are social acts, but they are probably not symbolically mediated. Touch, in theses cases, directly causes feelings. The social role of hands increases significantly when we consider symbolic or communicative gesturing. The voice is our primary instrument for symbolic communication, but hands come in second, and hand gestures may be an ancient form of signaling. Of course, some sign languages do use gesture as a primary medium, but those of us who don't sign spontaneously typically use gestures to supplement or replace speech.

Studies of gesture suggest that we use our hands to mark off items when making a verbal list; we also use gestures when expressing spatial information, as when we describe something as large or small. There are even metaphorical uses of spatial gestures. Nick Enfield has shown that people use gestures to trace tree structures in the air when listing the members of their family. Some of these gestures may serve as cognitive crutches for the speaker, to help with mental bookkeeping, but they may also help listeners in the same way.

Moving beyond these spontaneous gestures, there are also many more conventional uses of the hands in communication that are learned and symbolic but still fall short of a full-blown sign language. One striking thing about conventional hand gestures is the diversity of the communicative roles they can play. Hands play a range of roles akin to what we find in speech.

Gestures can serve a role comparable to that of the assertoric sentence, as when we use an okay sign or a "thumb's up" to indicate content, or when we raise a hand to answer a question. We may also answer numerical questions by holding up an appropriate number of fingers or answer size questions by spreading our digits or palms (e.g., "I love you this much" said with arms spread). These gestural assertions are usually answers to questions, but we can also make assertions without being asked anything, as when we engage in pantomime.

Gestures can also serve as demonstratives, when we point; as markers of emphasis, when we pound the table; and as salutations, when we wave. More notoriously, we use gestures as expletives, such as when we "flip someone off." These don't always translate cross-culturally; a North American okay sign means something much less polite in Brazil. Gestures can also be expressive of more positive feelings, as when we applaud, or a wider range of emotions, as in Indian dance. In the latter case, gestures might be said to play a poetic role, going beyond mere communication and adding an aesthetic dimension. A gesture can even function metaphorically, as when we snap to indicate a short period of time. Snapping can also be used to get attention (compare the interjection "hey!"). Other gestures serve as commands; a fetchingly curled finger may say "come hither," whereas a pointing sweep may say "be gone!" Gestures can be interrogative, too: a supplicating hand can be used to beg, and two hands can be pressed together in an act of prayer.

Given this range, it is no surprise that some people think language originated in gesture. Whether or not this is true, it is hard to deny that gesture is a major means of communication. And, it is also fingers that type and write for us, making the hand the source of most of the written word. Without hands, writing systems and literacy may have never emerged.

Typing: The Hand and Identity

Hands are in fact used for typing in two senses of the word: We use fingers to write words with keyboards, and we also use them as input for social classification. We can often tell what type of person we are viewing by looking at his or her hands. In this way, hands communicate identity.

Consider gender. One's hands can indicate whether a person is masculine or feminine by use of culturally specific gender markers such as long nails, nail polish, or gendered jewelry. In India, henna paintings (*mehndi*) are made on the hands of a bride who is about to be married. In many cultures, a ring indicates marital status. Rings may also indicate personal interests, taste, and subculture. A skull ring may say "rocker," a class ring may say "college graduate," and a cross on one's ring may say "Christian." Ostentatiously jeweled rings can also convey financial wealth.

Socioeconomic status can also be indicated by the physical condition of one's hands. Hands that labor hard are worn, dry, cracked, scuffed, and blistered. Hands of those wealthy enough to avoid manual labor may appear pale, clean, soft, and manicured. Hands can also divulge one's age, ethnicity, and the climate to which one has been exposed.

Hands even reveal individual identity, as opposed to group membership. We can recognize the hands of loved ones, and distinctive scars, tattoos, or marks can make one's hands unique. We each come to know our own hands, and we often use them to signal who we are or want to be.

Moving beyond casual perception, all hands are unique because fingerprints differ from person to person. This discovery, put to forensic use by Francis Galton, has shown that each person's hand leaves a distinctive trail that can be detected and traced. Taking fingerprints is common practice in the world's prisons, and it is increasingly used to track people who cross borders, or as a means of high-security identification. This aspect of individuation makes hands an instance of the cultural preoccupation with measurement and surveillance, which Michel Foucault links to the birth of the human sciences. In this sense, hands are part of what make us human beings, where being human means being an object of certain kinds of inquiry.

Salute the Hand

After this brief but probing tour, it should not seem hyperbolic to say that hands have a privileged place in human life. The range of functions that hands serve arguably exceeds any other part of the body. Hands sense, like eyes, but they also speak, sculpt, spar, and shape our world. They are both input systems and output systems. They allow us to survive as individuals, and they link us together socially. Hands are integral to who we are as a species, as members of groups, and as individuals. If any anatomical unit deserves a reverent salute, it is the hand. It is high time, then, that cognitive science and philosophy lift up this neglected appendage and attempt to learn more about its role in making us who we are.

Beforehand

Zdravko Radman

Philosophers love gaps. Gaps are invented and imposed wherever our imagination finds even the slightest opportunity to do so. And because they seem unbridgeable, the only way to deal with them is to respect splits and conform to oppositions. In order to explain, we discriminate; in order to discriminate, we delineate; delineations lead to oppositions; oppositions are understood as polarities; polarizations imply exclusiveness; exclusions lead to irreducibility. For instance, in philosophy of mind and the theory of consciousness, such bipolarities are found in the following contrasts: organic versus inorganic, mind versus matter, "I" (or self) versus "it" (brain), cognitive versus motor, internal versus external, subjective versus objective, perceptual versus conceptual, introspection versus detached observation, first-person versus third-person perspective, the view from "within" versus the view from "without," quale versus quantum, signification versus sensation, internalism versus externalism, pragmatism versus intellectualism, semantics versus syntax, and so on. And the dualistic schema can be extended with further elements of counterpointing that may include oppositions between fact and fiction, reason and intuition, logic and imagination, rationality and emotion, science and art, and so on.

No doubt, divisions of this sort may be useful, but they can also deceive. Indeed, if the gaps devised by mind designers were real, we, as living beings, would be profoundly dysfunctional. The exemplary case is the classical mind–body problem as established in the supposed exclusion of mind and matter, also of the mental and the motor, and of cognition and corporeality. It could be said that Descartes's curse still lives in the heads of theorists, yet not necessarily in the heads of those they are investigating, because from the very possibility of making some theoretical distinction, it does not follow that the proposed duality exists in the living cognitive organisms to which that duality is attributed. Fortunately, the divided man that haunts theoretical space is not to be found among living agents. Cartesian ideas inspire, but they also inhibit; they provide the optics for one kind of reading and at the same time blur our ability to see alternatives.

Evidently, we are more talented in creating fissures than in building bridges across them. Exclusion is somehow preferred over inclusion. The idea of counterpointing seems to be explanatorily more powerful than that of interconnectivity, that of particularization more telling than that of holism. It seems that we can clearly see what something *is* only when we (also) state what it is *not*—that is, when explanation is expressed in contrastive terms.

The dualistic methodology just described can work when applied to some domains, but it definitely fails to play a productive role in the major concern of this volume. Whichever aspect of the hand-phenomenon you choose, you are likely to meet serious difficulties in applying "gaps" to this organ or faculty, which refuses to conform to what has long been established as a comfortable habit of thought. Indeed, the application of the same dualistic methodology to our understanding of the mind and how it works may be equally unhelpful.

Not unrelated to this inclination to bipolar contrasting is our habit of *locating* the mental; a frequently posed question asks whether mind is *in* the head or *out* of it. If we were to respond to this misleading philosophical question from the manual perspective, probably the most adequate reply would be: it depends (which would then require extensive elaboration on the nature of mind and its relation to the world). Yet, rather than accepting the call for localization, we might be better off directing our interest toward figuring out a possible response to the question: How does the hand shape the head? Clearly, the hand does not act in a headless manner, and neither is the authority of the head built exclusively on disembodied contemplation. (Consider, for instance, the loss of a hand; the organism misses it so badly that the head literally invents the missing limb; yet the "phantom" is so real that it hurts—indeed in a place where there are no flesh and bones.)

However, gaps are not given but created in the heads of theorists, and it is plausible to suppose that they can be revised by the same means. It is not an easy task, but it's worth a try. This collection, with its focus on the hand, is, in a way, a challenge to the tradition. It questions, for instance, the habit of thought that has divorced the motor (destined to the bodily "periphery") from the cognitive (for which "higher" centers in our neural architecture are standardly held to be responsible) and that, in general, has separated movement from mind. It is a collective attempt to present another perspective on the role of the manual—one that largely ignores the boundaries between body, mind, and world, the triad that represents a proper context within which our investigations take place. The theoretical hands in which this volume is written vary, but all share a common general motivation to understand how the manual, mental, and mundane are interconnected and interdependent.

Transcribed in more commonsense terms, one can say: As handed beings we are not alone, and the world is not alien to us. Hands enable us to exceed our privacy, to join in social interactions, and to participate in cultural events. As "enhanded" agents we are not mere observers or detached thinkers. Because our participatory engagement in the world is based mainly on the art of manipulation, we leave traces of our activity, and creativity (both productive and destructive), that outgrow our physical existence. It is thanks to them that we have traceable memory and that the record of our deeds enables us to have our own histories. Only that which bears the fingerprints of our doings can be inherited; almost everything else is doomed to be forgotten.

I believe that further advancement of the kind of research outlined here will lead to the reconsideration and reconceptualization of some of the basic philosophical terms related to mind. For instance, it might help us realize that cognition is not only what cognitivists tell us it is; that thinking entails more than intellectualists grant; and that mind is so vitally anchored in action that it can itself be viewed as a *way of doing* (and thus is nothing like an "information-processing" system or a "language of thought").

In other words, what drives our minds is not only reflection, and what motivates our actions is not fully preconceived in thought; but neither is it exclusively the result of sensory dictate. The body is knowledgeable in its own terms and the hand possesses its authentic know-how. A "trivial" movement may thus give birth to mental happenings; a manual grasp may make you "see" a thing in a particular way, that is, understand it in pragmatic terms; bodily hints may stir ("internal") mental stories; fine feelings may be generated within the fingertips (*Fingerspitzengefühle*); touch may induce thoughts; reaching may engender modes of more complex relations to other bodies and minds; gesture may self-stimulate for mental leaps; manipulations may initiate the refurnishing of mental rooms. At every instance, manual intervening shapes and reshapes the mental and creates conditions for agents to act.

The idea implicit in this project is that we need a big, and above all more integrative and diversified, picture of the embodied mind (where "big" does not mean merely a quantitative increase, but rather an opening of fresh vistas within the broader context of human acting-in-the-world). I hope that the empirical evidence and theoretical arguments assembled in this volume suffice for the stimulation of novel insights that may encourage views beyond the orthodox. For instance, they may facilitate belief that the *mind is many ways*, by no means only that to which current mainstream theorizing has reduced it; that *thoughts are manifested also in movements*, for example, in the scripture of intercorporeal relations, participatory movements, and signifying gestures; that most of *acting is performed in disobedience to deliberation*; and, finally, that much of our coping in the world is in a significant (and for humans specific) way a manifestation of the "knowledge in the hands."

If the reader recognizes as relevant at least a portion of what we attempt to display in this collection, then the mission undertaken in this project has been well worth the effort.

Acknowledgments

An ancestor of this volume is the conference that took place at Interuniversity Center Dubrovnik (Croatia), May 1–3, 2009. I want to thank everyone who took part in that productive exchange: those who have also contributed to this collection, but also Frank Wilson, Eva Man Kit Wah, Evelyn Tribble, and Mădălina Diaconu, who, with their experience and innovative approaches, enriched our meeting yet were unable to contribute chapters here. However, as the publication project matured, new contributors joined to complement the thematic scope of the collection and to bring in new aspects of this still rather neglected topic.

I gratefully shake hands with: Andy Clark, for providing encouragement at the very early stages of this project; Jesse Prinz, for being sympathetic and supportive; Shaun Gallagher and Daniel Hutto, for their constructive assistance whenever I needed it; Michael Wheeler, who always finds the proper words; Nicholas Holmes and Massimiliano Cappuccio, for their helpful suggestions; Jonathan Cole, for his inspiring wit; Philip Laughlin, MIT Press Senior Acquisitions Editor, for his easygoing yet efficacious accompaniment through various stages of the publication-process; Judith M. Feldmann for her smart and detailed copyediting; the three anonymous reviewers, for a number of valuable observations and helpful ideas for improvements; and the members of editorial board of the MIT Press, who have the privilege to choose among the plentiful numbers of outstanding manuscripts and whose choice favored this collection.

Contributors

Matteo Baccarini obtained his Ph.D. in philosophy at the University of Ferrara. He has collaborated with the Institute of Human Physiology of the University of Parma, and with the Department of Philosophy of the University of Milan. He completed a postdoctoral fellowship at the INSERM unit Multisensory Space and Action, in Lyon. He is currently a postdoctoral fellow at the Neuroscience Center, University of Ferrara and National Institute of Neuroscience, Italy. His research focuses mainly on body representation, tool use, and spatial representation.

Andrew J. Bremner completed his D.Phil. at the University of Oxford in 2003. Following two postdoctoral appointments in London and Brussels, he took up an academic post in the Department of Psychology, Goldsmiths, University of London, where he is now a Senior Lecturer. His research interests include multisensory development, object recognition in infancy, the development of cognitive control, and more recently the development of representations of the body and limbs in early life. He is an Associate Editor of the *British Journal of Developmental Psychology*, and a member of the editorial board of *PLoS ONE*. He is currently conducting a research project, Human Embodied Multisensory Development, awarded by the European Research Council.

Massimiliano L. Cappuccio is Assistant Professor (philosophy of mind and cognitive science) at UAE University, Emirate of Abu Dhabi, and a member of the Neurophilosophy Lab of the State University of Milan. He is currently visiting the University of Memphis (Tenn.) under a Fulbright fellowship. His research combines enactive and extended approaches to social cognition, and focuses on empathy, motor intentionality, joint attention, gestures, and the origins of symbolic culture.

Andy Clark is Professor of Logic and Metaphysics in the School of Philosophy, Psychology and Language Sciences, at Edinburgh University in Scotland. His research interests include robotics and artificial life, the cognitive role of human-built structures; specialization and interactive dynamics in neural systems; and the interplay between language, thought, sociotechnological scaffolding, and action. He is the author of *Being There: Putting Brain, Body and World Together Again* (MIT

xxvi   Contributors

Press, 1997), *Mindware* (Oxford University Press, 2001), *Natural-Born Cyborgs: Minds, Technologies, and the Future of Human Intelligence* (Oxford University Press, 2003), and *Supersizing the Mind: Embodiment, Action, and Cognitive Extension* (Oxford University Press, 2008).

Jonathan Cole   is a consultant in clinical neurophysiology at Poole Hospital and a visiting professor at the University of Bournemouth. He trained in Oxford and London and has research interests in sensory loss, motor control, and pain. In addition to his empirical research he is also interested in portraying the subjective experience of neurological impairment. This has led to books on deafferentation, facial visible difference, living without movement or sensation (spinal cord injury), and living without facial animation. He is the author of *Pride and a Daily Marathon* (Duckworth 1991, MIT Press, 1995), *About Face* (MIT Press, 1999), *Still Lives: Narratives of Spinal Cord Injury* (MIT Press, 2004), and (with Henrietta Spalding) *The Invisible Smile* (Oxford University Press, 2008).

Dorothy Cowie   completed a D.Phil. at the University of Oxford in 2007. After that she moved to the UCL Institute of Neurology as a postdoctoral fellow. Since 2010, she has been a Lecturer in the Department of Psychology, Goldsmiths, University of London. Her research focuses on the visual control of action through the three parallel streams of typical behavior, development, and neural bases. She is particularly interested in locomotion and the development of visuomotor skills in young children.

Natalie Depraz   is Professor of Philosophy at the University of Rouen (ERIAC), Archives-Husserl (ENS-CNRS, Paris), and CREA (Paris). Her research interests include phenomenology, German philosophy, cognitive sciences, and spiritual traditions (Christianity, Buddhism). Current research themes include attention, surprise and the emotions, the first person, and phenomenology. Her publications include: *Lucidité du corps: De l'empirisme transcendantal en phénoménologie* (Kluwer, 2001); *Comprendre la phénoménologie: Une pratique concrète* (A. Colin, 2006); (together with F. J. Varela and P. Vermersch) *On Becoming Aware: A Pragmatics of Experiencing* (Benjamins Press, 2003) (in French: *À l'épreuve de l'expérience: Pour une pratique de la phénoménologie*, Zêta Books, 2011); and "Avatar: je 'te' vois: Une phénoménologie pratique" (*Ellipses*, autumn 2012).

Rosalyn Driscoll   investigates the sense of touch and the experience of the body through sculptures and installations. Her artwork has been exhibited in the United States, Europe, and Japan, and has received awards from the New England Foundation for the Arts and Massachusetts Cultural Council. She has had residencies at the Helene Wurlitzer Foundation of New Mexico and Dartington Hall Trust in the United Kingdom. She has presented at conferences worldwide on topics ranging from perception, haptics, enaction, and art history of touch, to neuroscience and cognitive science.

Harry Farmer   is currently studying for a Ph.D. in Psychology at Royal Holloway, University of London. His research focuses on how interactions with other people affect the experience of one's own body in sensation (e.g., giving rise to a sense of body-ownership) and action (e.g., giving rise to a sense of agency), using behavioral and neuroscientific techniques, in both normal and clinical populations. He is part of an interdisciplinary team funded by the European platform VW working on intersubjectivity and the body. He holds an M.Sc. in Psychological Research from the University of Oxford and a B.Sc. in Psychology and Philosophy from the University of Bristol.

Shaun Gallagher   holds the Moss Chair of Excellence in Philosophy at the University of Memphis. He also has secondary appointments as Research Professor of Philosophy and Cognitive Sciences at the University of Hertfordshire, Honorary Professor of Philosophy at the University of Copenhagen, and affiliated research faculty at the Institute for Simulation and Training, University of Central Florida. He has held visiting positions at the Cognition and Brain Sciences Unit, Cambridge, the Ecole Normale Supériure, Lyon and most recently at the Centre de Recherche en Epistémelogie Appliquée (CREA), Paris. His recent books include *How the Body Shapes the Mind* (Oxford University Press, 2005), *Brainstorming* (Imprint Academic, 2008), and (with Dan Zahavi) *The Phenomenological Mind* (Routledge, 2008). He is editor of *The Oxford Handbook of the Self* (2011) and editor-in-chief of the journal *Phenomenology and the Cognitive Sciences*.

Nicholas P. Holmes   studied psychology and neuroscience in Manchester and Oxford, then spent several years as a traveling postdoc in Lyon and Jerusalem before settling into a lectureship in Reading (Hand Laboratory, School of Psychology and Clinical Language Sciences). His research interests include perception and action in general, and more specifically hand perception and hand action. His current research projects focus on the motoric effects of the mere observation of hand actions, the rapid online control of reaching and grasping actions, and the integration of somatosensory stimuli applied to the two hands. To study these topics, he uses a broad range of behavioral, electrophysiological, and neuroimaging techniques, his favorites being psychophysics and transcranial magnetic stimulation.

Daniel D. Hutto   is Professor of Philosophical Psychology at the University of Hertfordshire. He is author of *The Presence of Mind* (John Benjamins, 1999), *Beyond Physicalism* (John Benjamins, 2000), *Wittgenstein and the End of Philosophy* (Palgrave Macmillan, 2006), and *Folk Psychological Narratives* (MIT Press, 2008). He is coeditor of *Folk-Psychology Re-Assessed* (Springer, 2007) and editor of the Royal Institute of Philosophy volume, *Narrative and Understanding Persons* (Cambridge University Press, 2007) and *Narrative and Folk Psychology* (Imprint Academic, 2009). A special yearbook issue of *Consciousness and Emotion*, entitled *Radical Enactivism*, which focuses on his philosophy of intentionality, phenomenology, and narrative, was published in 2006.

Angelo Maravita graduated in medicine from the University of Milan and completed his training in neurology at the University of Parma. He then obtained a Ph.D. in neuroscience at the University of Verona and held a Senior Research Fellow position at the Institute of Cognitive Neuroscience in London. He is now Associate Professor at the Department of Psychology, University of Milano-Bicocca. His research interests focus on body representation and multisensory integration in healthy and brain-damaged people. He has published articles on multisensory integration and the body schema, spatial disorders, and how tool use changes multimodal spatial interactions between vision and close-to-hand and within-reach experience.

Filip Mattens studied architecture and philosophy. In 2009, he obtained his doctorate at the Institute of Philosophy at the University of Leuven with a dissertation on sensory experience and perceptual space. Besides architectural theory, his research interests include philosophical issues related to consciousness, pictures, and the senses. He was a visiting fellow at the Humboldt-Universität zu Berlin and Harvard University. Currently, he is a Postdoctoral Fellow of the Research Foundation—Flanders (FWO).

Richard Menary read for a B.A. in philosophy at the University of Ulster, an M.Sc. in Cognitive Science at the University of Birmingham and then a Ph.D. in philosophy at King's College London. He has been a Senior Lecturer at the University of Hertfordshire and a Senior Lecturer and Head of the Department of Philosophy at the University of Wollongong. He now holds a joint position in the Philosophy Department and the Centre for Cognition and Its Disorders at Macquarie University. His research interests include embodied and extended cognition, neural plasticity and the enculturation of the brain, narrative and the self, expertise, and pragmatism. Major publications include *The Extended Mind* (MIT Press, 2010); *Cognitive Integration* (Palgrave Macmillan, 2007); and papers published in journals such as *Philosophical Psychology, Phenomenology and the Cognitive Sciences*, the *Journal of Consciousness Studies*, and *Mind and Language*.

Jesse J. Prinz is a Distinguished Professor of Philosophy at the City University of New York. He works primarily in the philosophy of psychology and has produced books and articles on emotion, moral psychology, aesthetics, and consciousness. He is the author of *The Conscious Brain* (Oxford University Press, 2012), *Beyond Human Nature: How Culture and Experience Shape Our Lives* (Allen Lane, 2012), *The Emotional Construction of Morals* (Oxford University Press, 2009), *Gut Reactions: A Perceptual Theory of Emotion* (Oxford University Press, 2006), and *Furnishing the Mind: Concepts and Their Perceptual Basis* (MIT Press, 2004).

Zdravko Radman is a Senior Researcher at the Institute of Philosophy, Zagreb, and teaches philosophy at the University of Split, Croatia. As an Alexander von Humboldt and a William J. Fulbright Fellow he was affiliated with the University of Konstanz and the University of California, Berkeley; as a visiting scholar he conducted research at the Australian National University, the University of Tokyo, and University College London, among others. He has published in the philosophy of mind, aesthetics, and the philosophy of language. He is the author of *Metaphors: Figures of the Mind* (Kluwer, 1997) and editor of *Knowing without Thinking: Mind, Action, Cognition, and the Phenomenon of the Background* (Palgrave Macmillan, 2012), *Horizons of Humanity* (Peter Lang, 1997), and *From a Metaphorical Point of View* (Walter de Gruyter, 1995).

Matthew Ratcliffe   is Professor of Philosophy at Durham University, UK. Most of his recent work addresses issues in phenomenology, philosophy of psychology, and philosophy of psychiatry. He is the author of *Rethinking Commonsense Psychology: A Critique of Folk Psychology, Theory of Mind, and Simulation* (Palgrave, 2007) and *Feelings of Being: Phenomenology, Psychiatry, and the Sense of Reality* (Oxford University Press, 2008).

Etiennne B. Roesch   is a Lecturer at the University of Reading, Centre for Integrative Neuroscience and Neurodynamics. He completed undergraduate studies in both software engineer and cognitive science and graduate studies in cognitive science, and received a Ph.D. in psychology from the University of Geneva. He was the recipient of a Fellowship for Prospective Researchers, awarded by the Swiss National Science Foundation, with which he joined the Cognitive Robotics group at Imperial College London. His research is situated at the core of cognitive science, as he combines computational neuroscience modeling to neuroimaging and robotics with a view to investigate general brain functioning, emotion, perception, attention, and consciousness.

Stephen V. Shepherd   is a postdoctoral researcher working in the Department of Psychology and Neuroscience Institute at Princeton University. He investigates the neural mechanisms of communication and coordination between individuals. His doctoral thesis, with Michael Platt of Duke University, focused on gaze-following behavior in lemurs, monkeys, and humans. His postdoctoral research has expanded to consider the processing and production of facial expressions.

Susan A. J. Stuart   is a Senior Lecturer at the University of Glasgow. Her research interests center on enkinesthesia, languaging and ethics, kinesthetic imagination as the underpinning for cognitive and creative imagination, and Kantian metaphysics and epistemology. Currently she chairs the Consciousness and Experiential Psychology subsection of the British Psychological Society.

Manos Tsakiris   is Reader in Neuropsychology at the Department of Psychology, Royal Holloway, University of London. He investigates the neurocognitive mechanisms that shape the experience of embodiment and self-identity using a wide range of research methods, from psychometrics and psychophysics to neuroimaging. He has published widely in neuroscientific and psychology journals, and his current research projects investigate the plasticity of self-identity.

Michael Wheeler is Professor of Philosophy at the University of Stirling. His primary research interests are in philosophy of science (especially cognitive science, psychology, biology, artificial intelligence, and artificial life) and philosophy of mind. He also works on Heidegger and is particularly interested in developing philosophical ideas at the interface between the analytic and continental traditions. His book, *Reconstructing the Cognitive World: The Next Step*, was published by MIT Press in 2005.

I HAND-CENTEREDNESS

1 "Capable of whatever man's ingenuity suggests": Agency, Deafferentation, and the Control of Movement

Jonathan Cole

Any theory of human intelligence which ignores the interdependence of hand and brain function, the historic origins of that relationship, or the impact of that history on developmental dynamics in modern humans, is grossly misleading and sterile.
—Frank R. Wilson

1 Introduction

Our use of the hand, and the complex neural control involved, is so given that we are hardly aware of it. The great nineteenth-century surgeon, anatomist, and neuroscientist Sir Charles Bell put it thus, in his astonishingly prescient book, *The Hand: Its Mechanism and Vital Endowments as Evincing Design*: "The human hand is so beautifully formed, it has so fine a sensitivity, the sensibility governs its motions so correctly, every effort of the will is answered so instantly, as if the hand itself were the seat of that will. . . . We use the limbs without being conscious, or at least, without any conception of the thousand parts which must conform to a single act" (Bell [1833] 1979, 13–14).[1]

One of the messages of Bell's book is obvious and yet profound, as Frank Wilson (1998) recognized, namely, that the hand has a central importance in our development and evolution. But the hand is not just an instrument. My wife tells me that one thing she liked about me when we first met was my hands. Hands are an important part of who we are; people with severe rheumatoid arthritis, which can cruelly disfigure the fingers and joints, may become ashamed of their hands and hide them—as best they can—in public. Similarly, those with Parkinson's disease, which causes tremor of the hands, will often also hide them from view. The hands, then, are not only highly evolved instruments of manipulation, they are thought to reveal something of ourselves, even beyond the amusing and outlandish claims of palmists.

In this chapter, I will give a brief account of some of the neuroscience underpinning hand function before going on to reflect on the hand's function given the consequences of some impairments of the hand, of sensation, and of movement, and—I hope—derive from these examples some understanding of the roles of the hand. First, however, I will sketch something of the evolution of the hand and of its role in our evolution as human beings.

2   Manual Evolution

When our ancestors stood and walked on their hind legs, this freed the forelimbs to become arms and the forepaws to become hands. Already, with the assumption of a life in trees there had been a move toward more complex uses of the forearms for brachiation (swinging from branch to branch using the arms and forepaws), and of the eyes facing forward in the skull for binocular vision. As Wilson suggests, these must have driven brain development to control enhanced arm and forelimb use: "brachiation placed an enormous burden on the brain's kinaesthetic monitoring and spatial computing power, since there were so many new places the hands could actually *be*" (Wilson 1998, 29).

Once begun, these trends continued, with enlargement of the thumb allowing better hand use with a more controlled opposable thumb, which is not exclusive to humans. These changes, which allowed development of the hand and tool use, were also crucial in allowing the evolution of the face. Once it no longer needed to be involved in the killing of prey and ripping its flesh, since the hands could be used for that, then, in Daniel McNeill's phrase, the face could be demilitarized (McNeill 1999). As smell and touch were less important, superseded by vision, so the hair on the face was no longer necessary, and the face could become more visible. Without need for crushing movements or for control of the vibrissae (bristlelike hairs), facial muscles could be used for other purposes, as display and expression of those more complex emotional states themselves evolved (Cole 1998). The hand and the face, then, are crucial to the development of the human being, and it is no coincidence that Bell's other great book is on the face (Bell 1824).

3   The Sensory and Motor Apparatus of the Human Hand

Bell also realized that the development of the hand required the development of the brain to control its increasingly fine movements. Movement is controlled through three main systems in the brain, the extrapyramidal system, affected in Parkinson's disease, the cerebellum, which leads to severe unsteadiness of movement, and the cerebral cortex, considered to be involved in more fine and conscious control of movement. It is the latter system, including the sensorimotor cortex, which has shown most development in mammalian and, in particular, primate evolution as the hand become more important. Though Penfield's (1958) pioneering work, stimulating the brain of awake patients directly during exploratory brain surgery, showed just how large an area of motor and sensory cortex is involved in control of the hand (and tongue), as Phillips related, it was Hughlings Jackson who first suggested such an idea: "The parts of whose muscles are represented in the greater number of different movements would be represented by the greater quantity of gray matter" (Jackson 1931; quoted in Phillips 1986, 93).

An important development within primate evolution was the direct connection between motor cortex neurons descending through the pyramidal tract and the alpha motoneurones of the ventral nuclei of the spinal cord, going to the muscles of body parts that require fine, rapid control of movement. Such connections are seen in some monkeys to the muscles of the feet and the prehensile tail, but they become more important still in control of the hand in our closer relatives. These corticomotoneurones are separated from the hand muscles by just one nerve synapse and the neuromuscular junction, thus allowing refined, rapid control of movement by the cortex in ways not available to other animals with slower, less pure connections between brain and hand (Phillips 1986, ch. 9). As refined cortical motor control required greater cortical areas dedicated to this, sensory feedback from the hand became more important to allow the brain to know where the hand and fingers were in relation to each other and to explore the world in new ways. The hand is as much a refined organ of exploration and manipulation as it is of pure movement; and this active touch requires an intimate relation between movement and resultant peripheral sensation. So, within the brain, the main sensory areas receiving feedback from the skin and the muscle spindles and tendon organs concerned with movement and position sense (proprioception) are found next to each other, and the area of sensory cortex devoted to the hand expands similarly to the motor cortex. Many studies, beginning in the nineteenth century, have shown how much more sensitive touch is, in terms for instance of two point discrimination and localization, on the skin of the hand and particularly on the fingertip than, say, on the forearm or back. By recording from single peripheral sensory nerve cells in the peripheral nerves of conscious, awake human subjects through a technique known as microneurography, researchers have mapped the small size of receptive fields on the fingers compared with other, less movement- and sensation-salient areas more proximally. In addition it has been possible to selectively stimulate single nerve fibers in humans and ask them what, if anything, they felt. Researchers have shown that a single discharge in a peripheral axon from a sensory receptor on the hand leads to a distinct, localized sensation in a way not seen in nerves from other body parts (Vallbo et al. 1984). Not only are the connections from the motor cortex to the hand highly developed, but this increased development has also required the evolution of an exquisite sensory apparatus. The sensory and motor connections between cortex and hand, therefore, allow the fast, fractionated, exploratory, feedback-dependent movement of our hands and fingers. But these evolutionary developments reflect the beginning of human hand use rather than its end. It is the uses and inventions to which the hand has been put that, arguably, define us, whether this is tool use, writing, gesture, musical notation, or painting.

Damage to various areas of the brain in front and behind the sensorimotor cortex leads to difficulties not in the hand's power or movement, which are usually preserved, but in its use, in a group of conditions termed the apraxias. A person finds himself unable to perform tasks or movements when asked, even though he understands the request and has already learned it, and has a hand that can move in simple tasks normally. Limb-kinetic apraxia results in difficulties in making precise complex movements; one cause of this, which will be discussed in detail, is due to a peripheral problem, but it can also be seen in brain disorders. People with ideomotor apraxia can no longer perform everyday tasks such as use a pen properly. They have lost their normal, unconscious, implicit but learned skills in tool use. Other subjects have difficulties in sequencing appropriate movements, such as dressing; this condition is called ideational apraxia. Both are associated with parietal lobe damage. One way of looking at these conditions is that, though the sensorimotor cortex is working normally, the areas involved in sequencing complex movements, which we all take for granted, are not. Many of the case studies in Oliver Sacks's *The Man Who Mistook His Wife for a Hat* (1985) describe apraxias of various sorts. Abnormalities in areas of cortex surrounding the main sensorimotor areas are also involved in the rare but fascinating alien or anarchic hand syndrome, when the hand moves unbidden (for review, see Frith, Blakemore, and Wolpert 2000). This is associated with damage to an area just anterior to the motor cortex, the supplementary motor cortex (SMA), which is a higher-order area that is active during selection of the appropriate sequence of movements during a task. The SMA is active during tasks that require precision in timing of an action sequence and during its preparation rather than during movement execution, and it is also very active during imagination of a task. Abnormal activity in this area may therefore unleash formed movements, for example, pulling at one's clothes, which are not intended (by more anterior regions in the brain).

Many other more prosaic abnormalities in hand control and use are seen in medicine, from weakness due to individual peripheral nerve lesions or more generalized neuropathy, to pain and deformity due to arthritis, fractures, or tendon damage. In the remainder of this chapter, though, I will explore the hand from the perspective of two main neurological impairments, sensory loss and spinal cord injury, to try to tease out some aspects of its function.

4 Consequences of Sensory Loss to the Hand

Charles Phillips suggested that "if acute withdrawal of proprioceptive feedback were practicable in man all parts of the act of reaching and grasping would become impossible immediately" (Phillips 1986, 24). He had been considering results from a study on a subject with (the very rare) cortical loss of movement and position sense by Jeannerod et al. (1984).

4.1 Neuronopathy Syndrome
At around this time, the mid-1980s, a few subjects with exactly this problem—acute loss of proprioception and, in most cases, touch—were beginning to emerge in the literature. The condition, known as the acute sensory neuronopathy syndrome, was first described by Herb Shaumburg's group (Sterman et al. 1980). It is the result of a cross-reaction to an antibody raised to a foreign infection that leads to an autoimmune attack on the large myelinated nerve cell bodies lying in the dorsal root ganglia just outside the spinal cord at segmental levels down the body. Subjects with the condition are left permanently without touch, movement, and position sense in the body. Other classes of nerve cells are unaffected, so subjects retain normal function in motor nerves, that is, there is no loss of power, and normal perceptions of pain and temperature remain.

The syndrome is very rare and only a handful of subjects are to be found in the medical literature, though there must be more not written about. Oliver Sacks described one woman, Christina, who had lost movement and position sense (proprioception) but had retained touch (Sacks 1985); most

people with this condition appear to lose both cutaneous touch and movement and position sense in the body and limbs, though there is always an
upper level to this loss on the neck or head. Of the two subjects best known
in the literature, and who have been the subjects in many papers, GL, a
woman from Quebec, was 19 or 20 when she became afflicted (see Forget
and Lamarre 1987). Her level of loss is high, from the lower face down, so
that she has no sensation in the lower lip down and no proprioception in
the neck. Though Ian Waterman (IW) was a similar age to GL when it
happened, his sensory level of loss was crucially lower than hers, at spinal
level Cervical 3. He has feeling in the face and in most of the neck; his
sensory level is at the collar line anteriorly, sweeping up the back of the
head behind. This means that, unlike GL, he does not have to think about
head control. Both GL and IW have been studied for over 25 years.

The crucial thing in all these subjects was that although they had
normal movement or motor nerves and so normal power in the muscles,
they had no proprioception, and without that their brains could not coordinate movement at all. As Philips suggested, without movement and
position sense, any controlled movement becomes impossible. Initially, all
these subjects needed full nursing care, not because they could not move,
but because they could not *control* any movement. They showed the phenomenon called sensory ataxia, or limb kinetic apraxia, in which any
attempt to move leads to uncoordinated and shaky movements that are
useless for activities of daily living. Sacks wrote about Christina: "When
she reached out for something, or tried to feed herself, her hands would
miss, or overshoot wildly" (Sacks 1985, 44). The subjects could not dress,
feed themselves, or stand. Their subsequent progress, however, can vary
according to various neurological and other reasons.

Most have not learned to stand or walk again independently and have
lived in a wheelchair. GL, with the most severe loss, does not try to lift a
cup to her mouth, preferring to use a beaker with a spout or straw; her
ataxia means that she may spill an open drink. Because she cannot feel
her lower face, mouth, and lip, she chews by counting and smokes using
a cigarette holder (she would crush any cigarette between the lips). Her
lack of feedback from the neck is a particular problem; stability of the head
is crucial for vision and eye control and for the vestibular system in the
inner ear, so crucial that some suggest that half the muscle spindles in
the body, the receptors that give information about proprioception, are in
the neck. As we will see, recovery of some movement control is possible,
by thinking about movement and observing visually how one has moved.
But to have to think the whole time about head posture must be a huge

cognitive burden. CF, with a similar syndrome but at a lower level, is affected equally clinically and remains ataxic, living in a wheelchair years later. His sensory level is lower, but he developed the syndrome at age 60, reducing his chances of a functional recovery.

The third subject, IW, I first met in the middle 1980s, a decade or so after the onset of his illness (Cole and Sedgwick 1992; Cole 1991, 1995). He, like the others, was completely disabled by his sensory loss. It came on over a few days associated with viral diarrhea. He was admitted via an acute hospital to a neurological center, but the doctors were unable to do anything; in a few days his nerve cells had been destroyed. He was discharged to his mother's care and he spent a few months at home living in a single room, completely dependent on others for all care. Then a physiotherapist suggested he be admitted for rehabilitation. He spent the next year and a half living in the hospital, concentrating on trying to move in new ways. This may be one reason for his greater functional recovery (none of these subjects has regained sensation). In contrast, GL had a husband and young family and decided early on to live in a chair and to care for her family as best she could. IW's sensory level is lower too, so he does not have to think about movement and posture of the head.

IW slowly realized that the useless, ataxic movements could be tamed, if he looked at the part he was trying to move and concentrated on, and thought about, each movement. His first controlled movement was to sit up in bed, by looking at his stomach and consciously contracting the muscles. When he sat up he was so elated that he forgot to think about the movement and fell backward again. The elation was also tempered by concern, since he knew then that he would be able to improve but that it would take a huge amount of mental effort and concentration to do so. He stood after a year and walked a few months later; he finally left the hospital after seventeen months. The story of his recovery of movement and return to independence has been told elsewhere (see Cole 1991, 1995), so here we will concentrate on how he improved hand function and how he uses his hands to this day. But first we will consider his walking, since this allows us some insight into his new way of moving.

4.2 Walking Anew

When IW walks down the street, few people turn to look. His gait is slow, deliberate, and a little ungainly, but having accompanied him on many occasions, I can say that he does not attract attention. But this is not to say that for him the problem of walking is similar in its solution to control subjects with proprioception. There are various constraints that have forced him to walk in a certain way. He has to watch his position and so has to keep his head bowed slightly forward; to compensate, his arms are slightly backward and out from the side. He cannot trust his knees to support his weight and so he always walks with a stiff, straight leg. He does not trust himself to rise up on the balls of his feet; his ankles might give way and in any case verticality is bad in general. So he lands on his heels and takes off flat-footed. To avoid tripping over his feet when moving his leg forward on the through phase, he abducts the leg at the hip, giving his gait a Chaplinesque aspect. Though his walking appears nearly normal, he has to think about it the whole time; if he sneezes he might fall, and if he daydreams he will stagger. (I have seen him do this once when he saw a pretty girl and forgot what he was doing.)

4.3   New Tricks

His use of his hands is similar. It may appear fairly normal, but it is far from it. He is at once more deliberate but never ataxic, and yet to see him sitting in a bar no one would look twice.

One of the first things to realize is that for hand function one needs arm control; Bell realized this: "The hand is not a distinct instrument; nor is it a superadded part. The whole frame must act in reference to it" (Bell [1833] 1979, 19). IW had to learn to control his arms and shoulders to allow his hand to be in the correct positions to be usable. He did this by first preventing ataxia by concentrating not simply on movement but also on not moving. Then he learned a series of movements at the shoulder and elbow. He soon learned that if he moved his arm outward he had to avoid falling over that way, since the arm weighed a fair amount, by a compensatory lean of the torso the other way. He also learned the biomechanical constraints of his limbs. He was often asked to touch his nose with his index finger with eyes shut, a common clinical test. He learned that if he had his arm up and out to the side, bending his elbow brought his hand to his face. Since he has feeling on his face, he can use this to guide his subsequent movement of finger to nose.

Once he had learned to position his arm correctly, he had to work out how to move and use his hand. He faced, and continues to face, several severe limitations. Jacques Paillard, the French neuroscientist who studied GL, made a useful distinction between two sorts of movements: morphokinetic and topokinetic. The first has to be accurate in shape but not position (e.g., tracing a circle in the air), while the latter has to be accurate in position in relation to the external world. IW soon learned that making a movement to pick up an object, where he had to position his hand accurately, was far harder than making a motion in the air. The second problem was the sheer mental concentration involved in moving five fingers, each with three joints. In fact, this problem proved insuperable for him. So, instead, he grasps with his thumb and first two fingers, and hardly uses his ring and little finger.[2] When he cannot see his hands or fingers, he does not know where they are. So, initially, if trying to pick up a mug, he might trap his fingers in something hidden behind the cup. He learned to inspect the back of whatever he grasps. Pottery cups are relatively easy, but plastic cups were a big problem. Once a woman at work brought him a cup of coffee in such a cup. He picked it up, and not being able to gauge how much pressure he should put on it, he overcompensated and crushed it, so that hot water spilled over his hand and desk. That night he took home a stash of disposable cups and stayed up all night learning how to pick them up. By sliding his hand around it low down, making a shape with his fingers and thumb and lifting the cup up as it got wider by resting it in his grasp, he was able to successfully disguise his limitations.

Finally, the hand is unique in that it is an exploratory organ, whose movements are determined by the feedback it receives from what it is touching and from its own movement. As Bell wrote: "Accompanying the exercise of touch, there is a desire of obtaining knowledge; in other words, a determination of the will toward the organ of the sense. Bichat says it is active when others are passive. . . . in the use of the hand there is a double sense exercised; we must not only feel the contact of the object, but we must be sensible to the muscular effort which is made to reach it, or to grasp it with the fingers" (Bell [1833] 1979, 150–151). "Sense of touch differs from other senses by this—an effort is propagated toward it as well as a sensation received from it . . . the sense of motion and of touch are necessarily combined" (ibid., 197–198).

When we take coins out of our pocket, our hands are moving and feeling the whole time, each part merging to make touch an active merging of action, peripheral touch, and proprioception. IW can no longer do this, and he has not found a substitute for it. He never has coins in his pocket; he usually buys something with a note and has the change put on the counter where he can sweep it into his palm and put it in a wallet pocket.

IW had to discover these limitations in his own terms and on his own, since his rehabilitation team, though supporting him, did not have a good idea of what was wrong or how to help. His improvement was also very slow. In the hospital he dedicated himself to relearning movement using visual supervision and mental effort. He preferred to dress himself even if it took an hour; he fed himself even if the food went cold. He devised games to play to help him learn to reuse his fingers. He made paper-clip chains to help himself learn to move more precisely. When visiting with CF, he taught him these games. The player sits still, resting his elbows and hands on a table to reduce limb ataxia. The paper clips are then placed in a pincer grip between thumb and forefinger and the other fingers folded out the way. Having been unable to contemplate such an action before, CF learned, in a few minutes, to do this after a fashion, once aware of the trick. His initial attempts, with his fingers and arms unsupported in the air were—for the task—hopelessly ataxic.

It was similar watching IW and GL sitting on a lawn playing catch. CF said that each time he tried to throw the ball he fell over backward. As he moved his arm behind him in preparation for the throw forward, it moved his weight distribution back and over he toppled. IW found this initially too, and so learned to throw by keeping the arm in front of him and throwing underarm. For CF this was revelatory; he had been trying to perform an action, throwing, by doing it the same way as before the neu-ronopathy. In contrast, IW showed him the same actions can be made using different movements with the same result; the way forward was not to try and repeat the movement patterns of before the neuronopathy, but to find new ways of doing and to invent new ways of being. (In throwing there are two components: making the throw and letting the ball go at the right time. CF became more successful at the former fairly quickly, but needed more practice in coordinating the latter.)

In the hospital, IW spent weeks doing jigsaw puzzles, learning to pick up and manipulate the pieces. It was a boring task, he knew, but the goal for him was to coordinate his hands and fingers. He saw a picture on the wall and asked what it was. Marquetry, he was told, with small chards of wood forming the image. "That's too difficult for you, you'll have your finger off." Such dismissive statements only served to motivate him more; he spent weeks and months learning how to carve and cut wood, and still has several examples of his own marquetry pictures.

4.4 Learning New Skills

IW can talk about how he learned to make new movements, by thought and with visual feedback, and how the movements afterward were easier if he had done them before the neuro. But he is less able to discuss how difficult this was now, years later. In fact, it is difficult to learn new move-ments as adults, and we rarely do it. We learn most movements as children and young adults, whether writing, feeding, or dressing ourselves. Learning to drive, as a young adult, can prove difficult not just because it involves learning the rules of the road, but because it involves mastering new movements.

One group who do have to learn new movements as adults are those with adult-acquired neurological impairment. One person, with a spinal cord injury affecting the neck, had no movement or sensation in his body and legs and had become incontinent as a result, but had retained sufficient power in his arms and hands—just—to maintain self-care. Now incontinent and with no ability to control his trunk, he described learning to self-catheterize his bladder thus:

Some things take a long time to learn. Say transfer from chair to bed; it used to take ages for me to use a board and move across. Now I do it without a board and almost without thinking. But it has taken nearly a year to get to that point.

When I first started doing intermittent catheterization I got a urinary tract infection and my nurse gave me a real bollocking. I was trying the best I could, so I asked why. She said, "Your technique is shit."

He had been shown how to do it and had followed instructions. But he was just not capable to doing all the new movements required, in the right order and with the right timing. For him to pass a catheter into his bladder while balanced on a chair, with weak hands and shoulders, and with the problems of seeing what he was doing, were beyond his best efforts. But slowly, after several months of doing it several times a day, he improved.

I now open the catheter, open a bit of sticky, split the packet, place it on the wall or loo, get a piece of tube, put that in my mouth while I get the catheter, shuffle forward on the chair, undo trousers and get everything out, connect, pass catheter and place over the loo. I do not know whether it [cognitive difficulty] is part of spinal cord injury but it has taken me 6 months to do this simply. I would forget the order to do things. I would withdraw the catheter and urine would still bubble out. Now I rarely wash my hand. It is such a simple technique—and I do not have infections. Bowel care is the same and is a voyage of discovery. (Quoted in Cole 2004)

New, complex movements are difficult to perform even when one is neurologically intact.

4.5 Skill Acquisition and Motor Programs

Early on in his rehabilitation, IW got drunk. He ended up with his head stuck in a bucket after he tried to throw up. He looks back at this with amusement, but for days afterward he found it difficult to move, even in the small ways he had taught himself. Even now, decades later, if he has a head cold he retires to bed since without good mental attention his ability to control his movement is substantially reduced. After two sleepless nights with an earache recently he found it difficult even to pick up a cup. The fragility of his motor recovery is a constant in the back of his mind.

It remains difficult to know how to quantify what training can achieve and why IW has been able to do more than others with the same condition, though as discussed above both neurological and social reasons play a role. IW relearned to make some movements and to avoid ataxia by applying mental concentration. He also appears to move rather more quickly than other similar subjects and, rather than use online visual feedback to guide his next movement, he says that he plans movements in his head and then uses visual supervision to make sure those movements are occurring. Athletic coaches talk of the importance of previsualizing events as a way of preparing for them. IW wishes he had been aware of that sooner, since that is what he discovered too. Even today, decades later, if faced with a new task, say, a new experiment, he likes to go the day before to see the lab and work out how he will walk through it so that he can then mentally rehearse overnight.

Two series of experiments shed some light on this. First, there is evidence that IW has a central awareness of what force he is applying, even without feedback; and second, there is some experimental evidence that he makes plans for movement and runs them forward in time.

In an early experiment, we asked IW to press a simple lever connected to a resistance force with one finger and match the force with a finger of the opposite hand, all without visual feedback and, importantly, without training (Cole and Sedgwick 1992). He could push down and maintain a given force, but he could not match forces between hands. We argued that he could maintain a level of force but without conscious awareness of it, that is, he did not have conscious awareness of the force he made from knowledge of his motor output—so-called efference copy. Subsequently, evidence from both GL and IW has shown that the situation may be more complex.

GL was asked to make a target force with one hand and then match this three seconds later, and she was able to do so (Lafargue et al. 2003). Since GL had no feedback, it was suggested that she perceived her applied muscular force through central effort, even though she had no feeling of fatigue, nor awareness of how hard she was trying. They concluded that consciousness of the size of the motor command may require interaction between internally generated signals and afferent input. One problem in interpretation of these experiments is the training effect. These subjects have to consciously choose and initiate movement, and so they may be able to, say, choose to apply 100 percent or 50 percent of a given force. Though they do not know if they have been successful, they do know what they intended. This intention may occur at a premotor-cortex level of action selection rather than a postmotor-cortex level of efference copy. In part to overcome this, more recently Luu et al. (2011) asked IW to make a strong sustained movement such that he fatigued his thumb flexor muscles of one hand to 50 percent of their maximum. Then he and one other deafferented subject were asked to judge how hard that thumb was pressing by matching its force with the other thumb. Control subjects in this experiment judged the weight as being 25 percent lighter than it was, possibly because of fatigue of muscle spindle activity in the fatigued muscle on which this peripherally dependent percept depends. In contrast, IW and the other subject judged the force to be twice as great. It was concluded that, without proprioception, they relied on a central sense of the motor effort being put into the weakened muscle; the muscle was half as strong, so they were having to put in twice as much effort. For some tasks involving force maintenance, IW and the others may have some awareness centrally of what they have done—the motor cortex efference output—as well as of what they have asked for—their command to the motor cortex. Further research may better distinguish these two features.

Such mechanisms may not be useful during dynamic, unfolding movements. Some years ago the Canadian group asked GL to trace her fingers around a Star of David placed on the table in front of her. The trick was that she only had mirror-reversed visual feedback. Control subjects have severe difficulties with the task, especially at the corners, when they see their fingers going one way and want them to go the other. GL was able to do the task easily first time (Lajoie et al. 1992). They concluded, logically, that this was because she had no mismatch between where she saw her fingers going and where she felt them to go, that is, no visual/proprioceptive mismatch. More recently, Chris Miall and I have repeated the experiments in a little more detail and shown that IW, though slightly faster than controls, also gets stuck at corners of objects; in other words, he still has a mismatch (Miall and Cole 2007). Since he lacks proprioception, we suggested that the mismatch is between where he sees his fingers moving to and where he wants them to go. This, in turn, suggests that he has a central plan, or motor program, predicting or planning his movements.

Introspectively, control subjects relate the same phenomenon; when they perform the task they feel they are stuck because of the mismatch between where they want to go and where they see themselves going, and not between where they feel themselves going and where they want to go. This suggests that one way IW has managed to achieve so much is by elaborating, in some way, however fragile, central motor programs or predictive models of action. He then seems to use vision not for direct feedback—it is too slow for that—but for reassurance that his plan is working, and in this he seems very different from the other subjects studied. GL did not get stuck, because she tends to use visual feedback for direct movement control and has not learned to elaborate motor programs in the same way as IW. IW certainly agrees that a lot of his recovery has been through imagining and rehearsing movements over and over again, presumably building up these models of action.

As we have learned, he also finds movements he made before his illness far easier than ones that are new, though it is difficult to investigate this scientifically. There is a long tradition that analyzes movement and motor learning in relation to schemas or central programming, not least to overcome the problem that movements unfurl faster than either proprioceptive or visual feedback can guide them (e.g., Bernstein 1967; Wolpert and Miall 1996). Movements are usually framed in relation to the learning of skills going from being cognitively driven as we learn, to automaticity or unconscious control; we learned to ride a bike or play piano, and then we can just do it (for some recent work on the computational complexity of this learning, see Körding and Wolpert 2004; Turnham et al. 2011). In IW's case, this is difficult to dissect; he has to think in order to move, so says he never develops unconscious skills, yet his ability to think about the way to do a task, in his head, improves with practice. He says that he has developed a small series of individual movements, like words in a vocabulary, and has learned to use them in various orders to create a sequence of movements, like sentences or paragraphs of speech.

What does seem clear is that his movements, especially in relation to the world and external objects, are comparatively short in temporal duration; he finds that he has to think a given movement through, perform it, and then stop to reassess. When he is walking, he will walk ten meters and then pause to look around and think his way through the next bit. Just as his walking has many constraints, so do his hand movements. His writing post illness was unrecognizable. He picks up cups and forks in a similar way each time; he types one key at a time on a keyboard (in which he is not alone). He appears to be able to hold a large number of movement sequences in his head for the tasks of daily living and to have sufficient control over his actions to perform these in near real time. To the frustration of some neuroscience experimenters, he does not always choose the same strategy for a task. His performance is also affected by how tired he is; his performance during experiments falls off before he is aware of being fatigued, though I can tell by just looking at his walk or reach. There is one movement, however, that may be an exception to some of these constraints: gesture.

4.6   Recapturing Gesture

It is fascinating to watch IW, GL, and CF sit and chat; they gesture as much or more than others. Where GL and CF's gestures are slightly ataxic and uncertain, IW's gestures are controlled and precise. Why and how do they gesture? GL and IW realized that when they learned movements to maintain self-care and live independently again, they had not relearned how to control gestures, and without these they thought they appeared stiff and inanimate. So both set about controlling gestures, consciously, to appear normal, conceal their problem, and appear expressive in and through the body. Their use of gesture was a conscious performance designed to appear natural.

Over the years this motor activity has become one of the most automatic movements they make, though constraints remain. IW knows that he will be unstable if he uses big gestures when standing, because his arms will be further from his body and their weight topples him forward. So he sets a safe workspace for his gestures. Both IW and GL usually gesture only when they can see their arms, so they can keep an eye on things. I once watched as they talked to each other. At the end of each of his responses IW would glance down and fold his arms on his lap, and then at the beginning of the next there was another short glance down to unfold the arms and release them for gesturing.

IW agrees he needs to pay only a little attention to most gestures now, under suitable conditions. Some gestures, however, he does have to think about synchronizing with speech. He has certain styles for what gestures he uses; recently he liked bringing his hand up to his ear, pointing at his head with a circling index finger, to accompany him describing someone as "crazy." This is complicated for him since he cannot see his hand during it and it has taken some practice to perfect. Other gestures that require two-hand coordination and accuracy, such as describing a square in space, also require attention.

David McNeill has made a lifetime's study of gesture in extraordinary and meticulous detail, analyzing the relation between gesture shape and timing, and speech in minute detail with complex computerized video techniques. For McNeill, gesture and language are inseparable parts of one whole: a thought-language-hand system (see McNeill 2005). He has studied

IW on two occasions, both with and without vision of his arms. The latter was done by sitting IW down and placing a tilted screen below his neck in front of him so that he could no longer see his arms and hands (and body and legs; see Cole, Gallagher, and McNeill 2002; Gallagher 2005, ch. 5, and McNeill 2005, ch. 8).

We analyzed IW's gestures in terms of timing with speech, morphokinesis (gesture shape and its position in space), topokinesis (location of the hands relative to each other), and from two spatial perspectives, that of the character being described and that of the narrator or observer. With vision, McNeill finds IW's gestures close to normal, though he found a reduced number of gestures and that they were also performed one by one, suggesting some conscious involvement. With vision obscured, timing with speech and morphokinetic accuracy remained, but topokinetic accuracy was reduced and use of a character viewpoint rare. But there was a clear contrast between IW's ability to gesture and his inability to perform instrumental actions, such as unscrew a thermos cap, without vision. Without vision, IW was still able to synchronize speech and gesture and to quicken and slow them in tandem. His gestures also continue for far longer than other movements; the need for rest and reassessment seems far less. These results, suggests McNeill, are evidence that IW's control of his hands during gesture is through a system that also controls thought and language in a way different from the systems he uses for instrumental action. Put another way, it is further evidence that gesture is not simply added on to language but instead is essentially coexistent with it and controlled through specific pathways in the brain, the thought-language-hand system.[3]

5   Observing the Body as Agent

Thus far we have considered sensory loss in relation to movement control, but its loss, in and of itself, has many consequences. I once saw IW run his hand down the curved side of an old computer. I asked why he did it, since he could no longer feel it. It is clear that his desire to feel and to explore objects, through touch as well as through vision, persists. He loves to pet dogs. Just as he is aware of the grace and enjoyment in movement that he has lost, and which he likes to see, for instance, in deer, so he is aware of the loss of immediate engagement with the world that the absence of touch leads to. His relation with his own body is more difficult to discern; after thirty years or so he lives with his own standard of what is normal. GL has talked about feeling more like a captain of a ship, as she thinks about moving her body the whole time; but with more potential for movement, IW does not feel quite so disembodied.

Normally we run, sit, eat, and talk, with little attention to the body that allows us to act in the world. It just does what we want it to do. Our bodies seem phenomenologically absent from our attention for much of the day. Merleau-Ponty wrote, "I observe . . . objects, but my body itself I do not observe" (Merleau-Ponty 1962, 91). IW now finds his relation with his body in these terms normal, having lived thus for decades. So, to look at how impairment can affect this relationship to our bodies, we turn, once more, to those who live with spinal cord injury. One of the interesting features of these people is that they *do* observe their own body. When paralyzed and insentient they have to look after their body, to avoid pressure sores and to maintain continence control, as they never did before. Their reduced function can also shed light on our otherwise unconscious sense of agency, our ability to own and initiate action.

Debbie Graham was a patient with a high spinal injury, which meant she had some weak hand movement but was only able to do a few movements that were useful. She then had a Freehand system, which involved electrodes under the skin of her forearm onto her muscles, with wires connected to a small receiver in her anterior chest wall, under the skin, driven by a radio transmitter lying on the skin over the receiver. The transmitter signals were controlled, in turn, by movement of the subject's shoulder via a shoulder position-sensor attached to her. Then, moving the shoulder backward or forward allowed her hand to open and close, like a mole wrench, giving her more hand function and hence more independence. Her use of this device also allows us to understand a little of the way she viewed her hand function and her own control of it, which is typically taken for granted in normal subjects.

She has two main grips, one a "pinch" with thumb apposition and another grip with wrist flexed and thumb and fingers acting together. One allows a fine grip for holding a fork or writing, the other a coarser grip for larger, heavier objects.

To use my Freehand [on the right] I move my left shoulder. To pick a pen up I first position the pen in front of me, then I put my left arm round the back of the chair to stop falling over [she has no postural control of her body]. I switch the system on by pressing a button on my chest and then I pull my left shoulder back and the right wrist is extended and the hand open. Then, as I move the shoulder forward— slowly and a very small amount—the wrist and hand close. Once my hand is gripping as I want, I flick my shoulder suddenly and it locks the system. Then I can hold with my clenched hand, without thought, and move my shoulder independently. When I want to release my hand I flick the shoulder again and the hand relaxes. Then I switch it off by pressing again. (Cole 2004, 136)

She learned to use the device through a sophisticated process of trial and error and a lot of fine-tuning from her engineers. It is so complicated that using it is a long way from being "second nature."

I always know what I am going to do. I decide to make a movement. I look at my hand but think about my shoulder. I am not thinking about muscles in my arm or hand, [but about] moving my shoulder and about getting the pen in the right place visually. Once locked I write from my whole arm, since I cannot use my fingers. My writing is not what I would like, but it is easier than before, because I have a proper grip. Before I could not press down or the pen would come out of my hand. To write I think about the whole arm. I cannot really think about anything else. I have to hold my breath.

If I am on the phone, at my brother's boat business, then that is slightly harder, since I am holding the phone with my left hand, with a hook on the back of the handset for my fingers, and trying to type a name with my right hand locked by the Freehand. I have got a metal pen with a rubber to type with. I am then careful not to move my shoulder, and I also lean onto my left elbow to give me the stability to type and have my left arm hooked behind my chair to avoid falling forward. (Ibid., 137–138)

Her answering the phone seems more complicated than most people's movements in their daily lives. She was doing it from a chair with a combination of the Freehand system, balance, and determination—coupled with mental concentration on various movements of her trunk, shoulder, and arm. Living in a wheelchair may look passive and devoid of action, but that is far from the truth. I wondered if now, two years on, she was more at home with the system and thought about it less.

First I thought about the system more than the goal. When I used it to pick up a fork, my attention was more on the shoulder movement and the fork's position in my grip than in using the fork to eat. My attention is now more on the action, and less on the hand. (Ibid., 139)

I was interested whether she considered her hand movements hers or not. It was evident that this is difficult for her to answer because, still, she does not feel completely in control of the system.

When I control it then it becomes me. Well, kind of me. When it is not working then it is the stupid bloody thing. When it is working it is me, but I do look at my hand differently when using the Freehand system. It is still not completely me, still alien, still something else, since it is being controlled by a computer in a bag on the back of my chair. Without the Freehand it is me completely, even though I cannot move it well. When it works it is the system and me controlling the system, but still not me completely. (Ibid.)

A crucial indicator of the success of any device is how often it is used in daily living:

At home I use it for eating, writing and doing my hair. In the shop I use it all the time to write. At home I may use it to put a video in. It can be fun to use, to see if I can open the fridge or cook a potato in the oven. I also find increasingly that I go to prepare a grip with my shoulder even when I have not connected the Freehand. (Ibid., 140)

Also, it had enabled one new movement: "I use it when I brush my hair." Previously she had had short hair since her injury; being able to brush her hair had allowed her to grow it long again.

Debbie's experience of her hand and the Freehand is therefore a little like our experience of tool use. It becomes part of us when it works without our thinking about it; otherwise it remains outside our perception of ourselves. It must be extraordinary, though, to feel this way about your own hand.[4]

We have considered hand function in relation to rare sensory loss and more common loss of hand function due to spinal cord injury. Another common cause of unilateral loss of hand function is stroke. People who have had a middle cerebral artery stroke, which affects the sensorimotor cortex, often experience paralysis down the other side of the body. Fortunately this usually improves, even more since new drugs have been developed, but the one function that does not recover so well is hand movement. The brain seems to be able to compensate for and restore movements of the leg and arm even when cortical recovery is incomplete, but hand movement, which is dependent more on sensorimotor cortex, does not recover as well. Someone may be able to move his leg to walk, though often not entirely normally, and may move his arm, but still have little or no movement of the fingers or hand. As a result, tasks requiring bimanual activity are difficult. Dressing, cutting up food, tying shoelaces, driving; many—most—movements of the hand are bilateral, and one needs to be resourceful to find ways around them. Some people find that, following severe stroke their fingers, once paralyzed, settle into flexion, and that hand becomes contracted into a useless fist. Not only do they lose the hand as a tool to manipulate other objects; they lose an expressive, hugely personally salient part of themselves. This is seen in the manner such people often sit caressing their immobile hands with their other hand, as though trying to bring it back to life. I have recently addressed something of the consequences of living with hand problems in a performative lecture entitled "The Articulate Hand" (http://thearticulatehand. com),[5] with help from an eminent actor, choreography, and theater director, Andrew Dawson.

6 Hand and Culture

David McNeill's (1992, 2005) works on gesture deliberately echo Vygotsky's great book on the social origins of language, *Thought and Language* (published in 1986 but written in the early twentieth century). All these works convey something of the relation between the evolution of the hand and of thought, and hence culture. Though this chapter has discussed hand function by reflecting on the consequences of loss of normal hand function, the hand itself in its everyday use seems crucial for human development. It is used for instrumental actions in a myriad of ways, and through it we have altered our development and even the environment in which we live. It is used in gesture, intimacy, and grooming. More importantly, though, our cultural development required the hand to write; without that our culture would have remained aural. Without the writing hand we would not have had Shakespeare's plays, Bach's Cello Suites (with no hands to play, nor notes to be written), no novels, no symphonies, no Velasquez, no Moon landing, no engineering, no science; in sum, no extended mind (see Clark, this vol.). As Bell suggested, "the hand [is] capable of executing whatever man's ingenuity suggests" ([1833] 1979, 209). And, of course, as other chapters relate, and as Bell realized, that ingenuity, imagination, and intelligence often are expressed through the hand but depend on development of the brain.

Notes

1. Much of Bell's book concerns the comparative anatomy of the hand, and though his account glorifies God and his creation, in retrospect, much of his book provides evidence refuting the biblical creation account. He clearly accepts that the anatomy of man's hand is similar in design to that of many other animals: "Bones of large animals are found imbedded in the surface of the earth . . . in beds of rivers, where no waters flow and they are dug up under solid limestone rock. . . . The principle of this great plan of creation was in operation previous to the revolutions which the earth has undergone; that the excellence of form seen in the skeleton of man was in the scheme of animal existence long previous to the formation of man and before the surface of the earth was prepared for and suited to him" (Bell [1833] 1979, 23–24).

2. We were once at NASA's the Johnson Space Center, in Houston, where they have a robot, called the DART, which has two arms with joints that move like a human's at the elbow, with wrist, hand, and fingers, and which is controlled by motion capture from a subject moving his or her arms and hands nearby. The subject looks down onto the robot arms from cameras installed in the robot's head shown in a

head-mounted display. The NASA scientists found that subjects could soon learn to move the hands and manipulate objects without difficulty; but, with vision of the robot but no peripherally originating proprioceptive feedback from it, people could only learn to control its thumb and two fingers, so they designed to hands without fingers 4 and 5. IW saw it and said he could have told them that, and so saved thousands of dollars of development money.

3. This is not to say that language and gesture add the same meaning to communication. McNeill considered this in an earlier study (McNeill 1992). He suggests that language segments and linearizes meaning in time. An instantaneous thought may be divided up into parts to describe and elongated in time: "Language can only vary along the single dimension of time—phonemes, words, phrases, sentences, discourse; at all levels language depends on variations along this axis of time" (ibid.). In contrast, gestures are "multidimensional and present meaning without segmentation or linearization" (ibid.). Gestures, for instance, are not combined in hierarchies to make other gestures. They also anchor expression in the body, and though they may convey complementary meanings to coexistent language, they also unfurl over subtly different durations and place communication in the body as well as in sound.

Such an account perhaps does not emphasize that language and gesture are not just parts of the same expression but also parts of the same creative process. Sometimes, in the "tip of the tongue" phenomenon, it seems that we use gesture to help ourselves remember a word or phrase. The same movement programs that we use to gesture also help us remember a word. One wonders in these deafferented subjects whether their reacquisition of gesture was entirely to appear normal, i.e., was just for us, or whether the subjects themselves gained any linguistic fluency from unfurling gesture with speech, over and above the enhanced facilitatory feedback from others in conversation they presumably receive when gesturing. If so, then since they do not get any proprioceptive feedback from gesture, does any advantage come from seeing the gesture, or from the internal brain activations from which it results?

4. One of the extraordinary perceptions during use of the DART robot is that within a minute of using it one becomes so immersed in it that one thinks of oneself as being in the robot, rather than in one's own body a few feet away. What we see and move we become, even when we may not feel (see Cole, Sacks, and Waterman 2000).

5. The title comes from Wilson's (1998) book.

References

Bell, Sir C. [1833] 1979. *The Hand: Its Mechanism and Vital Endowments as Evincing Design.* Cleveland: Pilgrim Press.

Bell, Sir C. 1824. *Essays on the Anatomy and Physiology of Facial Expression.* London: John Murray.

Bernstein, N. 1967. *The Co-Ordination and Regulation of Movements.* Oxford: Pergamon Press.

Cole, J. D. 1991, 1995. *Pride and a Daily Marathon.* London: Duckworth; Cambridge, MA: MIT Press.

Cole, J. D. 1998. *About Face.* Cambridge, MA: MIT Press.

Cole, J. D. 2004. *Still Lives.* Cambridge, MA: MIT Press.

Cole, J., S. Gallagher, and D. McNeill. 2002. Gesture following deafferentation: A phenomenologically informed experimental study. *Phenomenology and the Cognitive Sciences* 1 (1):49–67.

Cole, J., O. Sacks, and I. Waterman. 2000. On the immunity principle: A view from a robot. *Trends in Cognitive Sciences* 4 (5):167.

Cole, J. D., and E. M. Sedgwick. 1992. The perceptions of force and of movement in a man without large myelinated sensory afferents below the neck. *Journal of Physiology* 449:503–515.

Forget, R., and Y. Lamarre. 1987. Rapid elbow flexion in the absence of proprioceptive and cutaneous feedback. *Human Neurobiology* 6 (1):27–37.

Frith, C. D., S. J. Blakemore, and D. M. Wolpert. 2000. Abnormalities in the awareness and control of action. *Philosophical Transactions of the Royal Society of London, Series B* 355:1771–1788.

Gallagher, S. 2005. *How the Body Shapes the Mind.* Oxford: Oxford University Press.

Gordon-Taylor, G., and E. W. Walks. 1958. *Sir Charles Bell: His Life and Times.* Edinburgh: Livingstone.

Jackson, J. H. 1931. On the scientific and empirical investigation of epilepsies. In *Selected Writings of John Hughlings Jackson*, vol. 1, ed. J. Taylor, 162–273. London: Hodder & Stoughton.

Jeannerod, M., F. Michel, and C. Prablanc. 1984. The control of hand movements in a case of hemianaesthesia following a parietal lesion. *Brain* 107:899–920.

Körding, K. P., and D. M. Wolpert. 2004. Bayesian integration in sensorimotor learning. *Nature* 427 (6971):244–247.

Lafargue, G., J. Paillard, Y. Lamarre, and A. Sirigu. 2003. Production and perception of grip force without proprioception: Is there a sense of effort in deafferented subjects? *European Journal of Neuroscience* 17 (12):2741–2749.

Lajoie, Y., J. Paillard, N. Teasdale, C. Bard, M. Fleury, R. Forget, and Y. Lamarre. 1992. Mirror drawing in a deafferented patient and normal subjects: Visuoproprioceptive conflict. *Neurology* 42 (5):1104–1106.

Luu, B. L., B. L. Day, J. D. Cole, and C. R. Fitzpatrick. 2011. The fusimotor and referent origin of the sense of force and weight. *Journal of Physiology* 589 (13): 3135–3147.

McNeill, Daniel. 1999. *The Face: A Guided Tour*. London: Hamish Hamilton.

McNeill, David. 1992. *Hand and Mind*. Chicago: University of Chicago Press.

McNeill, David. 2005. *Gesture and Thought*. Chicago: University of Chicago Press.

Merleau-Ponty, M. 1962. *Phenomenology of Perception*. London: Routledge.

Miall, R. C., and J. Cole. 2007. Evidence for stronger visuo-motor than visuo-proprioceptive conflict during mirror drawing peformed by a deafferented subject and control subjects. *Experimental Brain Research* 176 (3):432–439.

Penfield, W. 1958. *The Excitable Cortex in Conscious Man: The Sherrington Lectures V*. Liverpool: Liverpool University Press.

Phillips, C. G. 1986. *Movements of the Hand: The Sherrington Lectures XVII*. Liverpool: Liverpool University Press.

Sacks, O. 1985. *The Man Who Mistook His Wife for a Hat*. London: Duckworth.

Sterman, A. B., H. H. Schaumburg, and A. K. Asbury. 1980. The acute sensory neuronopathy syndrome: a distinct clinical entity. *Annals of Neurology* 7 (4):354–358.

Turnham, E. J., D. A. Braun, and D. M. Wolpert. 2011. Inferring visuomotor priors for sensorimotor learning. *PLoS Computational Biology* 7 (3):e1001112. Epub March 31, 2011.

Vallbo, A. B., K. A. Olsson, K. G. Westberg, and F. J. Clark. 1984. Microstimulation of single tactile afferents from the human hand: Sensory Attributes related to unit type and properties of receptive fields. *Brain* 107 (3):727–749.

Vygotsky, L. 1986. *Thought and Language*. Cambridge, MA: MIT Press.

Wilson, F. R. 1998. *The Hand: How Its Use Shapes the Brain, Language, and Human Culture*. New York: Pantheon Books.

Wolpert, D. M., and R. C. Miall. 1996. Forward models for physiological motor control. *Neural Networks* 8:1265–1279.

2 Developmental Origins of the Hand in the Mind, and the Role of the Hand in the Development of the Mind

Andrew J. Bremner and Dorothy Cowie

1 Introduction

We look at and move our hands more than other limbs. This fascination with the hand is apparent from the first stages of life. While awake, and sometimes while asleep, newborns move their arms and hands rapidly and in ways that seem motorically complex. And yet the hands also appear to be a puzzle for infants. A four-month-old will stare at her hand with great intensity, only to misconstrue it as an external object and attempt to reach for it. The hand is an important focus of psychological activity in infancy, but its mastery (both in terms of its movement and in terms of the infant's understanding of it) poses an important developmental challenge. Yet when reviewing the research relevant to the development of representations of the hand or body more generally, it becomes apparent that this has remained restricted to a limited number of research questions.

Theories concerning the ontogeny of representations of the hand are also relatively obscure (at least for the developmental psychologist). The most obvious contributions come from Merleau-Ponty (1962), who acknowledges the importance of multisensory integration in the development of the "body schema" (the internal model of the body, which subserves action). Others have taken phantom limbs as the test case for the origins of body representations. Simmel, for instance (1966; see also Price 2006), shows that phantom limb experience is dependent on rather extensive prior experience of that limb in the first years of life, and argues that the body schema is constructed gradually (although see Gallagher and Meltzoff 1996; Melzack et al. 1997, for alternative accounts). It is perhaps surprising that Piaget, for whom the active hand was a key driver of cognitive development, devoted so little attention to the emerging representation of the hand itself.

In contrast, however, the question of what role the hand plays in development has received much more treatment. In this chapter, we begin by outlining the long-held but unfinalized debate concerning the role of active hands in cognitive development. We then go on to describe more research that has addressed the less considered but perhaps more fundamental question of how our perceptions and understandings of our own hands arise in early life. By addressing the question of how human infants learn about the interface between their hands, their body, themselves, and the environment, we may be able to shed more light on the unresolved debate concerning the role of the hand in human intelligence.

2   The Role of the Hand in Cognitive Development

A central question of this volume, and one that has also been on the minds of philosophers and psychologists from the outsets of their respective disciplines, concerns the relationship between the hand and intelligence. The concern is whether our intelligence arises through our interactions with our environment—our hands providing us with the information we need to act on and understand the world around us—or whether it is our knowledge of the world that drives us to use our hands in a competent manner (see Gallagher, this vol.). On first consideration, one would think that developmental psychology is well placed to provide an answer to this debate. If the hand drives our knowledge of the world, we would expect the manual limitations of younger age groups to have a significant impact on their cognitive abilities. Alternately, if cognitive abilities arise independently of manual competence, we might adopt the more rationalist position in which the hand is subordinate to the intelligent (visual) mind. Perhaps the best way to begin addressing these alternatives is by reiterating some of the key theoretical positions on the relationship between action and cognitive development.

2.1   The Hand Constructs Cognition in Human Infancy
The hand plays a central role in Piaget's sensorimotor theory of development in the first two years of life (Piaget 1952, 1954). In his investigation into the origins of concepts of objects and space, Piaget (1954) traced the development of object-directed behaviors in his three children from birth to two years of age. With these observations of manual, visual, and oral behaviors toward objects, he described how the young child actively adapts and combines her reflexes into increasingly complex forms of action

schema that allow her to interact with, and thus mentally represent, objects in increasingly more sophisticated ways.

For Piaget, it is the combination of various action schemas, among which manual schemas are placed in a position of importance, that leads to progress in cognitive development. He argued that the initial steps toward objectivity in infants' representations of their world were achieved in part through the integration of separate, modality-specific schemas (Piaget 1952). Take the following observation, in which Piaget notes what he termed the "organization of reciprocal assimilation," when his four-month-old daughter, for the first time, watches herself grasping an object:

> Lucienne makes no motion to grasp a rattle she is looking at. But then she subsequently brings to her mouth the rattle she has grasped independently of sight and sees the hand which holds the object, her visual attention results in immobilising the movement of her hands; however, her mouth was already open to receive the rattle which is 1cm away from her. Then Lucienne sucks the rattle, takes it out of her mouth, looks at it, sucks again and so on. (Piaget 1952, 102)

Piaget goes on to describe this as a new kind of schema in which infants can grasp what they see, and see what they grasp: a reciprocal relationship in which the visual and tactile universes are coordinated. For Lucienne, the object is no longer the thing of looking, nor the thing of grasping, but an entity existing in a more objective representation. So, for Piaget, an objective understanding of the world is promoted by the active exploration of the hand in relation to vision.

However, despite this central role of the hand (and indeed the body more widely) in the development of intelligence, it is very difficult to fit Piaget into the enactivist or embodied cognitive mold, in which cognitive processes are situated within the context of the body and hand (e.g., Bermudez et al. 1995; Gallagher 2005; Varela et al. 1991). Piaget almost characterizes the hands and body as the limiting factors in the emergence of an objective understanding of the environment. This was because he was primarily concerned with how infants and later children develop action schemas and ways of thinking that are independent of their own perceptions and actions—independent of their hands and bodies. The purpose of sensorimotor development was thus, in a sense, to escape the egocentric tyranny of the sense organs and the body. The abstraction of cognitive processes from the hands and body continued to be a key theme for Piaget in his study of development beyond infancy, as he described how young children gradually decenter their "imaginal" mental representations from their own egocentric perspectives.

Piaget's view faces certain problems. First, the argument that we require both hands and vision (i.e., visually guided reaching) in order to develop with typical intelligence runs into difficulty when we consider that the many instances of blind children or children without limbs who manage to do just this (e.g., Gouin Décarie and Ricard 1996).[1] Second, fifty years of empirical research utilizing measures of looking duration (e.g., Baillargeon 2004; Slater 1995) have shown that very young infants, before six months, are able to perceive objects and spatial arrays in ways that do not resemble the "egocentric" state that Piaget described. Observations such as these led to a quite different view on the role of the hand in cognitive development, which we shall now describe.

2.2 Cognition Precedes Representations of the Hand in Human Development

The view of the hand as secondary to the mind in cognitive development is perhaps most clearly articulated by Spelke and colleagues: "At three and four months of age, infants are not able to talk about objects, produce and understand object-directed gestures, locomote around objects, reach for and manipulate objects, or even see objects with high resolution. Nevertheless, such infants can represent an object that has left their view and make inferences about its occluded motion. . . . Infants represent objects and reason about object motions" (Spelke et al. 1992, 627). This perspective argues that knowledge of the environment and the objects it contains is provided by our phylogenetic inheritance in the form of "core knowledge" (e.g., Spelke and Kinzler 2007). Such knowledge is proposed to be modular in nature, but later elaborated on via ontogenetic interactions with the environment (e.g., Baillargeon 2004).

The proposal that such core knowledge exists rests primarily on the findings of experiments that measure patterns of looking duration in infants' visual inspection of objects and events. The success of visual inspection paradigms have made them the staple for investigating perception and knowledge over the first year and a half of life, yielding many surprising findings including demonstrations that young infants have an understanding of objects (their permanence, solidity, and constancy; e.g., Baillargeon 2004; Spelke et al. 1992), space (e.g., Bremner et al. 2007; Kaufman and Needham 2011), and even number, mathematics, and people's actions (see Spelke and Kinzler 2007). Such demonstrations have forced developmental scientists to revise the view inherited from Piaget that objective perception of the physical world develops slowly.

Even more than Piaget, the core knowledge approach downplays the importance of the hand in development. Whereas Piaget considered the hand (and the body) to be a tyrannical egocentric framework from which thought processes eventually become unfettered, he at least considered them to play an important role in the construction of objective knowledge. The core knowledge approach places a much lighter emphasis on the role of the hands and body in the development of cognitive abilities. In a sense, this is one of the main problems for this approach, to which even its exponents admit: it leaves out the question of how we come to be able to make use of our knowledge in order to deploy our actions appropriately in our environments.

2.3   Resolving Viewpoints: The Development of Hand Representations in Early Life

The problem of understanding how infants and young children come to be able to act on representations of objects and their spatial environments has received a great deal of attention, partly because it presents some strange paradoxes. While measures of looking duration have tended to indicate precocious abilities to represent objects in the extrapersonal environment, spatial orienting tasks, which require infants to locate objects in the environment relative to themselves (with manual responses, or visual orienting behavior), have provided a mixed picture of early spatial abilities. Typically, before eight months of age, infants do not even attempt to uncover hidden objects within reach (Piaget 1954). Such dissociations between looking responses and spatial orienting responses have now been shown in a variety of domains (see, e.g., Hood et al. 2000).

So, if infants can represent objects and spatial layouts, why then are they unable to make use of this knowledge to appropriately deploy actions until much later? Among the explanations that have been proposed are appeals to the delayed development of executive abilities (Diamond 1991), the different strengths of representation required to respond visually and manually to hidden objects (Munakata et al. 1997), and the different kinds of representation required (e.g., spatiotemporal vs. identity-based) to succeed in looking time tasks and action-based tasks (e.g., Mareschal and Johnson 2003). In the next section we provide the grounding for an alternative possibility, namely, that developmental lags between knowledge and action might be explained by a protracted development of an understanding of the hands and body relative to the early-acquired representations of the external environment described in section 2.2 (Bremner et al. 2008a).

3   Developing Representations of the Hand in the World

There is obvious scope for significant development in how we represent the hand and body during childhood. Any early ability to represent the layout of one's hand with respect to oneself would need to be continually retuned throughout development in order to cope with physical changes in the sizes, disposition, and movement capabilities of the hand and arms among the other parts of the body (Bayley 1969). Changes in body size, distribution, and movement give rise to a central difficulty for the developing infant or child, namely, how to combine sensory information about the body and the world across the multiple modalities through which that information arises (Bremner et al. 2010). As the body changes shape, size, and posture, the individual must reconfigure the ways in which the senses are combined. Indeed, as we describe in section 3.2, postural variation also presents a significant multisensory problem for the mature organism. Table 2.1 displays some of the most significant anatomical, perceptual, and motor milestones in the development of the hand.

3.1   The Hand from the Uterus to Infancy

The human fetus's upper limbs develop earlier than the lower limbs, appearing first as "limb buds" around four weeks of gestation.[2] The hands gradually develop from a paddle-like appearance at five weeks to having defined fingers, which remain joined at six and a half weeks' gestation. Fingers are entirely distinct and separate from around eight weeks. At six and a half weeks, the hands start moving for the first time relative to the torso. The fingers begin practicing the movements involved in the "palmar grasp reflex" (in which the fingers clasp down on the palm and alternately release) by the eleventh week of gestation.

Hand-to-mouth movements appear to be an important feature of the early manual behavior of the prenatal and newborn infant (Butterworth and Hopkins 1988; Castiello et al. 2010; Rochat and Hespos 1997; Rochat et al. 1988). The finding that a newborn's mouth will open in anticipation of the arrival of the hand at the mouth or the perioral region has been cited as evidence of intentionality of movement (Butterworth and Hopkins 1988). More recently, Castiello and colleagues (2010) have even suggested that by fourteen weeks' gestation, a fetus's directed hand movements toward the back and the head of a twin who is sharing the uterus indicate intentional interaction between co-twins.

Despite these pockets of competence, a casual observation of young infants' behavior with respect to their own hands suggests significant limitations both in their control and in their perception and understanding of their limbs. As suggested by the following anecdotal observation of a four-month-old, a naïve appreciation of the relation between tactile and visual spatial frames of reference could lead to a number of problems in controlling and executing appropriate actions: "Sometimes the hand would be stared at steadily, perhaps with growing intensity, until interest reached such a pitch that a grasping movement followed as if the infant tried by an automatic action of the motor hand to grasp the visual hand, and it was switched out of the centre of vision and lost as if it had magically vanished" (Hall 1898, 351).

What are the hurdles we have to overcome to gain mastery of our hands? In the next section, we shall argue that the principal difficulties concern learning how to combine stimuli impinging on the body across multiple sensory channels. The necessary multisensory information specifying the layout of our bodies with respect to the world is typically provided by touch, proprioception, vision, and occasionally audition. These senses convey information about the environment and body in different neural codes and reference frames (e.g., a tactile location arrives at the brain in body surface coordinates, whereas a visual location is initially coded retinocentrically). Do infants know how to combine these sensations?

3.2   The Development of Visual-Proprioceptive (V-P) and Visual-Tactile (V-T) Correspondence: Early Multisensory Hand Representations?

The question of whether infants and young children are able to perceive spatial correspondences between sensory inputs concerning the hand has fascinated philosophers and psychologists since the time of Molyneaux and Locke (Locke 1690). The first empirical approaches attempted to discern whether children, and later infants, could recognize an object in one sensory modality that had previously been presented only in another ("crossmodal transfer tasks"). Early pioneers of this technique in children (Birch and Lefford 1963) observed that the accuracy of children's crossmodal matching of stimuli between vision and touch on the hand increased across early childhood. However, later studies (Hulme et al. 1983) indicated that these early childhood crossmodal developments were fully explained by concurrent developments in unimodal perceptual accuracy.

More recently, evidence appears to suggest that very young infants are able to perceive matches between touches on the hand and visual sensations. Streri and colleagues (see Streri 2012; also Meltzoff and Borton 1979) have shown that an ability to transfer texture, and to some extent shape, between touch and vision is present at birth. Rose and her colleagues have

Table 2.1
The development of the hand: Key anatomical, motoric, and perceptual milestones (see Bremner et al. 2012)

Age	Anatomical development	Motor development	Perceptual development
Four weeks gest.	"Limb buds" appear Maturation of cutaneous receptors Visual anatomy matures		First responses to cutaneous stimulation
Six weeks gest.		Hands start moving relative to torso	
Eight weeks gest.	Fingers entirely separate on the hand Auditory receptors mature		
Eleven weeks gest.		Onset of "palmar grasp reflex"	
Fourteen weeks gest.		Hand-to-mouth movements apparent Intentional hand movement towards a co-twin (Castiello et al. 2010)	
Twenty-four to twenty-eight weeks gest.			Visual and auditory systems become functional
Birth (40 weeks gest.)	Substantial changes in the proportional size of the hand, and its distance relative to the body and sense organs continue right through to adulthood.	Mouth opens in anticipation of hand (Butterworth and Hopkins 1988; Rochat and Hespos 1997) Visually perceived objects elicit proto-reaches (Von Hofsten 1982) Coordination of hand movements towards visual locations (without sight of the hand; Van der Meer 1997)	Bidirectional visual-tactile transfer of texture (Sann and Streri 2007)
Three months		Proprioceptive guidance of reaching to visual or auditory targets (without sight of hand; Clifton et al. 1993)	Matching of visual movement of legs with proprioceptive movements (or efferent motor copies; Bahrick and Watson 1985)
Five months		Grasps which anticipate the orientation of an object (Von Hofsten and Fazel-Zandy. 1984)	

Table 2.1
(continued)

Age	Anatomical development	Motor development	Perceptual development
Six months			Successful manual orienting to tactile stimulus on the hand in familiar arm postures (Bremner et al. 2008b)
Nine months		Grasps which anticipate the size of an object (Von Hofsten and Rönnqvist 1988)	
Ten months			Successful visual orienting to tactile stimulus on hand (Bremner et al. 2008b) Successful manual orienting to tactile stimulus on the hand across familiar and unfamiliar arm postures (Bremner et al. 2008b)
Fifteen months		Infants' reaches become sensitive to visual feedback concerning the hand (Carrico and Berthier 2008)	
Four years		Increased reliance on vision in reaching	Increasing importance of vision in determining perceived hand position
Five and a half years			Crossed-hands effect emerges indicating the consolidation of an external frame of reference for locating tactile stimuli (Pagel et al. 2009)
Eight to ten years			Haptic and visual information about objects integrated optimally (Gori et al. 2008)
Adulthood			

also shown that more robust bidirectional crossmodal transfer of shape between touch and vision is available by at latest six months of age (Rose et al. 1981). Rose et al. also present evidence that, between six and twelve months, infants become progressively more efficient in their crossmodal recognition (see Rose 1994).

Researchers have also examined the early development of spatial correspondence between vision and proprioception. A series of studies conducted by several research groups has examined whether infants from as young as three months of age are able to respond to crossmodal visual-proprioceptive correspondences arising from the movement of their own limbs (e.g., Bahrick and Watson 1985; Rochat 1998; Schmuckler and Jewell 2007). Although these studies focus on multisensory representations of the legs—legs and feet are a source of fascination for infants around four months of age, especially when clothed with striped socks, as they were in many of these studies—the early competence displayed seems likely to be applicable to the hands also. In several of these studies, infants were presented with a visual display showing their own legs moving in real time. In other words, these displays presented spatiotemporal synchrony between visual and proprioceptive information about their leg movements. Infants' looking at these synchronous displays was compared with their looking at a visual presentation of another child's leg movements. The preferences infants showed for the unfamiliar asynchronous display across a number of different conditions of stimulus presentation have provided strong evidence that young infants are indeed able to detect proprioceptive-visual spatiotemporal correspondences.

Thus, the evidence seems fairly conclusive that, soon after birth, humans are able to register correspondences across the direct and distance receptors (at least for V-T and V-P correspondences). However, in considering the development of multisensory representations of the body and limbs, it is important to examine whether V-T and V-P spatial correspondences are perceived with respect to parts of the body (e.g., head-, eye-, body-, limb-centered spatial coordinates). Although this question has been tackled to some extent with respect to V-P correspondences (Rochat and Morgan 1995; Schmuckler 1996), there is a fundamental difficulty with addressing representations of body-centered space with this particular paradigm, as it requires that visual cues concerning the body be presented on a video display outside bodily and peripersonal space (i.e., beyond reach in extrapersonal space). It is thus quite possible that infants respond in this task on the basis of some other correspondence between these visual and proprioceptive inputs, such as their correlated movements (cf. Parise et al.

2012). Indeed, one of us has argued elsewhere (Bremner et al. 2010) that the findings from these studies may tell us more about infants' perceptions of contingencies between their own movements and the movements of objects or people in their extrapersonal environment than about representations of their own bodies.[3]

Finally, even if young infants do register V-T and V-P commonalities according to a representation of their own bodies, we have little information regarding the spatial precision of such representations. It is unclear to what extent the crossmodal perceptual abilities of young infants could provide the basis for useful sensorimotor coordination in the immediate personal and peripersonal environment. It is our assertion (see also Bremner et al. 2008a) that, in order to glean unambiguous information concerning the development of spatial representations of the body, spatial orienting paradigms are needed. Spatial orienting responses have the advantage of requiring coordination between the stimulus location and the intrinsic spatial framework within which the response is made. We will now describe the findings of a study of infants' orienting responses to tactile stimuli presented to their hands. As we shall see, there is a lot more to orienting to a location on the body than there is to registering V-T and V-P correspondences.

3.3   Locating the Hand across Changes in Body Posture

As we have related, information about the body arrives across numerous different neural codes and reference frames. For example, touch is coded with respect to the body, whereas visual stimuli are coded with respect to the retina. Crucially, it is not sufficient to learn one-to-one mappings between coordinates in these reference frames, because the relationships between sensory modalities frequently change, for example, when the arms or eyes move. To solve this kind of computational problem, adults' brains dynamically remap the ways in which sensory stimuli concerning the hand and body relate to the external environment in response to different circumstances, for example, across changes in posture or when using tools (see Holmes, this vol.; see also Azañón and Soto-Faraco 2008; Overvliet et al. 2011; Graziano et al. 2004). This dynamic remapping of the relationships between multisensory spatial cues occurs automatically in human adults (e.g., Kennett et al. 2002), but we cannot necessarily infer that the same is true for earlier stages in development.

Spatial orienting to tactile stimuli in early infancy is observable in a number of newborn reflexes, such as the grasp reflex. In a study of infants' orienting to tactile stimuli on the hands, which we next describe, Bremner et al. (2008b) addressed whether infants are able to update their orienting responses across changes in the posture of the hands. The challenge of forming the kinds of dynamic multisensory representations of the hand in the environment posed by changes body layout and posture is likely particularly difficult when considered across development. As well as changes in size and distribution of the body due to physical development, the number and variety of postural changes that a child can readily make increase substantially, particularly in the first years of life (Bayley 1969). An ability to put a hand on the other side of the body midline develops in the first year of life and continues to emerge across early childhood (Carlier et al. 2006; van Hof et al. 2002).

Bremner et al. (2008b) measured infants' spontaneous manual orienting responses to vibrotactile sensations presented to the infants' hands when placed both in the uncrossed-, and crossed-hands postures (see fig. 2.1). First, they found that visual orienting responses to tactile locations increase significantly between six and a half and ten months of age, suggesting that an ability to represent spatial colocation between touch on the hand and vision develops substantially in the first year. Manual orienting responses to touches were present in both age-groups. However, six-and-a-half-month-old infants demonstrated a bias to respond manually to the side of the body where the hand would typically rest (i.e., to the left of space for a vibration in the left hand), regardless of the posture (uncrossed or crossed) of the hands. This indicates a reliance on the typical location of the tactile stimulus with respect to the body. Later, at ten months, a greater proportion of manual responses were made appropriately in both postures, suggesting the development of an ability to take account of posture in remapping correspondences between visual and tactile stimulation.

One of us has argued elsewhere (Bremner et al. 2008a) that these developmental findings converge with the results of neuroscientific and behavioral research in adults in suggesting that representations of the hand and near space arise from two distinct mechanisms of sensory integration, which follow separate developmental trajectories. The first mechanism we propose, *canonical multisensory body representation*, integrates bodily and visual sensory information but relies substantially on the probable location of the hand, derived primarily from prior multisensory experience about the canonical layout of the body. We argue that this mechanism is present early in the first six months of life, explaining young infants' reliance on the typical location of touch with respect to the body. The second mechanism proposed by Bremner et al. (2008a), *postural remapping*, updates these multisensory spatial correspondences by dynamically incorporating information about the current posture of the hand and body. We argue that this mechanism develops after six and a half months of age, as a result of specific sensorimotor experiences (such as crossing the arms over the midline; see van Hof et al. 2000). We argue that this explains ten-month-olds' ability to orient correctly to touches across familiar and novel (crossed-hands) postures. We are not suggesting that the early mechanism of *canonical body representations* is wholly replaced by that of *postural remapping* but rather that they continue to work together (see fig. 2.1b).

This dual-mechanism framework proposes a way of understanding the development of spatial representations of the hands during the first year of life. However, the functional purpose of such hand representations is to provide a basis for manual action. While orienting responses can themselves be considered as exploratory actions with respect to the hand and the nearby environment, it is important to investigate whether our framework is also applicable to overt goal-directed actions with our hands within peripersonal space.

3.4   Hand Representations Underlying Action in Early Infancy

Of the measurable behaviors in early infancy, perhaps the most relevant ways to observe the development of representations of the hand in the world are in reaches and grasps made toward nearby objects. While newborn infants do not often manually contact objects, their reaches are more often directed toward an object if they are looking at it (von Hofsten 1982). Newborns have also been shown to change the position of their hand and initiate decelerations of movement in order to bring it into sight under the illumination of a spotlight, which alternates between two locations near their body (van der Meer 1997). Thus, at birth, there is at least some spatial communication between incoming visual information and the movements of the hands. This may be suggestive of the coordination of visual, proprioceptive, and kinesthetic information (purely visual guidance cannot explain the anticipatory adjustments, since the hand is invisible when outside of the spotlight). However, it is difficult to determine whether this indicates early crossmodal spatial correspondence between proprioceptive and visual space or rather operant conditioning of particular arm movements, contingent upon the reward of seeing one's own hand.

In adults, removing vision of the hand during a reach significantly impacts on the movement patterns made (Berthier et al. 1996; Churchill et al. 2000). This suggests that, for adults, reaching is guided by visual feedback of the hand, such that it suffers when it is under proprioceptive control alone. In young infants however, a different picture emerges. The

(a) An infant receiving tactile stimulation to the right hand

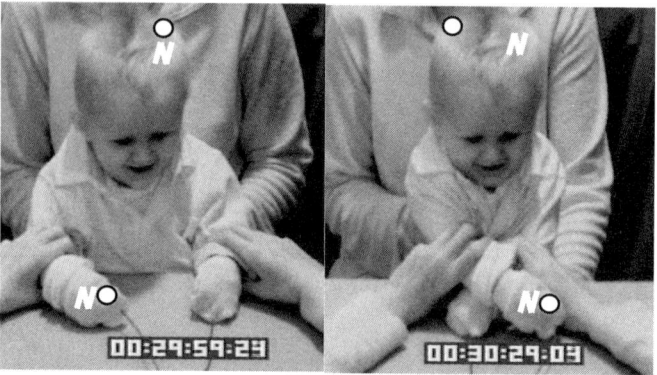

Uncrossed-hands   Crossed-hands

(b) Spatial information available for orienting

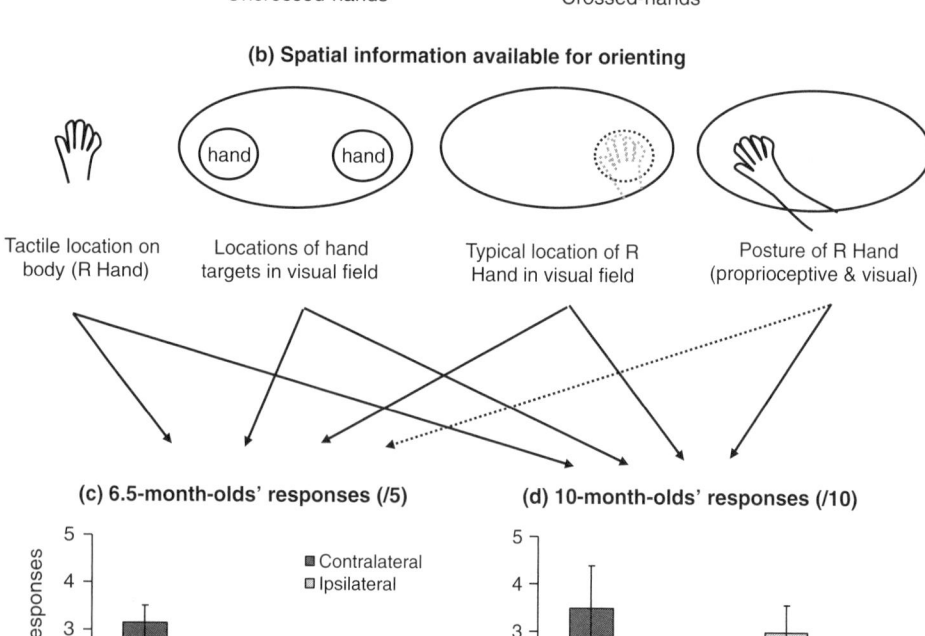

Tactile location on body (R Hand)

Locations of hand targets in visual field

Typical location of R Hand in visual field

Posture of R Hand (proprioceptive & visual)

(c) 6.5-month-olds' responses (/5)

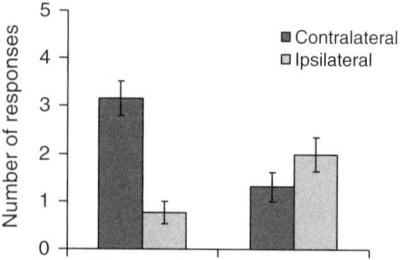

■ Contralateral
☐ Ipsilateral

Number of responses

Uncrossed-hands   Crossed-hands

(d) 10-month-olds' responses (/10)

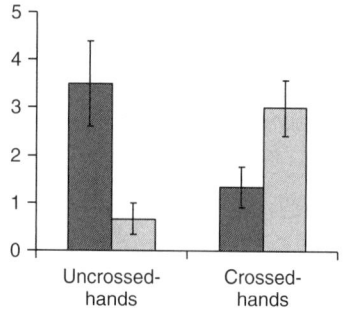

Uncrossed-hands   Crossed-hands

◀ Figure 2.1

Orienting to touch in familiar and unfamiliar postures. In the uncrossed-hands posture, both the visual information about the hand (circle) and a tactile stimulus on that hand (zig-zag pattern) arrive at the contralateral hemisphere (tactile stimuli were always delivered when the infants were fixating centrally). But with crossed hands, these signals initially arrive in opposite hemispheres (a). (b) shows the sources of information available to be integrated into a representation of stimulus location. We suggest that all sources of information are available to ten-month-olds, and all but current postural information is available to six-and-a-half-month-olds. (c) Six-and-a-half-month-old infants' manual responses to tactile stimuli. (d) Ten-month-old infants' manual responses to tactile stimuli. The infants' first responses on each trial were coded (from video-recordings) in terms of their direction in visual space with respect to the hemisphere receiving the tactile signal. Thus, contralateral responses are appropriate in the uncrossed-hands posture, and ipsilateral responses in the crossed-hands posture. The six-and-a-half-month-olds' manual responses (n=13; c) showed an overall contralateral bias, as predicted by a hypothesized reliance on the typical layout of their body relative to vision. The ten-month-olds (n=14; c) were able to respond manually in the appropriate direction in either posture, suggesting, in agreement with the proposed framework, that members of this age group are able to use information about current posture to remap their orienting responses (figure adapted from Bremner et al. 2008a). Asterisks represent significant comparisons. Solid arrows represent a strong contribution of a particular source of information to behavior. Dotted arrows represent a weak contribution of that particular source of information. Error bars show standard error of the mean.

uses and coordination of proprioceptive and visual information in the guidance of reaching has been investigated by comparing infants' target-directed reaches made in the light against reaches made in the dark (e.g., Robin et al. 1996). In this latter case, reaches are made toward sounding or glowing targets, without visual cues to the location of their hand. These studies have shown that successful reaching in the dark develops at the same age as in the light (three to four months). It seems that, in contrast to mature adults, young infants are no more reliant on visual than on proprioceptive information when controlling reaches toward objects.

Infants' reaches also provide clues to the multisensory nature of the spatial representations that they make use of in planning actions. Given that infants' first successful reaches toward visual targets can occur without any visual input about limb position, these actions are clearly generated by peripersonal spatial representations that integrate visual cues to the target with bodily (potentially proprioceptive) cues to the hand. However, the question that has not been resolved concerns whether the cues to hand position come from current proprioceptive information, or alternatively from a (potentially multisensory) representation of hand position that is strongly weighted toward the canonical body-centered location that the hand would normally occupy. In this second case, successful reaching in the dark could be accomplished by the integration of prior multisensory information of hand position with current visual information about the target. The framework described here in section 3.4 (and in Bremner et al. 2008a) would predict that if starting limb posture were to be varied, young infants' early reaches would be error-prone because they are more weighted to the typical location of the hand.

From four months of age, an infant's reaching movements gradually become more tuned to specific features of the goal of the reach. Grasps that anticipate the orientation of an object begin to emerge at around five months (von Hofsten and Fazel-Zandy 1984). By eight months, reorienting of the hand in anticipation of the orientation of a visual target also occurs independently of vision of the hand (McCarty et al. 2001), indicating that postural guidance is achieved proprioceptively at this age. Grasps that anticipate the size of an object are first observed from nine months (von Hofsten and Rönnqvist 1988). These developing abilities at producing more fine-grained ("goal-directed") postural adjustments toward objects (especially those made without sight of the hand; see McCarty et al. 2001) imply the use of current postural information to maintain spatial align-ment between different sensory inputs arising from the body and periper-sonal space. It is thus unlikely that such reaches could be achieved on the basis of a canonical representation of the body alone. In agreement with the research on tactile orienting responses described in the previous section, it seems reasonable to suggest that infants employ *postural remap-ping* in the planning and guidance of their reaches in the second half of the first year.

By fifteen months, infants' reaches are sensitive to the presence of visual information concerning the hand and arm (Carrico and Berthier 2008). Thus, it appears that the beginnings of the mature adultlike visually directed reaching emerge in the second year of life (although see Babinsky et al. 2012).

3.5 Developmental Changes in Hand Representations beyond Infancy
It is likely that the neural mechanisms underlying representations of the hand (and also, hand-centered space; see Holmes, this vol.) continue to develop beyond infancy. Children's representations of the layout of their hands and arms would need to be tuned and retuned throughout develop-ment in order to cope with physical changes in the disposition, sizes, and

movements of the limbs that continue even beyond adolescence. Such physical changes have important implications for the ways in which the sensory inputs concerning the body are weighted and integrated in order to provide the most reliable information about the body and its relation to the environment.

Some classic and more recent studies have indicated that the multisensory processes involved in representations of the hands and arms may change substantially in early childhood. For instance, several researchers have asserted that, across early childhood to adolescence, children come to rely more on vision relative to the other senses (Renshaw 1930; Warren and Pick 1970). Certainly, visual guidance of reaching movements of the hand increases from infancy into childhood (e.g., Carrico and Berthier 2008).

Pagel et al. (2009) have recently suggested that, across early childhood, vision also comes to play a more important role in the spatial representation of locations on the body surface. Pagel et al. examined the spatial coordinate systems children use when attributing tactile stimuli to their hands. Using a tactile temporal order judgment task, they demonstrated a crossed-hands deficit (poorer tactile temporal judgments in the crossed-hands posture) that emerged from around five and a half years of age (the same deficit has previously been documented in adults; e.g., Yamamoto and Kitazawa 2001). Pagel et al. argue that this crossed-hands effect is due to the conflict between external and anatomical frames of reference for locating tactile stimuli that arises when the hands are crossed. Given that this effect is not observed in congenitally blind adults, Röder (Röder et al. 2004; Röder 2012) has argued that the development of the crossed-hands effect is due to the acquisition of a visual external spatial frame of reference for representing tactile stimuli on the body.

One view concerning the functional purpose of an increased importance of vision in reaching (as observed in the findings reported above, and by Renshaw 1930) was put forward by both Renshaw (1930) and Warren and Pick (1970), who argued that children come to be dominated by visual cues because they derive from the most spatially accurate sensory organ. A more recent refinement of this argument has been proposed in response to the "optimal integration" account of multisensory integration in adults (e.g., Ernst and Banks 2002). This account proposes that all relevant senses contribute to spatial representations, but that the contributions of the individual senses are weighted in proportion to their relative task-specific reliabilities, thus yielding statistically optimal responses. Gori et al. (2008) have proposed that multisensory development converges in middle to late childhood onto an optimal sensory weighting. The evidence for this claim rests on recent demonstrations in which children develop from a state of either relying entirely on one sensory modality (which sensory modality this is depends on the task; see Gori et al. 2008) or alternating between reliance on one modality or the other (Nardini et al. 2008), to a situation in which they weight the senses in a statistically optimal fashion.

This account can provide a framework for understanding developmental shifts in middle to late childhood toward an increased visual influence observed across a number of spatial orienting behaviors (Pagel et al. 2009; Renshaw 1930; Smothergill 1973; Warren and Pick 1970). In addition, however, the optimal integration account can also explain developmental decreases in visual influence as due to convergence on an optimal weighting from a prior state in which vision was weighted disproportionately to its reliability. Indeed, we have recently observed such decreases in visual capture of hand position in a situation where vision and proprioception conflict later in childhood (see Bremner et al. 2012).

Thus, developmental changes in the role of vision in limb representations in early childhood may represent a process of convergence on an optimal weighting of the senses. Gori et al. (2008) propose an interesting account of why the development of optimal multisensory integration takes so long; they suggest that the optimal weighting of the senses is delayed so as to provide a point of reference against which to calibrate the senses in multisensory tasks. Because the body and limbs continue to grow over the course of childhood and adolescence, this recalibration is constantly required, and so, according to Gori et al. (2008), optimal weighting does not take place until the body has stopped growing substantially (although see Bremner et al. 2012, for an alternative explanation).

4 Developments in the Representation of the Hand (or Foot) in the Self?

The development of multisensory representations of the body and its movements is, of course, intrinsically tied up with the emerging representation of the self, and many researchers have derived conclusions concerning the emerging self-concept from studies investigating infants' and children's use of multisensory information about the body (see e.g., Rochat 2010). The apparent early ability of infants to detect multisensory synchrony between visual stimuli concerning the limbs and motor efferent outputs and/or proprioceptive signals has led some researchers to propose that multisensory contingency detection is an innate predisposition in humans that allows them to develop a "primary representation of the bodily self" (Gergely and Watson 1999).

However, the precise relationship between multisensory body (or hand) perception and perceptions of the self has been a subject of recent scrutiny by researchers studying body representations in mature adults. Specifically, researchers have investigated, in adults, the perceptual basis of our experience of body ownership and its neural correlates. Body ownership is defined as the "feeling" of ownership of, for example, a hand, which can be considered separately from the perception of it (see Tsakiris 2010).

A particular controversy in this debate has been whether body or limb ownership is derived—in a "bottom-up" fashion—from current spatiotemporal correlations between synchronous multisensory stimulation (e.g., the synchronous proprioceptive, tactile, auditory, and visual sensations that might arise when a baby bangs a hand or toy on a table) or whether additional sources of information picked out by an internal mental body model are used to derive experience of body-ownership in a "top-down" fashion (see Tsakiris 2010). Proponents of this debate have investigated the role of bottom-up and top-down cues to body ownership via the "rubber hand illusion" (RHI; Botvinick and Cohen 1998; Gallagher, this vol.). (See also the "mirror-hand illusion," described by Holmes, this vol.) In the RHI, a participant views a fake hand while his or her real hand is out of sight. When the real and fake hands are stroked synchronously, participants experience the fake hand as belonging to themselves, and their judgments concerning the location of their real hand are biased toward the fake hand.

Initial characterizations of body ownership in the RHI (Armel and Ramachandran 2003; Botvinick and Cohen 1998) suggested that bottom-up multisensory cues are both necessary and sufficient to support feelings of ownership (see also Makin et al. 2008). However, more recently authors have emphasized the additional importance of top-down influences from the internal body model (e.g., Tsakiris 2010), pointing to evidence that the strength of the RHI is also influenced by the visual appearance of a fake hand; the extent to which the fake hand resembles a hand (Tsakiris and Haggard 2005), the similarity of the fake hand's texture to that of the real hand (Haans et al. 2008), and the congruency of its posture with that of the real hand (Costantini and Haggard 2007; Pavani et al. 2000) have all been shown to affect the illusion.

What might we draw from this recent literature concerning the development of experience of body ownership in early life? Perhaps the most obvious implication is that infants' early abilities to detect the multisensory correlations between proprioception, touch, and vision from early in life (see sec. 3.3) may not necessarily imply that they experience of limb of body ownership under ecological conditions. Thus, we might ask whether young infants also respond to the cues that are purported to be used by the internal body model. At least two studies provide information that is relevant to this question.

Morgan and Rochat (1997), using the V-P technique described in section 5.3 with three-month-olds, found that changing the visual appearance of the legs, by bulking out the infants' socks with stuffing, disturbed the infants' ability to distinguish displays in which visual and proprioceptive signals about the leg were spatially congruent. Morgan and Rochat's (1997) evidence might be used to argue that infants of three months identify their own bodies by referring to an internal body model that specifies that their legs have a specific visual appearance, in addition to bottom-up visual-proprioceptive synchrony cues. A more recent study of visual-tactile cross-modal matching in infants prompts a similar conclusion. Zmyj et al. (2011) showed that ten-month-old infants are sensitive to temporal synchrony between visual touches to fake legs presented at a distance and tactile sensations arising from the infants' real legs. When the visually presented dolls legs were replaced with wooden blocks (this has been shown to disturb the RHI in adults; see Tsakiris and Haggard 2005), ten-month-olds showed a significant decline in their preference for visual-tactile synchrony, leading to the conclusion that the visual appearance of the limbs is important in detecting visual-tactile synchrony with respect to body parts.

Thus, these studies provide an indication that, in addition to bottom-up multisensory cues to body ownership (e.g., visual-tactile and visual-proprioceptive synchrony), an internal body model specifying the body's visual appearance may play an important role in identifying the self even from the first year of life. However, some important caveats to this conclusion remain. First, across both of these studies, the absence of crossmodal matching in the conditions in which the legs were of nonstandard appearance may have been driven by extraneous factors. Morgan and Rochat (1997) report that the infants' leg movements in their visual-propriocep-tive matching study also decreased when the appearance of the legs was bulked out (perhaps due to the impedance of their movement), and so the lack of differentiation in infants' looking behavior may have been due to there being simply not enough bottom-up multisensory synchrony to support crossmodal matching. In Zmyj et al.'s (2011) study, it is possible that synchrony matching declined as a result of the relative novelty of seeing blocks of wood in an infant seat across both conditions. Indeed, overall looking increased when the doll's legs were replaced with blocks of wood.

Perhaps most problematically for these studies of the body model in young infants, and as discussed in section 3.3, is that the visually presented legs were well beyond the spatial extent of the infants' bodies; both studies presented the legs at a distance of one meter from the infants' heads. This is well outside the bounds at which adults experience ownership over illusory limbs. Although adults experience the RHI across some spatial separation between the fake and real hands, the illusion of ownership is almost entirely removed when the hand is at a distance of sixty centimeters (Lloyd 2007). Relative to the body size of an infant in the first year of life, one meter is a particularly long distance over which to expect infants to perceive ownership over an illusory limb.

Thus, although it seems likely that young infants are able to perceive the multisensory synchrony cues that specify the layout of one's own body, it would seem premature to conclude that they perceive ownership of their limbs via the top-down influence of an internal body representation. Further research is needed to determine how sensory information arising from the body is used in a bottom-up or top-down fashion across development. As it is particularly difficult to assess subjective experiences of ownership in preverbal infants, young children may provide a more fruitful avenue of research with regard to this question. Intriguingly, Cowie et al. (in press) have recently shown that, before ten years of age, young children are more susceptible to the RHI than adults, even when bottom-up visual-tactile (stroking) cues are out of synchrony (i.e., inconsistent with the illusion).

5 Summary and Epilogue: The Role of the Hand in Cognitive Development

In this chapter, we have described how an ability to detect the multisensory information which specifies the hand and guides its actions undergoes significant developments, not just prenatally and in infancy but across early childhood. Research investigating crossmodal matching in infants (sec. 3.3) has shown that an ability to detect synchronous multisensory information concerning the hand, its movements, and interactions with objects is in place early in infancy.

However, it is unclear from crossmodal matching studies whether infants perceive sensations arising from the hand and body with reference to the spatial framework of the body. Indeed, studies of reaching and manual orienting responses to touch indicate that there is an extended period over which infants learn about how exactly to align their extrapersonal (visual, auditory) and corporeal (tactile, proprioceptive) spatial environments to form spatially coherent multisensory representations of their hands and body. Even beyond infancy, children undergo significant changes in the ways they weight sensory information in order to locate and act with their hands. We argue that these developmental changes in hand representation in infancy and childhood are driven in large part by physical changes in the body and the specific sensorimotor behaviors that emerge over early life.

Toward the end of the chapter we addressed the question of what we can infer from research on developing hand and body representations about the emerging self-concept. Research has demonstrated quite conclusively that infants, like adults, are able to use bottom-up multisensory cues (such as proprioceptive-visual synchrony) to the body. However, in adults, a sense of body ownership is considered to arise from a combination of both bottom-up multisensory cues, and the visual appearance of the hand as related to an internal model of the body. Although some studies have addressed this question, the jury is still out on whether such cues are used by infants and children, and so it is still difficult to characterize the sense of bodily self that we have in early life.

Finally, we return to the question with which we started this chapter: Does the hand play a role in cognitive development? Piaget answered strongly in the affirmative. However, his story does not seem to fit; remember (sec. 2.1) that he argued that the infant, who is initially egocentric in her appreciation of the environment, gradually constructs an objective representation of objects and the world from her sensorimotor interactions. The picture offered in this chapter suggests the opposite. Infants begin with knowledge of objective properties of the environment (see sec. 2.2), but the more difficult and gradual achievement discussed here seems to be the ability to represent the world in relation to the hand and body or egocentric space.

So can we accept that the hand plays little role in cognitive development? We should be careful in drawing this conclusion. In this chapter, we have placed some strong qualifications on the picture of early cognitive competence offered by developmentalists in the last two decades (see Spelke and Kinzler 2007; also, sec. 2.2). Whereas looking duration studies have tended to indicate precocious cognitive abilities, measures that require infants or children to represent the environment in relation to themselves or their own actions have provided a much more delayed picture of early perceptual and cognitive competence (see Bremner et al. 2008b), one that fits much better with Piaget's sensorimotor account.

A reasonable solution to this dilemma is to distinguish between the kinds of recognition knowledge expressed in looking time tasks and the practical knowledge required to take action on that knowledge (see Russell 1995). While numerous studies have shown that the former competencies seem to emerge independently of sensorimotor abilities, embodied knowledge of the world develops in a more protracted way, one that suggests that it is coupled with physical changes in the body and sensorimotor developments. Future research investigating the causal developmental relationship between sensorimotor skills and specifically *embodied* knowledge of the world will be crucial in confirming or denying the role of the hand in cognitive development.

Acknowledgments

A. J. Bremner is supported by an award from the ERC (241242). The authors would like to acknowledge helpful discussions with their colleagues, particularly J. Begum, F. Le Cornu Knight, N. P. Holmes, S. Rigato, and C. Spence.

Notes

1. There is reason to believe that Piaget considered visually guided reaching to be merely one example of how the reciprocal assimilation could be organized, and that he viewed evidence concerning infants without reaching or vision to be irrelevant to this matter (Gouin Décarie and Ricard 1996).

2. The lower limbs first appear as buds around one week after the upper limbs. Differentiation of fingers and the emergence of movement are similarly slower for the lower limbs.

3. Indeed, this problem is also present in a more recent study of infants' detection of visual-tactile synchrony of stimulation to the legs (Zmyj et al. 2011).

References

Armel, K. C., and V. S. Ramachandran. 2003. Projecting sensations to external objects: Evidence from skin conductance response. *Proceedings of the Royal Society of London, Series B: Biological Sciences* 270:1499–1506.

Azañón, E., and S. Soto-Faraco. 2008. Changing reference frames during the encoding of tactile events. *Current Biology* 18:1044–1049.

Babinsky, E., O. Braddick, and J. Atkinson. 2012. Infants and adults reaching in the dark. *Experimental Brain Research* 217:237–249.

Bahrick, L. E., and J. S. Watson. 1985. Detection of intermodal proprioceptive-visual contingency as a potential basis of self-perception in infancy. *Developmental Psychology* 21:963–973.

Baillargeon, R. 2004. Infants' reasoning about hidden objects: Evidence for event-general and event-specific expectations. *Developmental Science* 7:391–414.

Bayley, N. 1969. *Manual for the Bayley Scales of Infant Development*. New York: Psychological Corp.

Bermudez, J. L., A. J. Marcel, and N. Eilan. 1995. *The Body and the Self*. Cambridge, MA: MIT Press.

Berthier, N. E., R. K. Clifton, V. Gullapalli, D. D. McCall, and D. J. Robin. 1996. Visual information and object size in the control of reaching. *Journal of Motor Behavior* 28:187–197.

Birch, H. G., and A. Lefford. 1963. Intersensory development in children. *Monographs of the Society for Research in Child Development* 28 (5):1–48.

Botvinick, M., and J. Cohen. 1998. Rubber hands "feel" touch that eyes see. *Nature* 391:756.

Bremner, A. J., P. E. Bryant, D. Mareschal, and Á. Volein. 2007. Recognition of complex object-centred spatial configurations in early infancy. *Visual Cognition* 15:1–31.

Bremner, A. J., N. P. Holmes, and C. Spence. 2008a. Infants lost in (peripersonal) space? *Trends in Cognitive Sciences* 12:298–305.

Bremner, A. J., N. P. Holmes, and C. Spence. 2012. The development of multisensory representations of the body and of the space around the body. In *Multisensory Development*, 113–136, ed. A. J. Bremner, D. J. Lewkowicz, and C. Spence. Oxford: Oxford University Press.

Bremner, A. J., D. Mareschal, S. Lloyd-Fox, and C. Spence. 2008b. Spatial localization of touch in the first year of life: Early influence of a visual code, and the development of remapping across changes in limb position. *Journal of Experimental Psychology: General* 137:149–162.

Butterworth, G., and B. Hopkins. 1988. Hand-mouth coordination in the new-born baby. *British Journal of Developmental Psychology* 6:303–314.

Campbell, J. 1993. Objects and objectivity. *Proceedings of the British Academy* 83: 3–20.

Carlier, M., A.-L. Doyen, and C. Lamard. 2006. Midline crossing: Developmental trend from 3 to 10 years of age in a preferential card-reaching task. *Brain and Cognition* 61:255–261.

Carrico, R. L., and N. E. Berthier. 2008. Vision and precision reaching in 15 month old infants. *Infant Behavior and Development* 31:62–70.

Castiello, U., C. Becchio, S. Zoia, C. Nelini, L. Sartori, L. Blason, G. D'Ottavio, M. Bulgheroni, and V. Gallese. 2010. Wired to be social: The ontogeny of human interaction. *PLoS ONE* 5:e13199. doi:10.1371/journal.pone.0013199.

Churchill, A., B. Hopkins, L. Rönnqvist, and S. Vogt. 2000. Vision of the hand and environmental context in human prehension. *Experimental Brain Research* 134:81–89.

Clifton, R. K., D. W. Muir, D. H. Ashmead, and M. G. Clarkson. 1993. Is visually guided reaching in early infancy a myth? *Child Development* 64:1099–1110.

Costantini, M., and P. Haggard. 2007. The rubber hand illusion: Sensitivity and reference frame for body ownership. *Consciousness and Cognition* 16:229–240.

Cowie, D., T. Makin, and A. J. Bremner. In press. Children's responses to the Rubber Hand Illusion reveal dissociable pathways in body representation. *Psychological Science.*

Diamond, A. 1991. Developmental time course in human infants and infant monkeys, and the neural bases of inhibitory control in reaching. *Annals of the New York Academy of Sciences: pt. VII. Inhibition and Executive Control* 608:637–704.

Ernst, M. O., and M. S. Banks. 2002. Humans integrate visual and haptic information in a statistically optimal fashion. *Nature* 415:429–433.

Gallagher, S. 2005. *How the Body Shapes the Mind*. Oxford: Oxford University Press.

Gallagher, S., and A. N. Meltzoff. 1996. The earliest sense of self and others: Merleau-Ponty and recent developmental studies. *Philosophical Psychology* 9:211–233.

Gergely, G., and J. S. Watson. 1999. Early socio-emotional development: contingency perception and the social-biofeedback model. In *Early Social Cognition: Understanding Others in the First Months of Life*, 101–136, ed. P. Rochat. Mahwah, NJ: Erlbaum.

Gori, M., M. M. Del Viva, G. Sandini, and D. C. Burr. 2008. Young children do not integrate visual and haptic form information. *Current Biology* 18:694–698.

Gouin Décarie, T., and M. Ricard. 1996. Revisiting Piaget revisited, or The vulnerability of Piaget's infancy theory in the 1990s. In *Development and Vulnerability in Close Relationships*, 113–134, ed. G. G. Noam and K. W. Fischer. Mahwah, NJ: Erlbaum.

Graziano, M. S. A., C. G. Gross, C. S. R. Taylor, and T. Moore. 2004. A system of multimodal areas in the primate brain. In *Crossmodal Space and Crossmodal Attention*, 51–67, ed. C. Spence and J. Driver. Oxford: Oxford University Press.

Haans, A., W. A. Ijsselsteijn, and Y. A. de Kort. 2008. The effect of similarities in skin texture and hand shape on perceived ownership of a fake limb. *Body Image* 5:389–394.

Hall, G. S. 1898. Some aspects of the early sense of self. *American Journal of Psychology* 9:351–395.

Hood, B., S. Carey, and S. Prasada. 2000. Predicting the outcomes of physical events: Two-year-olds fail to reveal knowledge of solidity and support. *Child Development* 71:1540–1554.

Hulme, C., A. Smart, G. Moran, and A. Raine. 1983. Visual kinaesthetic and cross-modal development: Relationships to motor skill development. *Perception* 12:477–483.

Kaufman, J., and A. Needham. 2011. Spatial expectations of young infants, following passive movement. *Developmental Psychobiology* 53:23–36.

Kennett, S., C. Spence, and J. Driver. 2002. Visuo-tactile links in covert exogenous spatial attention remap across changes in unseen hand posture. *Perception & Psychophysics* 64:1083–1094.

Locke, J. 1690. *An Essay Concerning Human Understanding*. Oxford: Oxford University Press.

Lloyd, D. M. 2007. Spatial limits on referred touch to and alien limb may reflect boundaries of visuo-tactile peripersonal space surrounding the hand. *Brain and Cognition* 64:104–109.

Makin, T. R., N. P. Holmes, and H. H. Ehrsson. 2008. On the other hand: Dummy hands and peripersonal space. *Behavioural Brain Research* 191:1–10.

Mareschal, D., and M. H. Johnson. 2003. The "what" and "where" of infant object representations. *Cognition* 88:259–276.

McCarty, M. E., R. K. Clifton, D. H. Ashmead, P. Lee, and N. Goubet. 2001. How infants use vision for grasping objects. *Child Development* 72:973–987.

Meltzoff, A. N., and R. W. Borton. 1979. Intermodal matching by human neonates. *Nature* 282:403–404.

Melzack, R., R. Israel, R. Lacroix, and G. Schultz. 1997. Phantom limbs in people with congenital limb deficiency or amputation in early childhood. *Brain* 120:1603–1620.

Merleau-Ponty, M. 1962. *Phenomenology of Perception*. London: Routledge & Kegan Paul. (Translation of *Phénoménologie de la Perception* [Paris: Gallimard, 1945].)

Morgan, R., and P. Rochat. 1997. Intermodal calibration of the body in early infancy. *Ecological Psychology* 9:1–23.

Munakata, Y., J. L. McClelland, H. H. Johnson, and R. S. Siegler. 1997. Rethinking infant knowledge: Toward an adaptive process account of successes and failures in object permanence tasks. *Psychological Review* 104:686–713.

Nardini, M., P. Jones, R. Bedford, and O. Braddick. 2008. Development of cue integration in human navigation. *Current Biology* 18:689–693.

Overvliet, K. E., E. Azañón, and S. Soto-Faraco. 2011. Somatosensory saccades reveal the timing of tactile spatial remapping. *Neuropsychologia* 49:3046–3052.

Pagel, B., T. Heed, and B. Röder. 2009. Change of reference frame for tactile localization during child development. *Developmental Science* 12:929–937.

Parise, C. V., C. Spence, and M. Ernst. 2012. Multisensory integration: When correlation implies causation. *Current Biology*. doi:10.1016/j.cub.2011.11.039.

Pavani, F., C. Spence, and J. Driver. 2000. Visual capture of touch: Out-of-the-body experiences with rubber gloves. *Psychological Science* 11:353–359.

Piaget, J. 1952. *The Origins of Intelligence in the Child*. London: Routledge & Kegan Paul.

Piaget, J. 1954. *The Construction of Reality in the Child*. London: Routledge & Kegan Paul.

Price, E.H. 2006. A critical review of congenital phantom limb cases and a developmental theory for the basis of body image. *Consciousness & Cognition* 15:310–322.

Renshaw, S. 1930. The errors of cutaneous localization and the effect of practice on the localizing movement in children and adults. *Journal of Genetic Psychology* 28:223–238.

Robin, D. J., N. E. Berthier, and R. K. Clifton. 1996. Infants' predictive reaching for moving objects in the dark. *Developmental Psychology* 32:824–835.

Rochat, P. 1998. Self-perception and action in infancy. *Experimental Brain Research* 123:102–109.

Rochat, P. 2010. The innate sense of the body develops to become a public affair by 2–3 years. *Neuropsychologia* 48:738–745.

Rochat, P., E. M. Blass, and L. B. Hoffmeyer. 1988. Oropharyngeal control of hand-mouth coordination in newborn infants. *Developmental Psychology* 24:459–463.

Rochat, P., and S. J. Hespos. 1997. Differential rooting response by neonates: Evidence for an early sense of self. *Early Development & Parenting* 64:153–188.

Röder, B. 2012. Sensory deprivation and the development of multisensory integration. In *Multisensory Development*, 301–322, ed. A. J. Bremner, D. J. Lewkowicz, and C. Spence. Oxford: Oxford University Press.

Röder, B., F. Rösler, and C. Spence. 2004. Early vision impairs tactile perception in the blind. *Current Biology* 14:121–124.

Rose, S. A. 1994. From hand to eye: Findings and issues in infant cross-modal transfer. In *The Development of Intersensory Perception: Comparative Perspectives*, 265–284, ed. D. J. Lewkowicz and R. Lickliter. Hillsdale, NJ: Erlbaum.

Rose, S. A., A. W. Gottfried, and W. H. Bridger. 1981. Cross-modal transfer and information processing by the sense of touch in infancy. *Developmental Psychology* 17:90–98.

Russell, J. 1995. At two with nature: Agency and the development of self-world dualism. In *The Body and the Self*, 127–151, ed. J. L. Bermúdez, A. Marcel, and N. Eilan. Cambridge, MA: MIT Press.

Sann, C., and A. Streri. 2007. Perception of object shape and texture in human newborns: Evidence from cross-modal transfer tasks. *Developmental Science* 10: 398–409.

Schmuckler, M. 1996. Visual-proprioceptive intermodal perception in infancy. *Infant Behavior and Development* 19:221–232.

Schmuckler, M. A., and D. T. Jewell. 2007. Infants' visual-proprioceptive intermodal perception with imperfect contingency information. *Developmental Psychobiology* 48:387–398.

Simmel, M. L. 1966. Developmental aspects of the body scheme. *Child Development* 37:83–95.

Slater, A. M. 1995. Visual perception and memory at birth. In *Advances in Infancy Research*, vol. 9, 107–162, ed. C. Rovee-Collier and L. P. Lipsitt. Norwood, NJ: Ablex.

Smothergill, D. W. 1973. Accuracy and variability in the localization of spatial targets at three age levels. *Developmental Psychology* 8:62–66.

Spelke, E. S., K. Breinlinger, J. Macomber, and K. Jacobson. 1992. Origins of knowledge. *Psychological Review* 99:605–632.

Spelke, E. S., and K. D. Kinzler. 2007. Core knowledge. *Developmental Science* 10:89–96.

Streri, A. 2012. Crossmodal interactions in the human newborn: New answers to Molyneux's question. In *Multisensory Development*, 88–112, ed. A. J. Bremner, D. J. Lewkowicz, and C. Spence. Oxford: Oxford University Press.

Tsakiris, M. 2010. My body in the brain: A neurocognitive model of body-ownership. *Neuropsychologia* 48:703–712.

Tsakiris, M., and P. Haggard. 2005. The rubber hand illusion revisited: Visuotactile integration and self-attribution. *Journal of Experimental Psychology: Human Perception and Performance* 31:80–91.

Van der Meer, A. L. 1997. Keeping the arm in the limelight: Advanced visual control of arm movements in neonates. *European Journal of Paediatric Neurology* 1:103–108.

van Hof, P., J. van der Kamp, and G. J. P. Savelsbergh. 2002. The relation of unimanual and bimanual reaching to crossing the midline. *Child Development* 73: 1353–1362.

Varela, F. J., E. Thompson, and E. Rosch. 1991. *The Embodied Mind: Cognitive Science and Human Experience*. Cambridge, MA: MIT Press.

Von Hofsten, C. 1982. Eye-hand coordination in the newborn. *Developmental Psychology* 18:450–461.

Von Hofsten, C., and S. Fazel-Zandy. 1984. Development of visually guided hand orientation in reaching. *Journal of Experimental Child Psychology* 38:208–219.

Von Hofsten, C., and L. Rönnqvist. 1988. Preparation for grasping an object: A developmental study. *Journal of Experimental Psychology: Human Perception and Performance* 14:610–621.

Warren, D. H., and H. L. Pick, Jr. 1970. Intermodality relations in blind and sighted people. *Perception & Psychophysics* 8:430–432.

Yamamoto, S., and S. Kitazawa. 2001. Reversal of subjective temporal order due to arm crossing. *Nature Neuroscience* 4:759–765.

Zmyj, N., J. Jank, S. Schütz-Bosbach, and M. M. Daum. 2011. Detection of visual-tactile contingency in the first year after birth. *Cognition* 120:82–89.

3 Hand-Centered Space, Hand-Centered Attention, and the Control of Movement

Nicholas P. Holmes

1 Representing Visual Space

Two hundred and fifty million photoreceptors gather the light that enters your eye and transduce it into eight million action potentials every second (Perge et al. 2009), which are fed continuously to the thalamus and *superior colliculus*. In order to be useful to us, those action potentials must specify the size, shape, location, and distance of objects with respect to the parts of our body that move relative to them—most importantly the eyes, head, and hands. In this chapter, I will describe one way in which the brain might convert visual signals from the eye into muscle activity that guides our hand toward or away from visible objects. I will argue that the brain represents target objects and the hand in the same "map" of space, and that the hand is at the center of that map. Putting the hand at the center of a representation of visual space may allow rapid hand–object interactions to occur at a moment's notice, as well as provide a hand-centered focus of attention that guides multiple aspects of perception and action.

1.1 The Eye and the Hand

The retina contains the first eye-centered representation of the visible world, a blurred and filtered image focused by the lens onto a population of photoreceptor cells. Through about a 100:1 convergence, this image is further processed and projected onward to the thalamus and the visual cortex, and to other optical centers such as the superior colliculus. Many subsequent areas of the brain represent the visual world in an eye-centered fashion: when the eyes move, all such eye-centered representations are affected; after each new movement, each eye-centered neuron receives action potentials that carry information relating to a new portion of the world. How the brain maintains our sense of the visual stability of the world, despite large and frequent eye movements, has vexed scientists for decades, and it remains a key topic in neuroscience (Melcher 2011).

The general question of visual stability cannot be answered here, and I will not address it. Rather, I will argue more simply that there is, in some sense, a relatively stable hand-centered map of visual space that is created and used by the brain when needed in the service of object-directed hand actions. To create such a map, our visual system must have access to an estimate of where our hands are in space. This estimate could be derived in a number of ways. Hand position could be determined purely visually, by recognizing the hands from visual information alone, and by coding their locations in an eye-centered representation of space. Alternatively, estimates of hand position may be derived purely from somatosensory information, arising from the skin, muscles, tendons, and joints. For such an estimate to be used, for example, in commanding a hand movement toward a target object, it must then be combined with visual information about the target, and with information about the position of the eyes relative to the hands. This latter scheme is perhaps more complex than deriving an estimate of hand position purely visually, and might involve several sequential transformations of information between different reference frames. This would be achieved first by combining the eye-centered visual information with head- or body-centered somatosensory information, and eventually computing information about movement in a hand-centered reference frame (e.g., Andersen and Buneo 2002).

The brain likely uses both purely eye-centered and purely hand-centered schemes, as well as representing information in intermediate hand-and-eye-centered coordinate systems. All possible hand-related sensory signals, arising from multiple sensory modalities, in multiple reference frames, carry information that may be useful in bringing the hands toward or away from a target. These rich sources of information may be drawn upon directly or indirectly, with or without the intervening stages of sensorimotor transformation necessary to align signals in the same reference frame. In the remainder of this chapter, I will summarize the evidence for hand-centered representations of space in the primate brain, describe a number of experimental approaches that seem to tap into these representations of space in the human brain, and finally suggest ways in which these representations may be used in the control of hand movement.

2 Visual Receptive Fields Centered on the Hand

Hyvärinen and Poranen (1974) presented perhaps the first evidence for a neural representation of hand-centered space in the primate brain. They found that a number of neurons in macaque parietal area 7 responded to both visual and somatosensory inputs, and that these neurons responded during goal-directed movements. In particular, some of the cells responded selectively to near or approaching three-dimensional visual stimuli, while their somatosensory responses were restricted to stimulation of the hand, arms, shoulders, face, and head. Follow-up studies over the next twenty-five years (e.g., Graziano et al. 1997; Rizzolatti et al. 1981) investigated cells with visuospatial properties like these in greater detail, and discovered several such neuronal populations in portions of the ventral and dorsal premotor cortex, parietal area 7b, and subcortically in the putamen. Neurons that may be said to represent hand-centered space have the following properties: (1) visual selectivity for three-dimensional stimuli (objects) moving near to or approaching the hand; (2) visual responses organized in a hand-centered reference frame that moves with the hand as it moves, being relatively insensitive to eye or head position; (3) somatic selectivity to stimuli applied to the hand; and (4) motor activity related to object-directed or object-avoidance hand movements (for a more detailed treatment of the anatomical and physiological properties of these cells, see Brozzoli et al. 2011).

With Tamar Makin and a number of others, we recently described a series of experiments attempting to replicate, as closely as possible but in humans, the experimental situations described above for monkeys (Makin et al. 2009). Following on from Makin's earlier work (Makin et al. 2007), we devised an experiment in which five-centimeter diameter balls fell rapidly toward a human participant's right hand on either the left or right side of the body midline. During the ball's descent, we used transcranial magnetic stimulation over the left motor cortex to evoke motor potentials in the muscles of the right hand as an index of corticospinal excitability. By systematically manipulating the stimulus position, the hand position, and the eye position, and by cueing the participant's visual attention to one side or the other, we repeatedly demonstrated what we argued was a hand-centered representation of the stimulus position at a very early stage in processing—just 70 to 80 milliseconds after the onset of the visual stimulus. While we were unable to pin down the cortical or subcortical locus of this rapid visuomotor effect, we believe it is likely related to the hand-centered processing and selection of visual information for action, specifically for the control of hand and finger movements. Returning to what is known of the macaque monkey, because neurons with hand-centered visual receptive fields very often respond to somatosensory stimuli applied to the hand, as well as during hand movements, it is reasonable to suppose that such cells may represent the space around the hand in general, regardless of sensory modality. Although this notion is appealing, as yet we have very little evidence, for example, of auditory receptive fields centered on the hand. There is some evidence from single-unit studies in the macaque, of auditory receptive fields centered on the head and selective for stimulus distance (e.g., Graziano et al. 1999), but as yet there are no similar reports related to the hand, and the neural coding of auditory distance is complicated by a number of factors unrelated to distance (Moore and King 1999).

In the following, I describe a number of approaches to studying hand-centered representations of visual space in the human brain. These approaches can be considered under the umbrella term of "hand-centered attention," emphasizing the possible role of such representations in the selection, filtering, and processing of visual and motor information for the control of hand action.

3   Hand-Centered Attention

Because direct evidence for neuronal representations of hand-centered space is available only from invasive neurophysiology in the macaque brain, behavioral and neuroscientific studies of hand-centered space in humans require a different set of approaches. In general, to support the existence of hand-centered mechanisms, such evidence needs to show that the position of the hand, rather than of the eye, or head, or body, is the key modulating factor in a particular sensorimotor context. Here, I outline a number of experimental situations in which the visibility and position of a hand—not necessarily one's own hand—seem to be the critical factors in determining a participant's ability to detect, discriminate, or pay attention to visual or somatosensory stimuli, or to make targeted hand movements.

3.1   Hand Visibility

Perhaps the simplest evidence for hand-centered or hand-related mechanisms comes from studies showing that sensory or motor behaviors differ depending on whether the hand is visible or not. Our hands are probably the single most commonly viewed things throughout our lives, save perhaps for the sides of our nose (Mach [1886] 1959). It is difficult, for example, to position the hands in such a way that the arms can be seen but not the hands; when sitting at a table or desk, our legs may be hidden, but our hands are usually in sight. Given the near omnipresence of the hands, what happens when they are no longer visible?

When we lose sight of our hands, our estimate of where they are worsens, and more so if we also refrain from moving them. There is some debate over whether the perceived positions of our hands "drift" when we can no longer see them (e.g., Desmurget et al. 2000; Wann and Ibrahim 1992); however, it is clear that the accuracy of our sense of hand position diminishes substantially in the absence of visual information, and that providing such visual information, even if only a brief glimpse of the hand, can improve the accuracy and precision of subsequent reaching movements (e.g., Rossetti et al. 1994). Several recent studies have shown that simply placing one's hand next to visual stimuli (e.g., on a computer screen) affects the allocation of visual attention, both in healthy (Reed et al. 2006, 2010; Tseng and Bridgeman 2011) and in brain-damaged humans (di Pellegrino and Frassinetti, 2000; Schendel and Robertson 2004; though see Smith et al. 2008). Similar phenomena occur in monkeys (Thura et al. 2008), and such hand-related deficits or biases in attention may be absent in hand amputees (Makin et al. 2010). Suffice it to say here that the hand is an important and salient visual stimulus, and that both visual perception and attention are affected merely by the presence, position, and visibility of the hands.

3.2   The Mirror Hand Illusion

Just as seeing the position of our hands before a movement can improve the precision of that movement, so seeing the *incorrect* position of our hands can worsen movement precision. This has been demonstrated time and again in a number of ways, most often in studies of prism-adaptation, testing the effect of viewing the hand displaced by a small number of degrees to the left or right (e.g., Rossetti et al. 1995). More recently, this effect has been replicated, in a slightly cheaper way, by using a mirror (Holmes et al. 2004b, 2006; Holmes and Spence 2005). With a mirror oriented along the body's midline, one hand placed behind the mirror's reflective surface and the other in front, an experimental participant can see in the mirror a reflection of one hand that appears in a similar position to his or her other, unseen hand. This simple situation creates an illusion that one is looking directly at the hidden hand, an illusion that is strengthened by moving the two hands identically and in synchrony. By adjusting the distance between each hand and the mirror's surface, the experimenter can induce a multisensory conflict in the observer. For example, the hidden hand may actually be ten centimeters away from the mirror surface, but if the reflection of the other hand appears to be 16 centimeters from the mirror, such that a conflict between the veridical (proprioceptive) and apparent (visual) information is produced, where will the observer perceive her hand to be? Answer: approximately twelve centimeters from the mirror—about one-third of the way between the actual and the apparent hand positions. This effect arises after as little as six seconds of viewing the misplaced hand reflection and increases over time (Holmes and Spence 2005). Furthermore, this shift in perceived hand position has a strong biasing effect on subsequent hand movements: participants reach forward toward a target as if their initial hand position were one-third of the way between the real and the apparent hand positions.

One way to think of these data is that the proprioceptive representation of the hand "drifts" toward the visual representation over time. Another is to say that the precision of the proprioceptive sense degrades (but does not drift) over time, while the visual sense retains its reliability, and also does not drift. A weighted combination of a precise (but inaccurate) source of information with an increasingly imprecise (though accurate) source, will result in an apparent "drift" in hand position (Smeets et al. 2006; see also van Beers et al. 1999). For the experimental psychologist, a drift in proprioception or a change in the weighted combination of visual and somatosensory information will both produce differences in perceived hand position or reaching movement accuracy as time passes. The question for the neurophysiologist to answer, therefore, is: Is there a single representation of hand position in the brain that is constantly adjusted on the basis of multiple sources of information, or are there multiple representations of hand position that are drawn upon selectively in a task-dependent fashion? Given the messiness of neuronal representations and reference frames in the macaque brain, we are unlikely to reach an unambiguous answer to this question any time soon.

The mirror hand illusion described above has also been exploited in clinical settings in an attempt to ameliorate or even to "cure" hand perception problems, for example in phantom limb pain (Ramachandran et al. 2009). Although controversial (Moseley et al. 2008), the mirror illusion may provide a treatment, at least for some amputees. That a mirror-reflection of one hand can substitute surprisingly well for direct vision of the hand has led researchers to ask: Just how realistic must the viewed hand be in order to affect our sense of hand position? Could we be fooled into thinking *someone else's* hand was our own (Tastevin 1937), or someone else's foot (Horne 1931)? Yes, indeed we can.

3.3   The Rubber Hand Illusion or Fake Body Effect

If you wake up after lying awkwardly on your arm, you may feel that your arm and hand no longer belong to you, that they are limp, numb, and immovable (see, e.g., Gandevia and Phegan 1999; Sacks 1984). "Losing" a limb in this way is a relatively common yet thankfully brief and transitory experience, but gaining a limb from another person or from a prosthesis was, until relatively recently, a much rarer occurrence. The earliest version of this "fake body effect" that I am aware of comes from a Laurel and Hardy film, screened in 1931. In *Beau Hunks*, Stan and Ollie join the French Foreign Legion to draw their attention from the woman they both fell for. After a long march through the desert, they sit on a bed to take the weight off their feet. Laurel removes his boots, picks up a tired foot, rests it on his knee, and begins to massage away the day's pain. Closing his eyes and relishing the sensual relief of his foot massage, only after a minute or so does he realize that he has not his own, but Ollie's foot in his hands.

It is tempting to speculate that seeing Laurel and Hardy's intimate encounter in the French Foreign Legion encouraged the Frenchman J. Tastevin (1937) to experiment with the effects of such artificial displacement of body parts, thus bringing the gentle manipulation of fingers, feet, and other body parts into the psychological laboratory. But for a full-scale experimental assault on the fake body effect, psychologists would await Botvinick and Cohen (1998). This brief report single-handedly reanimated a field of research, allowing the many dozens of fake hands that had lain gathering dust for sixty years to be brought back into experimental service. Viewing a dummy hand while it is stroked in synchrony with strokes felt on one's own, but hidden, hand leads many participants to feel that the visible strokes on the dummy are directly causing the felt strokes on their real hand—that the dummy is, somehow, their own hand. Many dozens of papers have now been published on this "rubber hand illusion" or fake body effect (for reviews, see Aspell et al. 2011; Makin et al. 2008). What conclusions can be drawn from this literature about hand-centered space?

Viewing a dummy hand in a location near one's own hidden hand can bias reaching movements with that hand, almost as well as seeing a mirror reflection of one's own hand (Holmes et al. 2006; Zopf et al. 2010, 2011; see also Kammers et al. 2009). Such situations can affect tactile, proprioceptive, and visual perception of the hand and hand position (Pavani et al. 2000; Pavani and Zampini 2007; Tsakiris and Haggard 2005). If the neural mechanisms of hand-centered space rely at all on visual information to determine hand position, then it is reasonable to suppose that any visible handlike object may affect those mechanisms.

While collecting data for the series of "mirror-illusion" studies described above, I was surprised both at the effectiveness of the mirror reflection in biasing participants' reaching movements and at the sensitivity of the illusion to one particular experimental manipulation. Rotating the forearm and hand, so that one hand is palm-down and the other is palm-up, abolishes the mirror-illusion, and the effect of viewing a misaligned (i.e., rotated) *dummy* hand in the mirror is likewise diminished (Holmes et al. 2006). Despite this particular sensitivity, the rubber hand illusion is surprisingly robust to other experimental changes, for example in the size, shape, skin tone, hairiness, and even the species or plausibility of the visible hand (Austen et al. 2004; Pavani and Zampini 2007). This leads me to conclude that the rubber hand illusion may be most informative in that it provides a test of those features of visible hands that affect neural representations of hand-centered space. In particular, the position and orientation of a visible hand seem to be the most potent features, whereas the size, skin color, texture, and even the shape of the hand seem of less interest to the neural mechanisms of hand-centered space.

In the remainder of this section, I will examine the role of hand position and the effects of viewing artificial hands in the context of simultaneous visual and somatosensory stimulation, in two well-studied paradigms of crossmodal attention: crossmodal extinction and the crossmodal congruency effect.

3.4 Crossmodal Extinction

The finding that a visual stimulus presented near a brain-damaged participant's left hand could interfere with the detection of a simultaneously presented *tactile* stimulus on their right hand (i.e., crossmodal extinction, an attentional competition between modalities and/or sides of space; di Pellegrino et al. 1997; Mattingley et al. 1997) quickly led to studies on hand-centered space from the perspective of crossmodal extinction. Studies have since shown that, as the visual stimulus is moved farther away in space from the left hand, tactile detection at the right hand improves (Làdavas 2002). Whether the visual stimulus moves away from the left hand, or the left hand moves away from the visual stimulus, seems not to matter; rather, proximity to the left hand itself seems to be crucial. Further, proximity to just *any* left body part is not sufficient, since visual stimuli near the left side of the *face* are less effective at impairing tactile detection on the right *hand* than stimuli near the left hand (Farné et al. 2005). This suggests that the hands feature in a multisensory representation of space separately from the head and the face, and that this hand-centered representation can be damaged by lesions to the right hemisphere.

3.5 Crossmodal Congruency

Another experimental setting in which vision of the hand and hand position seem to be crucial determinants of behavior is the crossmodal congruency task (Spence et al. 2008). This task involves making speeded elevation discriminations ("upper" vs. "lower") of tactile targets presented to the index fingers and thumbs of the two hands, while trying to ignore simultaneously presented visual distractor stimuli positioned next to the four possible tactile target locations. When tactile targets and visual distractors are on the same digit (and thus require the same response), discrimination performance is fast and accurate. But when the visual distractor location is incongruent with respect to the tactile target (e.g., an "upper" visual distractor paired with a "lower" tactile target), participants are notably slower (by 50–100ms) and less accurate (with 5–25% more errors) at discriminating the location of the tactile targets. This difference in performance is the crossmodal congruency effect, a Stroop-like congruency effect in which the task-irrelevant distractor exerts a strong priming, perceptual, or biasing effect on responses to the task-relevant target.

What has made the crossmodal congruency task an attractive paradigm for research into hand-centered mechanisms of attention is that, as the visual distractor is moved further away from the tactile target, the congruency effect diminishes (Spence et al. 2001). A visual distractor on the left hand has much less of a distracting effect (about 30–50ms) on a tactile target presented on the right hand than a visual distractor near the right hand (about 60–100ms). This effect of distance also applies in the two other dimensions—move the visual distractor further from the hand in depth or vertically, but on the same side of space, and congruency effects decrease (e.g., Holmes et al. 2004a, 2007; Pavani et al. 2000).

Just like crossmodal extinction, the crossmodal congruency effect is also sensitive to experimental manipulations using mirrors (Maravita et al. 2002) and dummy hands (Pavani et al. 2000), as well as to their combination (Zopf et al. 2010, 2011). But questions remain about the crossmodal congruency effect. In particular, since crossmodal congruency effects are just as large for targets presented on the feet (Schicke et al. 2009), or even on the back (Aspell et al. 2009), as on the hands, one must ask whether this task is specific to *hand*-centered space. Most likely, crossmodal congruency effects (and similar considerations may apply to crossmodal extinction effects) arise from a number of sources (Shore et al. 2006), and determining exactly how these effects relate to neural representations of hand-centered space is likely to require neuroimaging evidence (e.g., Holmes et al. 2008), as well as more systematic study of their behavioral sources, as discussed in the next section.

3.6   Hand-Centered or Hand-Related Attention?

Whether the effects of hand visibility, hand position, the mirror illusion, the fake body effect, crossmodal extinction, and crossmodal congruency are primarily *eye-centered* or primarily *hand-centered* phenomena is not yet clear. The majority of studies discussed above have not manipulated hand position, eye position, or visual stimulus position sufficiently, for example, to tease apart the roles of different reference frames or to separate out any influence of (eye-centered) mechanisms from possible hand-centered mechanisms of attention. A conservative conclusion might therefore be to class all these phenomena as providing evidence of *hand-related* mechanisms of attention. While this should not be taken as trivializing the importance of such mechanisms, it does not help in explaining them either. It should by now be clear that the hand is perhaps the most salient visual stimulus that appears in our visual field, and it should no longer be surprising that the brain pays close attention to our hands, to anyone else's hands, and to any handlike object that should enter our visual field.

Neural representations of hand-centered space, as described above, have mostly been measured in human and nonhuman primates in relatively passive situations—presenting visual and somatosensory stimuli to the participant, and requiring him or her to keep visual fixation and, in humans, to make simple detection or discrimination responses. But if neural mechanisms of hand-centered space serve some behavioral purpose, it is perhaps unlikely to be simply for detecting, identifying, or discriminating sensory stimuli, and it is perhaps not even just for directing attention in space. Rather, hand-centered space may best serve the purpose of a sensorimotor interface for the control of hand movements. In the final section of this chapter, I will argue that the primary function of hand-centered and hand-related mechanisms of attention is the guidance of hand movements toward and away from desired or threatening objects. I will therefore review a number of experimental approaches that tap into hand-centered space closer to its roots—in action.

4   The Control of Hand Movement

Hand-centered representations in the macaque monkey have been studied by moving an object toward or away from the monkey's hand. Similarly, the multisensory attentional experiments described above all involve human participants keeping their hands still while static or moving stimuli are presented nearby. What about the converse of this—when the monkey or human moves its hand toward a target object? Do the same representations of space code the relative locations of hand and object when the former moves and the latter is static? Is it parsimonious to assume that they do—that hand-centered sensory representation of approaching or moving objects can be used very effectively to control hand movements toward those same objects? In this final section I describe a number of experimental situations of this sort.

4.1   Action-Centered Attention

On many occasions I have had to inhibit my tendency to grasp the fullest of beer glasses on the table, and reach instead for the one that I have paid for and partially drunk. Indeed, the fuller a glass is, the more we may have to inhibit our movements toward it (Nijnens 2010). Reaching for a glass of beer on a table with several other glasses, as Tipper and colleagues have pointed out, requires the selection of movement toward your target glass, as well as the inhibition of movements toward the other glasses (Tipper et al. 1992). These selective and inhibitory processes may operate in hand-centered reference frames. Similarly, if we have to hold the location of our beer glass in short-term memory, then that memory may be maintained in a hand-centered representation of space (Chieffi et al. 1999).

For the hand as for the eye, the trajectories of movements toward target objects may be influenced by the presence of distractors. Both the eye and the hand may deviate toward or away from distractors on their way toward their target (e.g., Chieffi et al. 2001; Doyle and Walker 2001). The distractor location is inhibited while the target location is selected. For the eye, it is easy to imagine how such phenomena are brought about in the brain. The superior colliculus contains an eye-centered motor map of saccade direction and amplitude. Stimulating a portion of this motor map evokes a saccade of amplitude and direction similar to that which is represented in the map during natural saccades (Stryker and Schiller 1975). Stimulating one portion of the map while inhibiting another portion may evoke saccades similar to those that occur when a target object must be looked at in the presence of distractor objects. The parallel effects of distractor objects on hand and eye movements suggest that a similar motor map may exist for the hand as for the eye, to guide movement toward its target. Indeed, the mammalian superior colliculus (SC) does show hand- and reaching-related activity (e.g., Werner 1993). The SC, however, is not best known for hand movement activity or for the spatial control of hand actions. A network of brain areas implicated in the visual and multisensory control of hand movements will be described below.

4.2   Hand-Object Interactions

Earlier, I described how magnetic stimulation was used to study the processing of visual information during simple hand movements (e.g., Makin et al. 2009). One can also use behavioral methods to examine the processing of visual and somatosensory stimuli during object-directed hand actions. Brozzoli and his colleagues (Brozzoli et al. 2009, 2010) used the crossmodal congruency task to measure the strength of visual-somatosensory interactions during a reach-to-grasp action. They found that the influence of a visual distractor on reaction times to discriminate a tactile target was increased shortly before and during a reaching movement as compared to longer before the movement started. Since the visual distractor stimuli were positioned on the target object itself, this result suggests that visual information is processed in an action-dependent fashion, and that, just before hand movements, visual stimuli on the target object are integrated with tactile stimuli felt in the hand. One interpretation is that the portion of distant space where the target lies is momentarily treated as nearby space; another is that the distant region of space is selected as a movement goal and that this attentional selection enhances the processing of visual information presented there. These two interpretations may in fact be identical—one describes the underlying mechanisms in terms of hand-centered space, the other in terms of hand-centered attention. Either way, this fascinating result deserves to be followed up.

4.3   Rapid Control of Hand Movements

When might we need to move our hands suddenly, or very rapidly? Either when unexpected things happen near us, when objects suddenly move or appear, or when objects are rapidly approaching us. It is probably an everyday occurrence that an object we are attempting to hold slips from our hand as we fumble in grasping it, or that something falls or topples off a table and we reach out, suddenly and almost instinctively, to catch it or else lessen the damage it may cause. I often find my hands making the most complex of intercepting, catching, and grasping movements with what seems like almost no deliberate or conscious control.

For example, I remember once accidentally knocking a half-full glass of water off a table with a heavy swipe from the back of my hand. Barely had it begun to topple as my hand jerked out, curved over the top of the glass, swooped around and under it, crafted a perfectly formed grip and caught the glass with at least 30 centimeters to spare before impact. Not a drop was spilled. This sort of rapid hand movement must require just as fine control as a carefully planned reach-to-grasp movement, yet the brain seems able to plan, initiate, and control such movements in the blink of an eye. How might it do that, and do hand-centered representations of space play a role?

When recording from neurons representing hand-centered space in the polysensory zone of the macaque ventral premotor cortex, Graziano and colleagues noticed that, if electrical stimulation was applied to this same area, the monkey's hand and arm would suddenly perform a complex movement and assume a fixed posture in space that was held until the stimulation was ended (Graziano et al. 2002). For many of these stimulations, the hand and arm assumed a "defensive" posture, as if to protect the face and body from some approaching object. Similar movements can be evoked either by stimulating the multisensory part of the ventral intraparietal area (Cooke et al. 2003), or by stimulating the monkey's body with an aversive air-puff (Cooke and Graziano 2003).

These results imply that hand-centered representations of space may be particularly useful for the rapid control of defensive hand movements. Extrapolating from and applying these results to humans, such representations of space may play a central role in the rapid, online control of hand movements both toward and away from objects (Makin et al. 2012). When reaching to grasp an object, if that object suddenly moves, the hand can react within just 100 milliseconds to follow the target object in space (e.g., Paulignan et al. 1991). To effect a visually driven change in hand movement in such a short time, the neural pathways involved must be relatively direct and simple. In the cat, about which most is known regarding the anatomical pathways mediating this rapid visual control of limb movements, the most rapid movements are likely mediated by a collicular, pontine, and cerebellar pathway, while slightly slower limb movement corrections may involve the cerebral cortex (Alstermark et al. 1984; Pettersson et al. 1997). It is possible that similar subcortical pathways are responsible for the very rapid visual control of hand movement in humans (Fautrelle et al. 2010).

It is also known from studies on optic ataxia, however, that the cerebral cortex is intimately involved in the online control of hand movements (Perenin and Vighetto 1988), and magnetic interference with neural processing in the parietal cortex may lead to disruptions of online movement control (Desmurget et al. 1999). I believe that the most interesting questions that remain in this domain relate to the involvement of the cerebral cortex in the control of rapid hand movements. Exactly when, where, and how do you need your cerebral cortex to control your hands? The answer to this question almost certainly involves multisensory hand-centered representations of space.

5 Conclusion

He pointed out . . . that the perfection of mechanical appliances must ultimately supersede limbs; the perfection of chemical devices, digestion; that such organs as hair, external nose, teeth, ears, and chin were no longer essential parts of the human being, and that the tendency of natural selection would lie in the direction of their steady diminution through the coming ages. The brain alone remained a cardinal necessity. Only one other part of the body had a strong case for survival, and that was the hand, "teacher and agent of the brain." While the rest of the body dwindled, the hands would grow larger.
—H. G. Wells, *The War of the Worlds*

In any future breed of *Homo sapiens*, I suspect that the hand and the brain will remain as important as H. G. Wells noted them to be. However, I would not hesitate to add the visual system as being indispensable to our future survival, so intimate is its involvement with the hand in the brain. The eye and the hand: teachers and agents of the brain, organs of the mind.

References

Alstermark, B., E. Eide, T. Górska, A. Lundberg, and L. G. Pettersson. 1984. Visually guided switching of forelimb target reaching in cats. *Acta Physiologica Scandinavica* 120 (1):151–153.

Andersen, R. A., and C. A. Buneo. 2002. Intentional maps in posterior parietal cortex. *Annual Review of Neuroscience* 25:189–220.

Aspell, J. E., B. Lenggenhager, and O. Blanke. 2009. Keeping in touch with one's self: Multisensory mechanisms of self-consciousness. *PLoS ONE* 4 (8):e6488.

Aspell, J. E., B. Lenggenhager, and O. Blanke. 2011. Multisensory perception and bodily self-consciousness: From out-of-body to inside-body experience. In *The Neural*

Bases of Multisensory Processes, 467–481, ed. M. M. Murray and M. T. Wallace. London: Taylor & Francis.

Austen, E. L., S. Soto-Faraco, J. T. Enns, and A. S. Kingstone. 2004. Mislocalizations of touch to a fake hand. *Cognitive, Affective & Behavioral Neuroscience* 4 (2):170–181.

Botvinick, M. M., and J. D. Cohen. 1998. Rubber hands "feel" touch that eyes see. *Nature* 391 (6669):756.

Brozzoli, C., F. Pavani, C. Urquizar, L. Cardinali, and A. Farné. 2009. Grasping actions remap peripersonal space. *Neuroreport* 20 (10):913–917.

Brozzoli, C., L. Cardinali, F. Pavani, and A. Farné. 2010. Action-specific remapping of peripersonal space. *Neuropsychologia* 48 (3):796–802.

Brozzoli, C., and T. R. Makin, L. Cardinali, N. P. Holmes, and A. Farné. 2011. Peripersonal space: A multisensory interface for body-object interactions. In *Frontiers in the Neural Basis of Multisensory Processes*, 449–466, ed. M. M. Murray and M. T. Wallace. London: Taylor & Francis.

Chieffi, S., D. A. Allport, and M. E. Woodin. 1999. Hand-centered coding of target location in visuo-spatial working memory. *Neuropsychologia* 37 (4):495–502.

Chieffi, S., M. Ricci, and S. Carlomagno. 2001. Influence of visual distractors on movement trajectory. *Cortex* 37 (3):389–405.

Cooke, D. F., and M. S. A. Graziano. 2003. Defensive movements evoked by air puff in monkeys. *Journal of Neurophysiology* 90 (5):3317–3329.

Cooke, D. F., C. S. R. Taylor, T. Moore, and M. S. A. Graziano. 2003. Complex movements evoked by microstimulation of the ventral intraparietal area. *Proceedings of the National Academy of Sciences of the United States of America* 100 (10):6163–6168.

Desmurget, M., C. M. Epstein, R. S. E. Turner, C. Prablanc, E. Alexander, and S. T. Grafton. 1999. Role of the posterior parietal cortex in updating reaching movements to a visual target. *Nature Neuroscience* 2 (6):563–567.

Desmurget, M., P. Vindras, H. Gréa, P. Viviani, and S. T. Grafton. 2000. Proprioception does not quickly drift during visual occlusion. *Experimental Brain Research* 134 (3):363–377.

di Pellegrino, G., and S. Frassinetti. 2000. Direct evidence from parietal extinction of enhancement of visual attention near a visible hand. *Current Biology* 10 (22):1475–1477.

di Pellegrino, G., E. Làdavas, and A. Farné. 1997. Seeing where your hands are. *Nature* 388 (6644):730.

Doyle, M., and R. Walker. 2001. Curved saccade trajectories: Voluntary and reflexive saccades curve away from irrelevant distractors. *Experimental Brain Research* 139 (3):333–344.

Farné, A., M. L. Demattè, and E. Làdavas. 2005. Neuropsychological evidence of modular organization of the near peripersonal space. *Neurology* 65 (11):1754–1758.

Fautrelle, L., C. Prablanc, B. Berret, Y. Ballay, and F. Bonnetblanc. 2010. Pointing to double-step visual stimuli from a standing position: Very short latency (express) corrections are observed in upper and lower limbs and may not require cortical involvement. *Neuroscience* 169 (2):697–705.

Gandevia, S. C., and C. M. L. Phegan. 1999. Perceptual distortions of the human body image produced by local anaesthesia, pain, and cutaneous stimulation. *Journal of Physiology* 514 (2):609–616.

Graziano, M. S. A., X. T. Hu, and C. G. Gross. 1997. Visuospatial properties of ventral premotor cortex. *Journal of Neurophysiology* 77 (5):2268–2292.

Graziano, M. S. A., L. A. Reiss, and C. G. Gross. 1999. A Neuronal representation of the location of nearby sounds. *Nature* 397 (6718):428–430.

Graziano, M. S. A., C. S. R. Taylor, and T. Moore. 2002. Complex movements evoked by microstimulation of precentral cortex. *Neuron* 34 (5):841–851.

Holmes, N. P., G. A. Calvert, and C. Spence. 2004a. Extending or projecting peripersonal space with tools? Multisensory interactions highlight only the distal and proximal ends of tools. *Neuroscience Letters* 372 (1–2):62–67.

Holmes, N. P., G. Crozier, and C. Spence. 2004b. When mirrors lie: "Visual capture" of arm position impairs reaching performance. *Cognitive, Affective & Behavioral Neuroscience* 4 (2):193–200.

Holmes, N. P., G. A. Calvert, and C. Spence. 2007. Tool use changes multisensory interactions in seconds: Evidence from the crossmodal congruency task. *Experimental Brain Research* 183 (4):465–476.

Holmes, N. P., H. J. Snijders, and C. Spence. 2006. Reaching with alien limbs: Visual exposure to prosthetic hands in a mirror biases proprioception without accompanying illusions of ownership. *Perception and Psychophysics* 68 (4):685–701.

Holmes, N. P., and C. Spence. 2005. Visual bias of unseen hand position with a mirror: Spatial and temporal factors. *Experimental Brain Research* 166 (3–4): 489–497.

Holmes, N. P., C. Spence, P. C. Hansen, C. E. Mackay, and G. A. Calvert. 2008. The multisensory attentional consequences of tool use: A functional magnetic resonance imaging study. *PLoS ONE* 3 (10):e3502.

Hyvärinen, J., and A. Poranen. 1974. Function of the parietal associative area 7 as revealed from cellular discharges in alert monkeys. *Brain* 97 (4):673–692.

Horne, J. W. (dir.). 1931. *Beau Hunks*. Written by H. M. Walker. Metro-Goldwyn-Meyer.

Kammers, M. P. M., F. de Vignemont, L. Verhagen, and H. C. Dijkerman. 2009. The rubber hand illusion in action. *Neuropsychologia* 47 (1):204–211.

Làdavas, E. 2002. Functional and dynamic properties of visual peripersonal space. *Trends in Cognitive Sciences* 6 (1):17–22.

Mach, E. [1886] 1959. *The Analysis of Sensations, and the Relation of the Physical to the Psychical*. New York: Dover.

Makin, T. R., N. P. Holmes, C. Brozzoli, and A. Farnè. 2012. Keeping the world at hand: Rapid visuomotor processing for hand-object interactions. *Experimental Brain Research* 219 (4):421–428.

Makin, T. R., N. P. Holmes, C. Brozzoli, Y. R. C. Rossetti, and A. Farné. 2009. Coding of visual space during motor preparation: Approaching objects rapidly modulate corticospinal excitability in hand-centered coordinates. *Journal of Neuroscience* 29 (38):11841–11851.

Makin, T. R., N. P. Holmes, and H. H. Ehrsson. 2008. On the other hand: Dummy hands and peripersonal space. *Behavioural Brain Research* 191 (1):1–10.

Makin, T. R., N. P. Holmes, and E. Zohary. 2007. Is that near my hand? Multisensory representation of peripersonal space in human intraparietal sulcus. *Journal of Neuroscience* 27 (4):731–740.

Makin, T. R., M. Wilf, I. Schwartz, and E. Zohary. 2010. Amputees "neglect" the space near their missing hand. *Psychological Science* 21 (1):55–57.

Maravita, A., C. Spence, C. Sergent, and J. Driver. 2002. Seeing your own touched hands in a mirror modulates cross-modal interactions. *Psychological Science* 13 (4):350–355.

Mattingley, J. B., J. Driver, N. Beschin, and I. H. Robertson. 1997. Attentional competition between modalities: Extinction between touch and vision after right hemisphere damage. *Neuropsychologia* 35 (6):867–880.

Melcher, D. 2011. Visual stability. *Philosophical Transactions of the Royal Society of London, Series B: Biological Sciences* 366 (1564):468–475.

Moore, D. R., and A. J. King. 1999. Auditory perception: The near and far of sound localization. *Current Biology* 9 (10):R361–R363.

Moseley, G. L., A. Gallace, and C. Spence. 2008. Is mirror therapy all it is cracked up to be? Current evidence and future directions. *Pain* 138 (1):7–10.

Nijnens, C. 2010. The influence of object identity on obstacle avoidance reaching behaviour. Paper presented at Body Representation Workshop, Goldsmiths University of London, March 29.

Pavani, F., C. Spence, and J. Driver. 2000. Visual capture of touch: Out-of-the-body experiences with rubber gloves. *Psychological Science* 11 (5):353–359.

Pavani, F., and M. Zampini. 2007. The role of hand size in the fake-hand illusion paradigm. *Perception* 36 (10):1547–1554.

Paulignan, Y., C. MacKenzie, R. G. Marteniuk, and M. Jeannerod. 1991. Selective perturbation of visual input during prehension movements. 1. The effects of changing object position. *Experimental Brain Research* 83 (3):502–512.

Perge, J. A., K. Koch, R. Miller, P. Sterling, and V. Balasubramanian. 2009. How the optic nerve allocates space, energy, capacity, and information. *Journal of Neuroscience* 29 (24):7917–7928.

Perenin, M., and A. Vighetto. 1988. Optic ataxia: A specific disruption in visuomotor mechanisms. I. Different aspects of the deficit in reaching for objects. *Brain* 111 (3):643–674.

Pettersson, L. G., A. Lundberg, B. Alstermark, T. Isa, and B. Tantisira. 1997. Effect of spinal cord lesions on forelimb target-reaching and on visually guided switching of target-reaching in the cat. *Neuroscience Research* 29 (3):241–256.

Ramachandran, V. S., D. Brang, and P. D. McGeoch. 2009. Size reduction using mirror visual feedback (MVF) reduces phantom pain. *Neurocase* 15 (5):357–360.

Reed, C. L., R. Betz, J. P. Garza, and R. J. J. Roberts. 2010. Grab it! Biased attention in functional hand and tool space. *Attention, Perception & Psychophysics* 72 (1):236–245.

Reed, C. L., J. D. Grubb, and C. Steele. 2006. Hands up: Attentional prioritization of space near the hand. *Journal of Experimental Psychology: Human Perception and Performance* 32 (1):166–177.

Rizzolatti, G., C. Scandolara, M. Matelli, and M. Gentilucci. 1981. Afferent properties of periarcuate neurons in macaque monkeys. II. Visual responses. *Behavioural Brain Research* 2 (2):147–163.

Rossetti, Y. R. C., M. Desmurget, and C. Prablanc. 1995. Vectorial coding of movement: Vision, proprioception, or both? *Journal of Neurophysiology* 74 (1):457–463.

Rossetti, Y. R. C., G. E. Stelmach, M. Desmurget, C. Prablanc, and M. Jeannerod. 1994. The effect of viewing the static hand prior to movement onset on pointing kinematics and variability. *Experimental Brain Research* 101 (2):323–330.

Sacks, O. 1984. *A Leg to Stand On*. New York: Summit Books.

Schendel, K. L., and L. C. Robertson. 2004. Reaching out to see: Arm position can attenuate human visual loss. *Journal of Cognitive Neuroscience* 16 (6):935–943.

Schicke, T., F. Bauer, and B. Röder. 2009. Interactions of different body parts in peripersonal space: How vision of the foot influences tactile perception at the hand. *Experimental Brain Research* 192 (4):703–715.

Shore, D. I., M. E. Barnes, and C. Spence. 2006. Temporal aspects of the visuotactile congruency effect. *Neuroscience Letters* 392 (1–2):96–100.

Smeets, J. B. J., J. J. van den Dobbelsteen, D. D. de Grave, R. J. van Beers, and E. Brenner. 2006. Sensory integration does not lead to sensory calibration. *Proceedings of the National Academy of Sciences of the United States of America* 103 (49):18781–18786.

Smith, D. T., A. R. Lane, and T. Schenk. 2008. Arm position does not attenuate visual loss in patients with homonymous field deficits. *Neuropsychologia* 46 (9):2320–2325.

Spence, C., A. F. Kingstone, D. I. Shore, and M. S. Gazzaniga. 2001. Representation of visuotactile space in the split brain. *Psychological Science* 12 (1):90–93.

Spence, C., F. Pavani, A. Maravita, and N. P. Holmes. 2008. Multisensory interactions. In *Haptic Rendering: Foundations, Algorithms, and Applications*, 21–52, ed. M. C. Lin and M. A. Otaduy. Wellesley, MA: AK Peters.

Stryker, M. P., and P. H. Schiller. 1975. Eye and head movements evoked by electrical stimulation of monkey superior colliculus. *Experimental Brain Research* 23 (1): 103–112.

Tastevin, J. 1937. En partant de l'expérience d'Aristotle: Les déplacements artificiels des parties du corps ne sont pas suivis par le sentiment de ces parties ni pas les sensations qu'on peut y produire. [Starting from Aristotle's experiment: The artificial displacements of parts of the body are not followed by feeling in these parts or by the sensations which can be produced there.] *L'Encéphale* 1:57–84.

Thura, D., D. Boussaoud, and M. Meunier. 2008. Hand position affects saccadic reaction times in monkeys and human. *Journal of Neurophysiology* 99 (5):2194–2202.

Tipper, S. P., C. Lortie, and G. C. Baylis. 1992. Selective reaching: Evidence for action-centered attention. *Journal of Experimental Psychology: Human Perception and Performance* 18 (4):891–905.

Tsakiris, M., and P. Haggard. 2005. The rubber hand illusion revisited: Visuotactile integration and self-attribution. *Journal of Experimental Psychology: Human Perception and Performance* 31 (1):80–91.

Tseng, P., and B. Bridgeman. 2011. Improved change detection with nearby hands. *Experimental Brain Research* 209 (2):257–269.

van Beers, R. J., A. C. Sittig, and J. J. Denier van der Gon. 1999. Integration of proprioceptive and visual position-information: An experimentally supported model. *Journal of Neurophysiology* 81 (3):1355–1364.

Wann, J. P., and S. F. Ibrahim. 1992. Does limb proprioception drift? *Experimental Brain Research* 91 (1):162–166.

Wells, H. G. [1897] 1946. *The War of the Worlds.* Harmondsworth: Penguin.

Werner, W. 1993. Neurons in the primate superior colliculus are active before and during arm movements to visual targets. *Experimental Brain Research* 5 (4):335–340.

Zopf, R., G. Savage, and M. A. Williams. 2010. Crossmodal congruency measures of lateral distance effects on the rubber hand illusion. *Neuropsychologia* 48 (3):713–725.

Zopf, R., S. Truong, M. Finkbeiner, J. Friedman, and M. A. Williams. 2011. Viewing and feeling touch modulates hand position for reaching. *Neuropsychologia* 49 (5):1287–1293.

4 Beyond the Boundaries of the Hand: Plasticity of Body–Space Interactions Following Tool Use

Matteo Baccarini and Angelo Maravita

1 Introduction

One of the most important steps in the phylogenetic evolution of human-kind is the acquisition of erect posture. Thanks to this postural change, humans developed a full capacity of vocalization, which evolved into a proper capacity for language. Although linguistic production is the main effect of bipedalism, it is not the only important one. Indeed, the acquisition of erect posture also led to the attribution of a key role for locomotion to the lower limbs. As a consequence of this specialization, the upper limbs and, more specifically, the hands have been made available for other tasks, such as gesture production and purposeful goal-directed object manipulation. Although several studies have shown that tool use is peculiar to neither humans nor—more generally—primates (e.g., see Bird and Emery 2009; Peeters et al. 2009; Weir et al. 2002), surely human beings have reached an incomparable proficiency in using external objects as tools. Unlike other species, human individuals are able to use tools assembled not only by themselves but also by other humans. For example, while only a restricted group of persons is able to build a fork, a much larger number of humans know how to use it. In humans, then, using a tool implies a more general, cultural, and symbolic knowledge (Johnson-Frey 2004) of its function that is shared among conspecifics regardless of their ability to assemble that same tool.

Moreover, while other animals use tools only in situations of immediate problem solving, humans are unique in their attitude of providing tool use with a cultural—as well as social—value. For example, as noted by Cardinali et al. (2011a), in most Western cultures it is acceptable for babies to eat with their hands, while this same behavior is less acceptable in the adults, who are expected to use cutlery. Finally, while other forms of animal tool use are basically stereotyped and relatively simple, only humans can use them in complex and flexible ways, so that the very same tool can be used in different contexts and for accomplishing different tasks. Tool use has become such a common practice in everyday life that there are relatively few activities that we perform without any tool, even when tool use would not be strictly necessary.

Importantly, as observed by Spence (2011), the most intriguing characteristic of tool use is that it can produce changes in tool-users themselves. More precisely, tool use can change the way in which people integrate different sensory information and act on the external environment (e.g., Cardinali et al. 2009a,b). In a broader sense, tool use is thus supposed to affect the way in which tool-users interact with the space around them. However, the precise mechanism by which this occurs is still a matter of debate, as we will discuss later.

In the present chapter, we will, first, offer a definition of what tool use is in our viewpoint. Second, we will outline how tool use can affect tool-users by changing the way in which they interact with the surrounding environment. Third, we will discuss possible mechanisms that may account for the effects of tool use, namely a dynamic modulation of the tool-user's body/space representation or modulations of spatial attention.

2   Tool Use: A Definition

In everyday life, we intuitively classify a large number of commonly used objects under the label of "tools." However, although commonsense judgments can work at an intuitive, coarse-grained level, they cease to be useful when a more exhaustive analysis is required. A pivotal problem in this respect is that a univocally shared definition of what should be considered a tool is lacking in the literature. Consequently, the notion of tool use is somewhat vague.

The first step in any serious reasoning concerning tool use consists thus in indicating a clear definition of what tool use properly is. In this chapter, we substantially adopt the definition formulated by Beck around thirty years ago. According to Beck (1980), tool use should be defined as "the external employment of an unattached environmental object to alter more efficiently the form, position, or condition of another object, another organism, or the user itself when the user holds or carries the tool during or just prior to use and is responsible for the proper and effective orientation of the tool" (10).

The relevance of this definition relies in its capacity to clearly highlight some pivotal characteristics that an object must not lack in order to be classified as a proper tool. First of all, this definition requires that a proper tool should be physically held during use. This would exclude, for example, treating mirrors as proper tools (e.g., Làdavas 2002), even if mirrors show measurable behavioral effects on the way people represent the surrounding space (Maravita et al. 2002).

Another important feature highlighted by Beck (1980) is that a proper tool should be an object that is used directly to alter our relation with the external environment. The function of a tool is indeed to produce a certain modification in one's subjective performance, and this can occur only if the tool plays an active role in these dynamics. For example, if our intention is to grasp something with a precision grip, the ring on our index finger should not be considered a tool even though we are moving exactly that finger—and the ring attached to it—in order to accomplish the task. Therefore, contact or, more generally, some relation with an acting body part is a necessary condition for considering an object a tool, even though it cannot be taken to be a sufficient condition. On the contrary, as it emerges from the above considerations, an object can be considered a tool if and only if it is relevant in the context of the action in which it is involved. Accordingly, a tool can be defined as a device that has a determined set of functional characteristics and is used according to its functional value.

Importantly, as recently noted by Cardinali et al. (2011a), this reference to the tool's functional characteristics should not be intended as limited to the functional value for which the tool was originally intended, since an object designed for a specific use can also be used in a different context. Consider, for example, a very commonly used object, such as a fork. Everybody (at least in Western cultures) knows that its original function is to allow us to pick up food and bring it to the mouth without touching it with our hands. According to the definition by Beck (1980), there is no doubt that the fork has all the attributes of a tool, including its capacity for modifying the external environment. However, this is not the only way in which we can use a fork as a tool. For example, we could use it to open a jar, or even to point to something on the table. In using the fork in these ways, we are not using it according to its original function, although we are still using it to act upon the environment—opening a jar or directing someone else's spatial attention. By contrast, if we hold the very same object while gesturing during a discussion with someone else, even if the fork may be indirectly useful for emphasizing our gestures, it would have no intentional effects on the external environment. It follows that the very same item could be considered either a tool or a simple object, depending on the contingent situation in which it is involved. On this perspective, the emphasis should not be only on the intrinsic features of a given object, but also on the way in which the object is used.

However, to define tool use in an even more stringent way, it becomes crucial to analyze why, with regard to our paradigmatic examples, using a fork for grasping food or opening a jar is different from holding the same fork while gesticulating. In this respect, unfortunately, the classic definition from Beck (1980) is not as exhaustive as it has been thought to be. Indeed, in the above quoted definition, Beck emphasized that, in a tool-use dynamic, the acting subject is responsible for the way in which the tool is used and that it is crucial that the use of the object affects the environment in some way. However, in our examples, all three activities imply the execution of some voluntary movement by the participant and may have an effect on the environment—in our example, food, a jar, or an interlocutor. Therefore, these effects on the environment cannot be considered as the only elements characterizing an object as a tool.

In our view, what makes the first two activities different from the third is that the former imply the use of the fork in an intentional fashion, while in the latter the object is wielded without clear intentional involvement in the context of the action. For this reason, in the present chapter we adopt a definition of tool use that adheres to the definition provided by Beck (1980), with the sole critical addendum of an explicit requirement of the involvement of the object in an intentionally goal-directed activity. In a recent work, Spence (2011) has elegantly underlined the necessity of this theoretical passage by proposing to use "tool use" to mean the "deliberate and purposeful manipulation of an independent (usually) hand-held object in order to act upon another object in achieving a specific behavioral goal" (Spence 2011, 221). Accordingly, we became tool-users neither when we passively hold an object nor when we use it in an aimless and/or unintentional way, even if the object is, by itself, charged with a functional value.

This revised version of the Beck's definition is also in line with some recent relevant data from primate electrophysiology (Umiltà et al. 2008) that allow drawing a clear distinction between movements and actions. In their study, Umiltà and colleagues registered the activation of premotor cortex neurons of macaque monkeys performing a task using a pair of pliers. The authors observed that the majority of recorded neurons did not fire following the mere act of opening or closing the pliers, but fired only when the tool was used to grasp bits of food. These results suggest that the definition of "tool" we propose here not only responds to functional criteria, but may also hold specific neural underpinnings.

3   Effects of Tool Use

According to the definition provided in the previous section, tools are objects with particular functional properties that are used intentionally with the aim of improving, or even making possible, the execution of a given task. Notably, humans can use tools to accomplish a wide variety of tasks. Furthermore, the ability in using tools is so flexible that we can often use the very same tool in different contexts. As a consequence, it is quite evident that the category of tools is a large and heterogeneous class with respect to the complexity of its elements. Some tools are very easy to use and thus commonly employed without any particular ability. By contrast, some tools have a very complex structure and function, thus requiring specific skills to be used correctly. In fact, almost every aspect of our life, from everyday experience to highly specialized jobs, benefits significantly from the use of dedicated tools.

While such a wide range of tool-use skills is a clear index of how sophisticated our cognitive abilities are, it also makes it difficult to categorize all tool-use activities under a single label. Accordingly, it is important to define the kind of tool-use activities we will address in the present chapter. In particular, we will focus on those tool-use activities that may be considered as the most basilar forms of tool-use, namely, those that can directly expand our action space.

In order to define the qualities of this kind of tool-use, we should first consider that we live in a complex spatial framework that includes both ourselves and the multitude of objects around us. The way in which our body/space interactions are mapped in the brain is a matter of current investigation among scientists in various disciplines (see, e.g., Vallar and Maravita 2009; Rizzolatti et al. 1997).

Crucially, the objects contained in the space around us are all but mere icons. On the contrary, we are always interacting with some of them and, interestingly, we can do it in different ways: We can grasp them, we can push or move them, and, if we are skillful enough, we can manipulate or even use them according to a given purpose. However, none of these interactions would be possible if we were not able, in the first place, to reach our target(s).

This assumption has two relevant theoretical consequences. First, given the impossibility of interacting with objects without contacting them in some way, the act of reaching something should be considered as the first basilar step from which any subsequent goal-directed behavior can be performed. Second, considering that surrounding objects can be viewed as at least potential targets of our reaching actions, it follows that the space around us can intuitively be conceived as near and far, depending on the distance of a given spatial position from our body. However, the nature of such "near" and "far" sectors must be carefully considered. The body, just like any other extended item, has some morphological properties, and, crucially, it is extended in length. The idea adopted in this chapter is that the external space is mainly represented in a bodily constrained—or body-centered—way (Vallar and Maravita 2009; see also Holmes, this vol.). Because body holds specific metric features (height, arm length, etc.), the simple fact that the body of each individual is the pivotal reference of the surrounding space implies the division of such a spatial horizon in a near—peripersonal—and a far—extrapersonal—sector (Rizzolatti et al. 1981; Rizzolatti et al. 1997). In this framework, external objects can be classified into two categories: Some objects are *sufficiently near* to us to be literally at hand and other objects are *far enough* to be considered out of reach. In agreement with these considerations, the boundary between near space and far space has been defined on the behavioral criterion of hand reach-ability (Berti and Frassinetti 2000; Berti et al. 2001; Maravita et al. 2003; Làdavas 2002), eventually implemented by the possibility of moving the trunk (Longo and Lourenco 2006).

Importantly, if an object is "out of reach," a local movement of the whole body is not the only way to reach it. Indeed, and critically to the present chapter, an alternative strategy can be adopted, namely, the use of a tool that is long enough to cover the distance between the subject and the object. Importantly, during the last three decades, the idea that we can modify our relationship with the external space in terms of body/space representation by using a tool has gained a progressively larger consensus in the neurophysiological, neuropsychological, and psychological litera-ture (see, e.g., Maravita and Iriki 2004). In the following sections we review the most relevant empirical findings that support this hypothesis, report-ing data from single-cell recordings in primates and behavioral experimen-tal paradigms in brain-damaged patients affected by unilateral spatial neglect or extinction as well as in neurologically intact humans.

Multisensory Peripersonal Space and Tool Use: Seminal Evidence from Electrophysiological Recordings in Primates

Faraway objects can typically be perceived through a limited number of senses, namely, olfaction, audition, and vision. By contrast, objects located near the body—or in direct contact with it—can be detected by all our sensory modalities, including somatosensation. If this simple and intuitive statement is true, it is reasonable to claim that the peripersonal space is more complex in structure than extrapersonal space. This asymmetry in complexity may be related to the activation of multimodal neurons that integrate different sensory inputs into a multimodal format. Of particular interest, for the aim of this chapter, is the discovery (in the monkey's brain) of a class of bimodal neurons that integrate visual and tactile information. While unimodal sensory neurons have a single receptive field, visuotactile neurons show two receptive fields. One is the somatosensory field (sRF hereafter) and responds to stimuli delivered to a certain portion of the body surface. The other is the visual field (vRF hereafter) and consists in a tridimensional region of the visual space immediately surrounding the sRF (e.g., Rizzolatti et al. 1997).

Importantly, these receptive fields show a certain degree of spatial relation between each other. First, a visual stimulus can activate a visuotactile receptive field only if it is presented near the body surface. Indeed, if the visual event is moved progressively away from the skin where the sRF of a given neuron is located, the response of that neuron is reduced or even abolished. Second, these two receptive fields are reciprocally anchored. Graziano et al. (1994) tested some visuotactile neurons electrophysiologically with sRFs located on the arm by varying the spatial position of the arm (see also Fogassi et al. 1996; Graziano et al. 1994). As a result, they observed that most of the registered neurons shifted the location of their vRFs according to the position assumed by the arm, even though the monkey's gaze remained fixed.

Interestingly, on the basis of the idea that monkeys can be trained to be skillful tool-users (see Ishibashi et al. 2000 for a thorough review of this topic), Iriki et al. (1996) registered the activity of parietal visuotactile neurons to investigate whether tool use could change the spatial properties of their visual responses. In their groundbreaking study they distinguished between two types of visuotactile neurons, neurons of distal type and neurons of proximal type. The distal neurons were characterized by sRFs normally located on the hand, while the proximal neurons usually showed sRFs located on the shoulder. Crucially, Iriki et al. (1996) noted that after a short period of active goal-directed tool use—retrieving bits of food placed at a distance on the table, using a long rake, for about five minutes— these neurons expanded their visual responses. After tool use, they became responsive to visual stimuli presented far away from the skin but near the tip of the tool—in the case of distal-type neurons—or in the spatial region physically reachable by the handheld tool—in the case of proximal-type neurons. In other words, a remapping of the crossmodal representation of spatial objects occurred following prolonged tool use. Of course, tool use did not modify the physical length of the monkey's arm, only its range of action. Following such evidence, the idea that tool use induces a spatial remapping by acting on the extension of the internal representation of the body was introduced in literature. More properly, since some pragmatic aspects of the bodily dimension—such as posture and spatial extension—are largely acknowledged by the classic—though controversial—notion of "body schema" (Head and Holmes 1911, 1912; Maravita 2006; Holmes and Spence 2006), Iriki et al. (1996) provided the speculative intuition that the body schema could include external objects, with a plausible neurobiological basis. This idea thus became a source of inspiration for several succeeding studies concerning the spatial effects of tool use on multisensory integration.

Effects of Tool Use in Brain-Damaged Patients Affected by Unilateral Spatial Neglect

Unilateral spatial neglect (USN) is a complex neuropsychological deficit consisting of a reduction, if not total absence, of responses toward stimuli coming from the region or body parts contralateral to the affected hemisphere (Bisiach and Vallar 2000; Heilman et al. 2003; Husain 2008; Vallar and Maravita 2009). Since the brain damage responsible for this syndrome typically involves the right hemisphere, the symptoms are typically observed in the left side of the body. USN cannot be considered a primary sensory deficit (e.g., Bisiach and Luzzatti 1978), although it is generally reckoned a disturbance of the space representation and awareness at a higher level (Heilman et al. 2003). A common standard test for assessing neglect is the line bisection task, in which patients are asked to mark the midpoint of horizontal lines of different lengths that are presented in front of them. Because of the brain damage, neglect patients tend to shift the subjective midpoint of the line toward the right (i.e., ipsilesional) side. As emerges from this description, bisection is a simple test and therefore a very handy and popular way of testing neglect, as well as a useful tool for studying extrapersonal (Jewell and McCourt 2000) and personal (Sposito et al. 2010) space representation in neurologically intact and brain-damaged humans.

For a long time, line-bisection tasks were only performed on segments presented immediately in front of the patient, thus reachable using a pencil with a simple hand movement. Only more recently have researchers started to use line bisection as a tool to look for possible differences in the amount of neglect between near and far space representation. In particular, Halligan and Marshall (1991) asked a neglect patient to bisect lines presented either in near or far space. Since the far segments were not reachable by hand, the patient was required to bisect far lines by means of a laser pointer or even by dart-throwing. The authors reported a more accurate performance in the bisection of far than near lines, regardless of the tool used to perform the task (laser or pen in near space, dart or laser in far space). In a similar study, Cowey et al. (1994) described an opposite dissociation, in which neglect patients made greater errors in bisecting far than near lines, thus pointing to a double dissociation between the representations of the two sectors of space. Finally, in light of the findings collected by Iriki and colleagues (1996) described in the previous section, Berti and Frassinetti (2000) used the traditional bisection task by introducing an additional condition in which a long stick was used to bisect far lines. In this seminal case study, the right-brain-damaged (RBD) patient P. P. was first required to bisect lines in near and far space by using her right hand or a laser pointer, respectively. Data form these two conditions showed the same tendency described by Halligan and Marshall (1991), namely, a larger rightward bias for near than far lines. However, when the laser pointer was replaced by a long stick that allowed P. P. to physically reach the far targets, the rightward bisection bias significantly increased for the far lines, reaching a level of severity comparable to that shown for the near lines. This pattern of results (see also Berti et al. 2001; Neppi-Modòna et al. 2007; Pegna et al., 2001, for logically related results) led the authors to hypothesize that, compatibly with the electrophysiological data from Iriki and colleagues (1996), the far space previously coded as extrapersonal may acquire the same mechanisms of representation that are proper to the near space, following tool use.

Effects of Tool Use in Brain-Damaged Patients Affected by Cross-Modal Extinction

Extinction is a neuropsychological condition occurring after brain damage in which the patient fails to detect stimuli presented to the contralesional side of space or body. Unlike neglect, contralesional omissions occur only, or mainly, when a competing stimulus is presented at the same time, or with a minimum delay, to the ipsilesional side (Bisiach and Vallar 2000; Heilman et al. 2003; Vallar and Maravita 2009). Although extinction often co-occurs with primary sensory deficits, in analogy with neglect, extinction cannot be considered a disorder of primary sensory processing. Indeed, contralesional stimuli can be perceived, by definition, if delivered in isolation, and there is evidence of unconscious brain activations for the extinguished events (e.g., Berti et al. 1992; Driver and Mattingley 1998; Eimer et al. 2002). Extinction is now regarded as a disorder of spatial processing due to an exaggerated competition between simultaneous events. Crucially, following brain damage, the ipsilesional and the contralesional stimuli would undergo an asymmetrical weighting for their spatial representation, with the result that the contralesional stimulus would often go undetected (Driver et al. 1997).

In the clinical as well as the scientific environment, extinction has been often regarded as a typical unimodal syndrome in which only stimuli of the same sensory format could lead to reciprocal interference (Inhoff et al. 1992). However, Mattingley et al. (1997) unequivocally showed that such a conviction was radically misleading, by providing evidence that a right ipsilesional tactile (or visual) stimulus can extinguish a contralesional left event delivered through the visual (or tactile) sensory modality. At the same time, Di Pellegrino and coworkers (1997) independently observed that a left tactile stimulus can be extinguished following the delivery of a visual event, provided that such a visual stimulation is delivered close to the hand. Indeed, the authors also showed a reduction of extinction when the distance between the right hand and the right visual stimulus was increased. In other words, the farther the visual event from the ipsilesional hand, the smaller the extinction of the contralesional touch. The idea underlying these intriguing results was that, in order to interfere with the contralesional tactile event, the ipsilesional visual stimulus activates a multimodal visual-tactile representation. This, in turn, would only occur within the peripersonal space, as suggested by neurophysiology.

Given such a precise spatial modulation, crossmodal visual-tactile extinction has been fruitfully used as a paradigm to assess the plasticity of representation of peripersonal space following tool use (see Maravita and Iriki 2004). Farné and Làdavas (2000) asked a group of extinction patients to retrieve objects from far space using a long rake. Before and after a short period of practice, crossmodal extinction was tested by delivering touches on the left, contralesional, hand and visual stimuli close to the tip of the rake held with the right, ipsilesional hand. In this situation the stimulus was far from the hand, and thus was less likely to fall in the range of putative crossmodal neurons with their RF centered on the peripersonal space, although it was close to the tool tip. Crucially, the visual stimulus induced more contralesional tactile extinction after tool use than before, suggesting an expansion of the boundaries of optimal crossmodal integration from the space near the hand to the tip of the tool. Additionally, and in close analogy with physiological data collected by Iriki and colleagues (1996), this effect was only temporary, since it faded and then totally disappeared after more or less ten minutes of inactivity. In an independent yet closely related study, Maravita et al. (2001) similarly found that crossmodal extinction of a left touch by a far right visual stimulus was more severe when the patient touched the visual stimulus with the tip of a stick held with the right hand than in the absence of the stick or when the tool was passively placed on the table but physically disconnected from the hand. Finally, Maravita et al. (2002a) showed that training an RBD extinction patient to use a rake with his nonplegic left hand to retrieve objects in the right ipsilesional space effectively reduced crossmodal extinction of left touches by right visual stimuli close to the tool tip. In this respect, prolonged tool use linking right visual events to the left hand had a positive effect on extinction, possibly by creating a common multimodal representation around the tool, in which opposite spatial stimuli did not compete anymore.

Other studies further sought for the effect of tool use on crossmodal extinction and showed the precise link between tool length and the expansion of the multisensory integrative interactions from near peripersonal to far space (Farné et al. 2005a,b).

Effects of Tool Use in Healthy Individuals

While the neuropsychological picture of crossmodal extinction allows us to examine the reciprocal modulations of sensory information coming from different sensory modalities in peripersonal space, the crossmodal congruency effect (CCE) has been widely used to assess crossmodal effects in healthy humans, including their spatial modulation following tool use (see Spence et al. 2004 for a review). In a typical crossmodal congruency study, participants are required to hold two little cubes, one in either hand, comprising two LEDs, while two tactile stimulators are placed in spatial proximity, one on an upper corner and the other in the corresponding lower corner of the cube. Participants keep their index finger on the upper stimulator and their thumb on the lower stimulator. On each trial, a vibrotactile stimulus is presented randomly from one of the four possible locations, simultaneously with an independent visual event, consisting of a LED flash from one of the four locations. On each trial, participants are required to make quick elevation discrimination responses, deciding whether the tactile target is presented at the thumb or index finger of either hand, while trying to ignore the visual distracter presented at the same time.

Critically, distracters are just as likely to be presented from the same elevation as the target, as from a different elevation. In this paradigm, participants are normally slower and less accurate in the tactile elevation discrimination when the distractor is presented from a different (incongruent) elevation than when it is presented from the same (congruent) elevation. This difference is the crossmodal congruency effect and can be measured in response times or accuracy.

Analogously to what we previously suggested for crossmodal extinction, the CCE is usually larger when spatially incongruent distracters are close to the tactually stimulated hand and declines when they are close to the opposite hand (Spence et al. 2004), again supporting the importance of spatial proximity for the integration of multisensory events.

Maravita et al. (2002b) conducted the first study in which a condition of tool use was included in the classical crossmodal congruency paradigm. Whereas in a typical crossmodal congruency task, the visual stimuli are presented on the hands, close to the tactile stimuli, Maravita and coworkers placed the visual distracters (LEDs) at the tip of a golf club held in each hand, again in an up/down position. As usual, participants were asked to discriminate the elevation of the vibrotactile stimulus delivered to the hands holding the golf club, while trying to ignore the LED flashes. During the experiment, the posture of the golf clubs was alternatively changed by crossing or uncrossing them, so that the visual distracters at the tip of each tool were alternatively placed in the ipsilateral or contralateral side as compared to the hand holding them and receiving the tactile stimuli. As expected, in the standard, uncrossed condition, visual distracters on the same hemispace exerted a greater interference on the tactile discrimination task performed by the ipsilateral hand, which was on the same spatial side and connected to the same tool. Crucially, however, when tools were crossed, the interference of visual distracters was now stronger on the hand placed in the contralateral space. This result was interpreted as a clue that the presence of the tool was effective in remapping space representation by linking near and far space. In the crossed-tools situation, although the visual distracters were spatially closer to the tactile stimuli on the same side, and therefore one might have expected a stronger interference on them due to spatial proximity, they actually showed stronger interference over the contralateral touches, to which they were functionally connected via the tool.

Other effects of spatial remapping following tool use on neurologically intact participants were recently provided by studies using the line bisection task (see above). Longo and Lourenco (2006) have recently shown that, when performing a line bisection task on lines placed at different distances from the body, participants committed a rightward error that was progressively larger going from near (30cm) to far (120cm) space. However, these same authors have shown that the aforementioned progressive increase in line bisection errors with increasing stimulus distance was abolished if participants used, instead of a laser pointer, a long stick that could physically reach the far space. This evidence supports a gradual transition from near to far space that affects the way in which egocentric space is represented in the observer (see also Cowey et al. 1999). Furthermore, it shows that egocentric space coding can be dynamically affected and reshaped following tool use by extending the features of the near space representation to the far space.

4 Body Schema, Peripersonal Space, Crossmodal Attention: What Is Modified by Tool Use?

Most of the literature reviewed in the present chapter is based theoretically on the idea that peripersonal space is affected by tool use. This view is mainly inspired by the seminal study by Iriki and colleagues (1996), from which many studies of normal observers or brain-damaged people descend (Maravita and Iriki 2004). Evidence that multisensory visuotactile neurons expand their visual receptive field to include the whole space reachable by the tool suggests a direct expansion of the so-called peripersonal space (Rizzolatti et al. 1981), as discussed previously. Such an expansion critically occurs when we actively use the tool to reach or act on distant objects, thus requiring an intentional and effective use of the tool (also in agreement with the definition of a tool discussed previously) and not simply a passive observation of the tool extending into far space (Farné et al. 2005b). Since, in the original study by Iriki and colleagues (1996), the expansion of vRFs was found in bimodal visuotactile neurons that had their tRFs centered on the skin, one idea is that not only the peripersonal field but also some aspects of body representation expand to include the tool itself as a new functional part of the body (Iriki et al. 1996; Maravita and Iriki 2004; Cardinali et al. 2009a). This idea has been present since the early conceptualization of the term "body schema" (see discussions of this term in Holmes and Spence 2004; Maravita 2006) by Head and Holmes (1911). In their seminal paper, when describing the features of the body schema, they proposed that "It is to the existence of these 'schemata' that we owe the power of projecting our recognition of posture, movement and locality beyond the limits of our own bodies to the end of some instrument held in the hand. Without them we could not probe with a stick, nor use a spoon unless our eyes were fixed upon the plate. Anything which participates in the conscious movement of our bodies is added to the model of ourselves and becomes part of these schemata" (Head and Holmes 1911, 188). They then conclude that "a woman's power of localization may extend to the feather in her hat" (ibid.), a symbolically powerful statement, although probably more explicative for the idea that body schema can be shaped by external objects that habitually were thought to lie in passive contact with the body (Aglioti et al. 1996), than by intentional tool use. However, Head and Holmes (1911) emphasized the efficiency of the human sensorimotor system in coding the presence, motion, and spatial position of objects that are in contact with the body as if they were true physical extensions of it. In the authors' words, external objects would be somehow "added to the model of ourselves" (ibid., 188), suggesting some sort of automatic, likely implicit incorporation into the representation of one's own body.

In recent years, after having been hypothesized for a long time, clues suggesting some degree of reshaping of the body schema following tool use have been progressively accumulated. In a recent study by Cardinali et al. (2009b), the kinematic profile of free hand movements was recorded before and after training with a long mechanical tool to grasp a target object. As a result, the authors found that the kinematic profile of post-tool-use movements was different from the profile of pre-tool-use movements. More precisely, they observed reduced maximal amplitude, longer latencies of kinematic parameters, and shorter movement times following tool use. All these data would be compatible with the representation of a longer arm by the participant.

In support of these kinematic data, further experiments have reported that tool use also induces an increased subjective arm elongation. Again, Cardinali and colleagues (2009b) asked blindfolded participants to point with their left hand to the location of vibrotactile stimuli randomly delivered to one out of three skin locations—the elbow, the wrist, and the middle finger—on the arm trained in tool use. By measuring the distance between the subjective positions of the vibrators indicated by the participants, it was found that the subjective mean distance between the elbow and the finger significantly increased after tool use, thus suggesting that some degree of extension of the arm's representation had occurred. Further data in this direction have been collected by Sposito and colleagues (submitted), who show that the ability to implicitly assess the length of one's own arm through an arm-bisection task (Sposito et al. 2010) was biased following tool use. More precisely, when performing a bisection task on their arm placed in a radial posture, participants indicated the midpoint between the tip of their middle finger and the elbow more distally, if tested following a twenty-minute training with a rake to operate on objects placed far from them, as compared to a baseline assessment. Interestingly, this effect was only found using a tool that extended the reaching space considerably (by 60cm) and not by a functionally useless (20cm long) tool, and could be induced in both the dominant and the nondominant arm. This suggests that bodily effects of tool use are not critically affected by the general skills of tool use—which are typically higher for the dominant hand—and only follow the use of functionally effective tools.

Given the experimental results discussed above, there seems to be enough evidence to support the view that tool use effectively alters the body/space relationship in two main ways: expanding the boundaries of peripersonal space and increasing the metric representation of the trained body part. An intriguing and open question in this respect is to better understand the close relationship between the concepts of body schema and peripersonal space. Whether these may be considered as two aspects of the same concept or two separate concepts is a matter of current discussion, although theoretical and experimental aspects exist in favor of their independence (Cardinali et al. 2009a). The very same evidence that spatial crossmodal effects of tool use can be found several centimeters far away from the hand (see examples in Maravita and Iriki 2004) while the subjective elongation of body metrics is typically of a few millimeters (Cardinali et al. 2009b; Cardinali et al. 2011b; Sposito et al. submitted) suggests that the relationship between body and peripersonal space representation, although present, is not strictly linear. However, it is conceivable that tool use may affect both body and peripersonal space representation to some extent. Further studies are needed to clarify the interdependence or even any causal effect between these two mechanisms. One suggestion could come, as previously mentioned, from the study by Iriki and colleagues (1996) in which activations of bimodal parietal neurons were recorded following tool use. In particular, those neurons holding a tRF centered on the hand expanded their vRF around the full extension of the tool, while those with a vRF centered on the shoulder expanded their vRF to include the whole space reachable by the tool. One may speculate that tool use would update, at the same time, the coding of both extrapersonal space, by extending the representation of the reachable space, and of the body boundaries, by expanding the visual responses near the hand. To what extent the expansion of the former may indeed correspond to or follow an update of body image/schema is still an open question (see discussion in Cardinali et al. 2009a).

Another theoretical aspect of tool use, which has recently been under investigation, is the idea that tool use would simply modulate the allocation of spatial attention toward the distant spatial locations made reachable by the tool. Broadly speaking, this hypothesis is based on the idea that tool use would provide the tip of the tool with the same attentive relevance normally attributed to the biological hand. Although we are multitasking agents, we have access only to a restricted amount of attentive resources. Therefore, in the case of tool use, the main focus of attention would be directed to the part of the tool that is most relevant to the task, namely its tip. On this view, it is reasonable to conclude that the tip of the tool substitutes for the hand in its role of attentive fovea, with the consequence that the optimal region of multisensory integration would simply shift there (Holmes et al. 2007). Critically, this sort of distal projection of attention would not require an extension of the tool-user's body/space representation.

For example, Holmes and colleagues (2004) asked participants to respond to far distant stimuli by pressing buttons either with the tip of hand-held tools or directly with the tool handle. They subsequently tested the cross-modal interference exerted by visual distracters placed either at the hand, at the tip of the tool, or halfway between the hand and the tool-tip, over somatosensory stimuli delivered at the hand. Critically, following tool use, the authors found no modulation of crossmodal interference from visual distracters placed halfway along the tool but only an increased interference from distracters at the tool-tip. These findings were taken as proof that no expansion of peripersonal space representation had occurred; otherwise, visual distracters presented at any position between the hand and the tool-tip should have increased their crossmodal interference on touches delivered at the hand, following tool use. By contrast, the authors hypothesized that only a shift in the focus of multisensory attention toward the place where the effector of the action—the tool-tip—was used had occurred. Logically related conclusions were reached from another study assessing the ability of distinguishing between single or double vibrations—perceived through a seventy-five-centimeter-long tool—while trying to ignore single or double visual stimuli (Holmes et al. 2007). Although the general task always remained the same throughout the experiment, participants were required to hold the tool in different positions. More precisely, during the first three experimental conditions, they were requested to hold a single tool with their right hand while the vibrations were presented at the distal end of the tool, from either the right or the left side of space. Crucially, the tip of the tool was held either uncrossed on the right side—thus connecting the vibrator on the hand with the ipsilateral visual space—or crossed over the subject's midline—thus connecting the vibrator on the hand with the contralateral visual space. In the remaining two experimental conditions, participants were required to hold two tools—one in each hand—in either an uncrossed or a crossed position. As a result of such manipulations, Holmes and colleagues found that holding a tool on one side of space significantly increased the magnitude of visuotactile interactions on the side of the tool, compared to the side opposite the tool. By contrast, actively crossing the tool sensibly reduced this tool-dependent effect after a few trials. Finally, the authors found that the difference between left- and right-sided distracters was also reduced in the last two conditions, where subjects were asked to hold—and use—two tools instead of a single one.

The findings from the above studies surely call for the importance of attentional factors in the genesis of tool-dependent crossmodal effects. On the other hand, however, there are also many good reasons in support of the idea of a remapping of peripersonal space. Farné et al. (2005a), for example, reached opposite conclusions from those obtained by Holmes and colleagues (2004), by testing the effect of tool use in a patient with crossmodal extinction (see previous paragraph). In their study, tool use increased the effectiveness of visual stimuli presented both at the tip and at midway along the tool in modulating the perception of contralesional touches (Bonifazi et al. 2007), compatibly with an extension of the boundaries of peripersonal space, and its crossmodal properties, along the whole extension of the tool. Another interesting point in this debate is the previously mentioned finding by Maravita and colleagues (Maravita et al. 2002) of an amelioration of crossmodal extinction following tool use. In this case, while the tip of the tool was constantly used in the ipsilesional right side of space during training, its beneficial effect was in improving the detection of contralesional left tactile targets. In other words, while the training required a sustained attentional focusing on the right side, its effects were found on the left side, that is, far away from the focus of attention during the task. This again seems, in our view, more in line with a resizing of peripersonal space and its crossmodal features than with a shift of the attentional focus.

There may be different explanations of the apparently contrasting results found in the studies reported above, such as differences in the tasks used (e.g., touching a vibrator versus retrieving an object), the attentional requests of the task, the experimental population (brain-damaged patients versus neurologically intact participants), and so on. However, even disregarding the methodological differences, this chapter defends the idea that the findings collected by the sustainers of the attentive framework are surely evidence for a role of attention in the crossmodal effects of tool use, though not necessarily a counterargument for a body/peripersonal space remapping mechanism. Broadly speaking, this means that no argument in favor of one kind of mechanism should be used as a direct counterargument for another.

In line with this perspective, the first theoretical conclusion we want to underline is that the attentional and the body/space hypotheses should be better conceptualized as complementary than as exclusive alternatives. However, to say that these theoretical frameworks are complementary does not imply—at least in our viewpoint—that these are exactly equivalent. On the contrary, our second conclusion suggests that they are not provided with the same heuristic power. More precisely, this chapter defends the idea that such an explicative asymmetry is imbalanced in favor of the body/space representation hypothesis.

First, it should be noted that many behavioral effects of tool use reviewed in this chapter, and relative to the increased crossmodal effects exerted by visual stimuli in far space over somatosensory targets, can be interpreted both as a cue of an attention reorientation and as a cue of a modification of body/space representation. By contrast, recent data presented here, and relative to the update of the kinematic parameters of reaching (Cardinali et al. 2009b) and the subjective metric estimation of the body (Cardinali et al. 2009b; Cardinali et al. 2011b; Sposito et al. submitted) following tool use, can be coherently explained only by an extension of the body schema.

In particular, the evidence collected in the kinematic study of Cardinali et al. (2009b) seems intriguing in this respect. In this work, unlike what was typically done in previous studies of tool use, the critical post-tool-use test did not require any perceptual or evaluative judgment and was even performed without the tool present at the scene. Accordingly, if the core element of tool-use plasticity does rely in an attentional shift, one may wonder where the attentive focus would exactly be reoriented. Furthermore, it is possible to reasonably exclude that these kinematic variations are primarily due to a modification in the perceived object distance that might be in turn related to a shift in spatial attention. Indeed, as reported by Witt et al. (2005), a reduction of the perceived subject-object distance affects kinematics in reducing both the peaks and the latencies of movement parameters, whereas according to the results collected by Cardinali et al. (2009b) the post-tool-use movements had, in fact, longer latencies. By contrast, these results are, again, in line with the idea that the post-tool-use free-hand performances reflect a change in the body representation for action.

To conclude, tool use proves to be an extremely useful framework to assess the way in which the human brain copes with the representation of space and body for action. Such sophisticated brain representation can be dynamically updated during the use of a tool, in order to increase our range of actions while maintaining our performance. A large body of evidence supports the idea that using tools that expand our range of actions can dynamically shape the tool-user's behavior through a multicomponential shaping of body/space interactions.

Acknowledgments

The preparation of this chapter was supported by a local grant (FAR) from the University of Milano-Bicocca awarded to Angelo Maravita. We are grateful to Francesco Marini for his helpful suggestions.

References

Agliotti, S., N. Smania, M. Manfredi, and G. Berlucchi. 1996. Disownership of left hand and objects related to it in a patient with right brain damage. *Neuroreport* 8 (1):293–296.

Beck, B. B. 1980. *Animal Tool-Use: The Use and Manufacture of Tools by Animals*. New York: Garland STPM Press.

Berti, A., A. Allport, J. Driver, Z. Dienes, J. Oxbury, and S. Oxbury. 1992. Levels of processing for visual stimuli in an extinguished field. *Neuropsychologia* 30 (5): 403–415.

Berti, A., and F. Frassinetti. 2000. When far becomes near: Remapping of space by tool-use. *Journal of Cognitive Neuroscience* 12 (3):415–420.

Berti, A., N. Smania, and A. Allport. 2001. Coding of far and near in neglect patients. *NeuroImage* 14 (1):S98–S102.

Bisiach, E., and C. Luzzatti. 1978. Unilateral neglect of representational space. *Cortex* 14 (1):129–133.

Bird, C. D., and N. J. Emery. 2009. Insightful problem solving and creative tool modification by captive nontool-using rooks. *Proceedings of the National Academy of Sciences of the United States of America* 106 (25):10370–10375.

Bisiach, E., and G. Vallar. 2000. Unilateral neglect in humans. In *Handbook of Neuropsychology*, 459–502, ed. F. Boller and J. Grafman. Amsterdam: Elsevier.

Bonifazi, S., A. Farné, L. Rinaldesi, and E. Làdavas. 2007. Dynamic size-change of peri-hand space through tool-use: Spatial extension or shift of the multi-sensory area. *Journal of Neuropsychology* 1 (1):101–114.

Cardinali, L., C. Brozzoli, and A. Farné. 2009a. Peripersonal space and body schema: Two labels for the same concepts? *Brain Topography* 21 (3–4):252–260.

Cardinali, L., C. Brozzoli, F. Frassinetti, A. C. Roy, and A. Farné. 2011a. Human tool use: A causal role in plasticity of bodily and spatial representations. In, *Tool Use and Causal Cognition*, 202–219, ed. T. McCormack, C. Hoerl, and S. Butterfill. New York: Oxford University Press.

Cardinali, L., C. Brozzoli, C. Urquizar, R. Salemme, A. C. Roy, and A. Farné. 2011b. When action is not enough: Tool-use reveals tactile-dependent access to body schema. *Neuropsychologia* 49 (13):3750–3757.

Cardinali, L., F. Frassinetti, C. Brozzoli, C. Urquizar, A. C. Roy, and A. Farné. 2009b. Tool-use induces morphological updating of the body schema. *Current Biology* 19: R478–R479.

Cowey, A., M. Small, and S. Ellis. 1994. Left visuo-spatial neglect can be worse in far than in near space. *Neuropsychologia* 32:1059–1066.

Cowey, A., M. Small, and S. Ellis. 1999. No abrupt change in visual hemineglect from near to far space. *Neuropsychologia* 37 (1):1–6.

di Pellegrino, G., G. Basso, and F. Frassinetti. 1997. Spatial extinction on double asynchronous stimulation. *Neuropsychologia* 35 (9):1215–1223.

Driver, J., and J. B. Mattingley. 1998. Parietal neglect and visual awareness. *Nature Neuroscience* 1:17–22.

Driver, J., J. B. Mattingley, C. Rorden, and G. Davis. G. 1997. Extinction as a paradigm measure of attentional bias and restricted capacity following brain injury. In *Parietal Lobe Contributions to Orientation in 3D Space*, 401–429, ed. P. Their and H.-O. Karnath. Heidelberg: Springer.

Eimer, M., A. Maravita, J. van Velzen, M. Husain, and J. Driver. 2002. The electrophysiology of tactile extinction: ERP correlates of unconscious somatosensory processing. *Neuropsychologia* 40 (13):2438–2447.

Farné, A., S. Bonifazi, and E. Làdavas. 2005a. The role played by tool-use and tool-length on the plastic elongation of peri-hand space: A single case study. *Cognitive Neuropsychology* 22 (3):408–418.

Farné, A., A. Iriki, and E. Làdavas. 2005b. Shaping multisensory action-space with tools: Evidence from patients with cross-modal extinction. *Neuropsychologia* 43 (2):238–248.

Farné, A., and E. Làdavas. 2000. Dynamic size-change of hand peripersonal space following tool use. *Neuroreport* 11:1645–1649.

Fogassi, L., V. Gallese, L. Fadiga, G. Luppino, M. Matelli, and G. Rizzolatti. 1996. Coding of peripersonal space in inferior premotor cortex (area F4). *Journal of Neurophysiology* 76:141–157.

Graziano, M. S., G. S. Yap, and C. G. Gross. 1994. Coding of visual space by premotor neurons. *Science* 266 (5187):1054–1057.

Halligan, P. W., and J. C. Marshall. 1991. Left neglect for near but not far space in man. *Science* 350 (6318):498–500.

Head, H., and G. Holmes. 1911. Sensory disturbances from cerebral lesions. *Brain* 34:102–254.

Head, H., and G. Holmes. 1912. Researches into sensory disturbances from cerebral lesions. *Lancet* 179:1–4.

Heilman, K., R. T. Watson, and E. Valenstein. 2003. Neglect and related disorders. In *Clinical Neuropsychology*, 296–346, ed. K. Heilman and E. Valenstein. New York: Oxford University Press.

Holmes, N. P., G. A. Calvert, and C. Spence. 2004. Extending or projecting peripersonal space with tools? Multisensory interactions highlight only the distal and proximal ends of tools. *Neuroscience Letters* 372 (1–2):62–67.

Holmes, N. P., G. A. Calvert, and C. Spence. 2007. Tool-use: Capturing multisensory spatial attention or extending multisensory peripersonal space? *Cortex* 43 (3): 469–489.

Holmes, N. P., and C. Spence. 2004. The body schema and the multisensory representation(s) of peripersonal space. *Cognitive Processing* 2:94–105.

Holmes, N. P., and C. Spence. 2006. Beyond the body schema: Visual, prosthetic and technological contributions to bodily perception and awareness. In *Human Body Perception from the Inside-Out*, 15–64, ed. G. Knoblich, M. Grossjean, and M. Shiffrar. Oxford: Oxford University Press.

Husain, M. 2008. Hemispatial neglect. In *Handbook of Clinical Neurology*, 359–372, ed. G. Goldenberg and B. Miller. Amsterdam: Elsevier.

Inhoff, A. W., R. D. Rafal, and M. J. Posner. 1992. Bimodal extinction without crossmodal extinction. *Journal of Neurology, Neurosurgery, and Psychiatry* 55 (1):36–39.

Ishibashi, H., S. Hihara, and A. Iriki. 2000. Acquisition and development of monkey tool-use: Behavioral and kinematic analyses. *Canadian Journal of Physiology and Pharmacology* 78 (11):958–966.

Iriki, A., M. Tanaka, and Y. Iwamura. 1996. Coding of modified body schema during tool use by macaque postcentral neurons. *Neuroreport* 7 (14):2325–2330.

Jewell, G., and M. E. McCourt. 2000. Pseudoneglect: A review and meta-analysis of performance factors in line bisection tasks. *Neuropsychologia* 38 (1):93–110.

Johnson-Frey, S. 2004. The neural bases of complex tool use in humans. *Trends in Cognitive Sciences* 8 (2):71–78.

Làdavas, E. 2002. Functional and dynamic properties of visual peripersonal space. *Trends in Cognitive Sciences* 6 (1):17–22.

Longo, M. R., and S. Lourenco. 2006. On the nature of near space: Effects of tool use and the transition to far space. *Neuropsychologia* 44 (6):977–981.

Maravita, A. 2006. From "body in the brain" to "body in space": Sensory and intentional components of body representation. In *Human Body Perception from the Inside-Out*, 65–88, ed. G. Knoblich, I. Thornton, M. Grosjean, and M. Shiffrar. New York: Oxford University Press.

Maravita, A., K. Clarke, M. Husain, and J. Driver. 2002a. Active tool use with contralesional hand can reduce crossmodal extinction of touch on that hand. *Neurocase* 8 (6):411–416.

Maravita, A., M. Husain, K. Clarke, and J. Driver. 2001. Reaching with a tool extends visual-tactile interactions into far space: Evidence from crossmodal extinction. *Neuropsychologia* 39 (6):580–585.

Maravita, A., and A. Iriki. 2004. Tools for the body schema. *Trends in Cognitive Sciences* 8 (2):79–86.

Maravita, A., C. Spence, and J. Driver. 2003. Multisensory integration and the body schema: Close to hand and within reach. *Current Biology* 13 (13):531–539.

Maravita, A., C. Spence, S. Kennett, and J. Driver. 2002b. Tool use changes multimodal spatial interactions between vision and touch in normal humans. *Cognition* 83 (2):B25–B34.

Maravita, A., C. Spence, C. Sergent, and J. Driver. 2002. Seeing your own touched hands in a mirror modulates cross-modal interactions. *Psychological Science* 13 (4):350–355.

Mattingley, J. B., J. Driver, N. Beschin, and I. H. Robertson. 1997. Attentional competition between modalities: Extinction between touch and vision after right hemisphere damage. *Neuropsychologia* 35 (6):867–880.

Neppi-Modòna, M., M. Rabuffetti, A. Folegatti, R. Ricci, L. Spinazzola, F. Schiavone, M. Ferrarin, and A. Berti. 2007. Bisecting lines with different tools in right brain damaged patients: The role of action programming and sensory feedback in modulating spatial remapping. *Cortex* 43 (3):397–410.

Pegna, A. J., L. Petit, A. S. Caldara-Schnetzer, A. Khateb, J. M. Annoni, R. Sztajzel, and T. Landis. 2001. So near yet so far: Neglect in far or near space depends on tool use. *Annals of Neurology* 50 (6):820–822.

Peeters, R., L. Simone, K. Nelissen, M. Fabbri-Destro, W. Vanduffel, G. Rizzolatti, and G. A. Orban. 2009. The representation of tool-use in humans and monkeys: Common and uniquely human features. *Journal of Neuroscience* 29 (37):11523–11539.

Rizzolatti, G., L. Fadiga, L. Fogassi, and V. Gallese. 1997. The space around us. *Science* 277 (5323):190–191.

Rizzolatti, G., C. Scandolara, M. Matelli, and M. Gentilucci. 1981. Afferent properties of periarcuate neurons in macaque monkeys. I. Somatosensory responses. *Behavioural Brain Research* 2:125–146.

Spence, C. 2011. Tool use and the representation of peripersonal space in humans. In *Tool Use and Causal Cognition*, 220–247, ed. T. McCormack, C. Hoerl, and S. Butterfill. New York: Oxford University Press.

Spence, C., F. Pavani, A. Maravita, and N. Holmes. 2004. Multisensory contributions to the 3-d representation of visuotactile peripersonal space in humans: Evidence from the crossmodal congruency task. *Journal of Physiology* 98 (1–3):171–189.

Sposito, A. V., N. Bolognini, G. Vallar, and A. Maravita. Submitted. Extension of perceived arm length following tool-use: Clues to plasticity of body metrics.

Sposito, A. V., N. Bolognini, G. Vallar, L. Posteraro, and A. Maravita. 2010. The spatial encoding of body parts in patients with neglect and neurologically unimpaired participants. *Neuropsychologia* 48 (1):334–340.

Umiltà, M. A., L. Escola, I. Intskirveli, F. Grammont, M. Rochat, F. Caruana, A. Jezzini, V. Gallese, and G. Rizzolatti. 2008. When pliers become fingers in the monkey motor system. *Proceedings of the National Academy of Sciences of the United States of America* 105 (6):2209–2213.

Vallar, G., and A. Maravita. 2009. Personal and extra-personal spatial perception. In *Handbook of Neuroscience for the Behavioral Sciences*, 322–336, ed. G. G. Berntson and J. C. Cacioppo. New York: John Wiley.

Weir, A. A. S., J. Chappel, and A. Kacelnik. 2002. Shaping of hooks in new Caledonian crows. *Science* 297:981.

Witt, J. K., D. R. Proffitt, and W. Epstein. 2005. Tool-use affects perceived distance, but only when you intend to use it. *Journal of Experimental Psychology: Human Perception and Performance* 31:880–888.

II TOGETHERNESS IN TOUCH

5 Touching Hands: A Neurocognitive Review of Intersubjective Touch

Harry Farmer and Manos Tsakiris

I can feel the twinkle of his eye in his handshake.
—Helen Keller

1 Introduction

As the above quote indicates, the hands play a vital role in our interaction not only with the world (Barnett 1972) but also with each other. It is through the hands that we can most directly reach out to others; the importance of touch in social interaction can be seen in metaphors such as "a soft touch," "deeply touched," and "the common touch." Historically, the scientific study of the social aspect of touch has been neglected, compared to other senses such as vision and hearing (Gallace and Spence 2010). In recent years, however, advances in the understanding of the neuroscientific basis of intersubjective touch (Morrison, Löken, and Olausson 2010; Keysers, Kaas, and Gazzola 2010) combined with a new emphasis in cognitive science on the role of the body in cognition in general (see, e.g., Shapiro 2010) and social cognition in particular (Grafton 2009; Goldman and de Vignemont 2009) have led to increased interest in the important role that intersubjective touch plays in both infant (Hertenstein 2002; Duhn 2010) and adult (Hertenstein et al. 2006b; Field 2011) relationships.

That touch should be so important in human relationships is not surprising. The skin is both the largest (Field 2001) and developmentally the most fundamental of all our sense organs (Atkinson and Braddick 1982). Intersubjective touch has also been claimed to be the evolutionarily most basic form of human communication (De Thomas 1971; Frank 1957). In this chapter, we review the current evolutionary, psychological, and the neuroscientific research into intersubjective touch. We conclude by reviewing recent work showing the role that touch can have in allowing us to empathize with others.

2   The Evolutionary Basis of Intersubjective Touch

One way of understanding the importance of intersubjective touch in our social relationships is by tracing the evolutionary history of intersubjective touch by comparing intersubjective touch in humans with intersubjective touch in other primates. The role of touch in other primates can be seen in the importance of grooming (Dunbar 2010). While grooming obviously has a hygienic function, the amount of time that primates spend grooming others (allo-grooming) is far greater than is needed for keeping fur healthy. This has led several researchers (e.g., de Waal 1989; Dunbar 2010) to argue that the primary function of allo-grooming is the establishment and maintenance of social relationships.

Two pieces of evidence support this hypothesis. First, the amount of time spent grooming is strongly correlated with social group size (Lehmann et al. 2007). Second, within social groups, grooming partnerships tend to be consistent and persistent across time, especially among females. This consistency of grooming partnerships allows for the establishment of important relationships between group members and has been observed to aid in the maintenance of alliances (Dunbar 1980) and to enhance reproductive success (Silk et al. 2003). In humans, the importance of grooming and touch in cementing social bonds has declined, possibly as a result of the fact that in the large group sizes of human communities the time required to build up social relationships by grooming is prohibitively high (Dunbar 1993). Nevertheless, the importance of allo-grooming in the forming of social bonds in other primates provides us with insight into the role of touch in human relationships and the evolutionary development of the mechanisms responsible for the effectiveness of touch in initiating and building up social relationships.

3   Factors Modulating Intersubjective Touch

A considerable amount of research has demonstrated that the sociocultural, situational, personal, and physical aspects of touch can modulate both the frequency and interpretation of intersubjective touch.

3.1   Gender

The relationship between touch and gender has been investigated in a large number of studies (Henley 1973; Jones 1986; for a review, see Stier and Hall 1984). Many of these studies have focused on gender differences in the initiation of touch. Henley (1973) observed that men were more likely to initiate cross-gender touch than women and explained this using Goffman's (1956) theory that touch is a "status privilege," meaning that those with high status are more likely to touch others than those with low status. On Henley's account, therefore, the difference in touching behavior between men and women arises from the higher social status of men compared to women.

Subsequent research has indicated, however, that the relationship between touch, gender, and status is less obvious than Henley's explanation predicts. Jones (1986) asked participants to keep a record of their touching behavior over seven days and found that women were both touched and touched others more frequently than men and that, contra Henley, women were more likely to initiate opposite sex touches than men. Jones's explanation of these results is based on Forden's (1981) finding that after watching videos of cross-gender touch participants rated a woman initiating touch as most dominant while a man receiving touch was seen as least dominant. Jones thus argued that, while men had other means of expressing dominance, for women touch was a key way in which they could assert control over a situation. On this interpretation of the role of status in touch, touch is not so much a "status privilege" as a means of raising status (see also Goldstein and Jeffords 1981).

However, Hall (1996) found no difference in the amount of touch initiation between high- and low-status individuals, although she did find that higher-status individuals were more likely to use more affectionate touches than lower-status individuals. The inconsistent findings of various studies on gender and touch suggest that Henley's (1973) status theory is at best an incomplete account of the relationship between gender and touch and that to fully explain the relationship between gender and touch it is necessary to recognize that factors other than status can interact with gender in determining the amount and initiation of touch.

3.2 Age

Age is one such factor. Hall and Veccia (1990) observed intentional touches in 4,500 pairs of individuals and found greater male-to-female touch in younger dyads and greater female-to-male touch in older dyads. A similar interaction was found by Willis and Dodds (1998), who observed greater male-initiated touch in couples where the male was younger than 20 and greater female-initiated touch in couples where the male was over 40. They interpreted this finding from an evolutionary perspective, arguing that in courting couples touch is used by men to increase intimacy in order to obtain sex while in couples in an established relationship women initiate touch as a means of maintaining resources and parental investment. Further supporting this account is the finding by Guerro and Andersen (1994) that in casual romantic relationships men were more likely than women to initiate touch, while among married couples the opposite was true.

3.3 Setting

The location and context of the situation can also modulate the frequency of intersubjective touch. For example, in an airport setting in which people were meeting or saying farewell to friends or family, Heslin and Boss (1980) observed a comparatively higher number and rate of intersubjective touch than in other public places.

Setting may also provide an explanation for Jones's (1986) finding of a lack of gender difference in touch initiation. Major, Schmidlin, and Williams (1990) observed intentional touch in three different settings: public nonintimate, greeting and leave taking, and recreational. They found greater male initiation of touch in public nonintimate and recreational settings. In contrast, for greeting and leave taking settings, in which touchers are likely to have a close relationship, no effect of gender on touch initiation was found. If women are more likely to initiate touch in more intimate, private situations, then using only data from observational studies in public settings will underestimate the amount of female-initiated touch when compared to studies, like Jones's, which use self-reported data and include touches initiated in private.

3.4 Culture

Cultural expectations can also play an important role in determining how people interpret and respond to intersubjective touch. One key distinction is between "contact" cultures (e.g., Arabs, Latin Americans, southern Europeans) and noncontact cultures (e.g., North Americans, Asians, northern Europeans) (Hall 1966). Jourard (1966) observed that couples in coffee shops in Puerto Rico touched each other an average of 180 times an hour, while couples in London cafes had an average of zero touches an hour. Remland, Jones, and Brinkman (1995) found more touching between Italian and Greek dyads than among English, French, and Dutch dyads. Culture can also affect which types of touch are considered acceptable. For example, in Mediterranean cultures public hand-holding is seen as an accepted display of friendship when exhibited by same-sex dyads but is unacceptable in mixed dyads, while in the Anglo-Saxon world the opposite is true (Burgoon and Hubbard 2005).

As the above example shows, cultural expectation of touch can also interact with gender. DiBiase and Gunnoe (2004) further demonstrated this in a study observing touch among young people from the United States, Italy, and the Czech Republic. They found that, as expected, people from the high-contact cultures of Italy and the Czech Republic touched more than those in the United States. They also found that, for hand touches, in the Czech Republic men touched women significantly more than women touched men, which was not the case in the United States or Italy. Since the Czech Republic was the culture with the most traditional gender roles (and therefore the greatest difference in status between men and women), this interaction between gender and culture supports Henley's (1973) status theory of gender touching, at least for hand touches.

3.5 Location and Type of Touch

Several studies have investigated the different signals sent by touch to different body areas (Hall and Veccia 1990; DiBiase and Gunnoe 2004; Lee and Guerrero 2001). Lee and Guerrero (2001) had participants watch video of intersubjective touch between coworkers and rate them on the strength of their relational and emotional message, their inappropriateness, and their level of harassment. They found a strong correlation between the ratings on these conditions, with a touch to the face being seen as both signaling the most affection and love and being the most inappropriate and harassing. Unsurprisingly, lacing an arm around another's waist was also rated as being highly inappropriate and harassing. In contrast, a handshake was rated as the most formal form of touch. The results of this study underline the fact that how touch is perceived is highly dependent on the relationship between the initiator and recipient; what can be felt as the most loving touch in one context can also be the most violating in another.

Type of touch can also modulate the emotional message that the touch conveys. Hertenstein et al. (2006a) examined this issue in a study in which participants in both the United States and Spain were given the role of either encoder or decoder of emotional touch. Participants sat on opposite sides of an opaque curtain, and the encoder used touch to the arm to convey one of twelve emotions to the decoder. The decoder then decided which of thirteen emotion words best described the touch he or she felt. Hertenstein et al. found that touch could be reliably used to communicate six emotions, namely, anger, fear, disgust, love, gratitude, and sympathy, at level of accuracy comparable to those found in the transmission and decoding of facial displays and vocal communication (e.g., Elfenbein and Ambady 2002; Scherer et al. 2003). Further research has found that touch to the whole body allowed happiness and sadness to also be reliably decoded (Hertenstein et al. 2009). Notably, all three "prosocial" emotions tested were transmitted reliably through touch, but none of the tested "self-focused" emotions were.

Interestingly, Hertenstein and Keltner (2011) observed a significant gender difference in the successful decoding of the affective meaning of touch. Decoding of sympathy was communicated at above-chance level only when one of the dyad was female, while anger was decoded at above-chance level only when one of the dyad was male. Happiness was only decoded above chance when both the encoder and decoder were female. These findings further illustrate the gender-specific aspects of intersubjective touch.

4   The Effects of Intersubjective Touch

Moving from the factors modulating intersubjective touch, a large amount of research has also been carried out on the effects of touch on both development and social relationships.

4.1   The Role of Touch in Infant Development

One early indicator of the importance of intersubjective touch on infants was the (in)famous research conducted by Harlow (e.g., Harlow and Zimmerman 1959). Harlow showed that infant monkeys prefer to cling to surrogate mothers made of terrycloth than of wire independently of whether the terrycloth mother dispensed food or not. The observed pattern highlighted the importance of comforting physical contact. Research on children who were raised in substandard orphanages and therefore experienced inadequate care, including minimal tactile contact with caregivers, has found that these children's cognitive and social abilities are below average compared to children raised in either normal family homes or better standard institutions (see Maclean 2003 for a review). Given the many other forms of sensory deprivation these children experienced, it is not possible to definitively state that these deficits are the result of deprivation of tactile stimuli (Nelson 2007; Gallace and Spence 2010). However, the lack of nurturing touch from caregivers has been directly linked to infant feeding disorders (Feldman et al. 2004) and to infant angry mood and acts of aggression later in development (Main and Stadtman 1981).

Interestingly, it is not merely the amount of touch that impacts on the infant's well-being. Fergus, Schmidth, and Pickens (1991) found that mothers with more symptoms of depression actually touched their infants more than nondepressed mothers. Analysis of the pattern of interaction between the more depressed mothers and their children revealed that the forms of touch used by these mothers were often intrusive and overstimulatory, for example, poking and tickling. A similar pattern was observed by Cohn and Tronick (1989), who also found that depressed mothers' touching behavior elicited negative affect and gaze aversion in their infants. These studies further emphasize the importance of the manner of touch in intersubjective touch.

Additional evidence for the importance of intersubjective touch in infant development comes from studies showing that close physical contact between infant and caregiver leads to more secure attachment between infants and parents (Anisfeld et al. 1990; for a review, see Duhn 2010), increases weight gain and caloric intake in premature infants (Helders, Cats, and Debast 1989; Scafidi et al. 1990), can control the state of arousal in the infant (Montagu 1971), improves the development of visuomotor skills (Weiss et al. 2004), and plays an important role in communication between infants and caregivers (for a review, see Hertenstein 2002).

4.2 Touch and the Development of Intersubjectivity

The importance of touch in communication between infant and caregiver can be seen a study by Gusella, Muir, and Tronick (1988) that utilized the still-face procedure (Tronick et al. 1978). In this procedure, following a period of normal interaction between mother and infant, the mother assumes a still, nonresponsive face and ceases to provide tactile and vocal stimulation to the infant before again interacting with the infant normally. The period of nonresponsiveness typically leads to a neutral and negative affect in the infant (Ellsworth, Muir, and Hains 1993).

In Gusella, Muir, and Tronick's (1988) study, half the mother–infant dyads conducted normal interaction with tactile as well as visual and auditory stimulation, while the other half used only visual and auditory stimulation. During the still-face period, only infants who received tactile as well as visual and vocal stimulation during normal interaction displayed reduced gaze and smiling. This finding suggests that tactile stimulation plays a key role in guiding and holding the infants' attention onto their mothers. Considering the importance that has been placed on the maintenance of shared attention between infant and parent in terms of the development of an early form of intersubjectivity (Trevarthen and Hubley 1978; Rochat 2009), these results indicate an important role for intersubjective touch in the establishment of social interaction.

The importance of touch in the development of social abilities is also suggested by studies on people suffering from autism. It has been shown that people with autism and Asperger's syndrome are hypersensitive to tactile stimulation (Zwaigenbaum et al. 2007; Blakemore et al. 2006) and infants with autism are often touch averse (Baranek 1999). Given the above evidence for the role of tactile stimulation in establishing intersubjective links between infants and caregivers, it is possible that the aversion to touch shown by autistic infants is a factor in the social deficits characteristic of autistic spectrum disorder. Careful research will be required, however, to tease out the relationship between touch aversion as an infant and social difficulties later in life in people with autistic spectrum disorders.

4.3   Effects of Intersubjective Touch in Adults

As well as playing a vital role in development, intersubjective touch also has powerful effects on adults. The most thoroughly researched of these is the ability of touch to increase compliance. This effect was first noted by Fischer et al. (1976), who found that if students were touched by a librarian while their library card was returned to them they showed greater satisfaction with the library than if they were not touched.

Intersubjective touch can also increase the likelihood of participants returning money left in a phone booth (Brockner et al. 1982) and of giving both time and money to strangers in the street (Kleinke 1977; Willis and Hamm 1980; Guéguen 2002).

In a similar manner, touch has been shown to lead to greater compliance from shoppers to a request to sample products and to an increased likelihood of purchasing the tested product (Smith, Gier, and Willis 1982; Hornik 1992). Finally, touch from a waiter or waitress has been shown to lead to higher tips in restaurants, an effect known as "the Midas touch" (Crusco and Wetzel 1984; Guéguen and Jacob 2005), and, in the context of a pub, to higher alcohol consumption (Kaufman and Mahoney 1999). Quantity of touch can have an effect on compliance, with two touches leading to greater compliance than one (Vaidis and Halimi-Falkowicz 2008).

Two different hypotheses have been developed to explain the effect of touch on increased compliance with requests. One is that touch makes the recipient feel more positively toward the toucher and view their request more favorably (Rose, 1990). In support of this view, it has been found that touch leads the recipient to perceive the toucher in a more positive manner (Alagna et al. 1979; Hubble, Noble, and Robinson 1981), which may then affect the recipient's decision whether or not to agree with the request.

This theory depends on the recipient's conscious awareness that he or she has been touched. However, many studies that have investigated the effects of touch have found that most participants report being completely unaware that they had been touched (Fischer et al. 1976; Guéguen 2002). An alternative explanation offered by Reite (1990) argues that the link between touch and stress reduction established in infancy explains why touch increases both positive appraisal and compliance in adulthood, even when participants are unaware of being touched. On this account, the positive effects of intersubjective touch are due to affective rather than cognitive processes and rely on automatic and implicit mechanisms (see also Gallace and Spence 2010).

5   The Neural Basis of Intersubjective Touch

Is there a distinct sensory and neural system involved in the processing of intersubjective touch as compared to other forms of tactile stimulation?

5.1   C-Touch Afferents: A Receptor for Social Touch?

The cutaneous system has traditionally been thought of as comprising four different modalities, which relay tactile, painful, thermal, and pruritic (itch) information to the central nervous system (McGlone et al. 2007). However, recent advances in our knowledge of the cutaneous system have led a number of researchers (Hollins 2010; McGlone and Reilly 2010; Morrison, Löken, and Olausson 2010) to argue for the inclusion of another distinct modality—that of pleasant touch.

The empirical motivation for this claim comes from the identification of receptors, which respond preferentially to low-force, slow moving stimulation. These unmyelinated C-fibers, known as C-tactile (CT) afferents, were first identified by Johansson et al. (1988) and are found in hairy skin with particular abundance on the face and arms (Nordin 1990; Löken et al. 2009). In contrast, no evidence for CT afferents has been found in nonhairy (glaborous) skin areas, such as the palms of hands and soles of feet (Johansson and Vallbo 1979; Vallbo et al. 1999). These CT afferents have a slow rate of conduction that renders them suboptimal for sensory discrimination (Morrison, Löken, and Olausson 2010). They do, however, have qualities that suggest they play a role in the encoding of affective touch. Löken et al. (2009) investigated the link between activation of CT afferents and the pleasantness of touch by using microneurography to record single afferent activity in participants' forearms while the arm was stroked mechanically and found that CT afferents responded most strongly to touch at a velocity of between 1 and 10 cms^{-1}. A separate group of participants then rated the pleasantness of the different velocities of stroking. There was strong correlation between the velocities of stroking that most strongly activated CT afferents and those rated as most pleasant. Finally, subjective ratings of pleasantness of touch when applied to the forearm or palm (which does not contain CT afferents) were compared, and it was found that participants rated stroking to the forearm as more pleasant than stroking to the palm. Recently, the same researchers studied a group of patients who, because of a rare heritable disorder, have a reduced number of CT afferents (Morrison et al. 2011). They found that these patients judged stroking on the forearm at the velocities that excite CT afferents as less pleasant than healthy controls did. Complementary research in patients who lack myelinated mechanoreceptors, and whose sense of touch is presumably primarily relayed via CT afferents, has found that tactile stimuli to skin areas containing CT afferents elicit weak pleasant sensations that are difficult to localize, while stroking on nonglaborous skin lacking CT afferents produces no reported tactile sensation at all (Olausson et al. 2002, 2008). This dissociation between the two groups of patients provides further evidence that CT afferents are linked to the affective, rather than discriminative, aspects of touch.

It is interesting to note that the type of stimulation that these CT fibers respond to and their distribution across the surface of the body strongly corresponds with the areas of the body most commonly involved in primate grooming (Dunbar 2010). This suggests that CT afferents may have evolved specifically to generate positive sensation in response to grooming and are therefore tuned to react to intersubjective touch. It is also notable that in Hertenstein et al.'s (2006a) study on the recognition of emotion through social touch, love was usually communicated as a slow stroking of the skin—exactly the form of stimulation that CT afferents are most receptive to. Determining the way that the forms of cutaneous stimulation used in the expression of the other emotions documented by Hertenstein et al. activate the skin's sensory receptors would be a valuable addition to our understanding of intersubjective touch.

5.2 Brain Areas Involved in the Processing of Intersubjective Touch

Current research suggests that CT afferents project into the posterior ventral medial and medial dorsal nuclei of the thalamus and then to the dorsal posterior insula and the anterior cingulate (Craig 2009). These areas are linked to the sensory and emotional processing of touch, pain, and other cutaneous sensations (Craig et al. 2000; Craig 2009; Rolls et al. 2003; Vogt 2005). The pathway taken by CT afferents is distinct from that taken by the Aβ fibers that support discriminative touch and which project from the ventral post lateral region of the thalamus into the primary somato-sensory cortex (SI) (Morrison, Löken, and Olausson 2010).

Precisely mapping the cortical areas that process incoming signals from CT afferents has proved difficult because stimulation of glabrous skin activates both CT afferents and Aβ fibers. Olausson et al. (2002, 2008) used fMRI to investigate the brain areas activated by touch in patients lacking Aβ fibers, which meant that the resulting brain activation could be attributed to the activation of CT afferents alone. Tactile stimulation to the forearm led to the activation of the dorsal insula cortex but not to the somatosensory cortices, further demonstrating that CT afferents play little role in the discriminative aspects of touch.

For normal participants, stimulation of glabrous skin, containing only Aβ fibers, and of skin that contains both CT afferents and Aβ fibers, activates both the posterior insula and the somatosensory cortices (Friedman et al. 1986; Disbrow et al. 1998; Olausson et al. 2002). This asymmetry between activation in the insula and in the somatosensory cortices suggests that the insula plays a more general role in the representation of touch, although in the case of signals from Aβ fibers this may be mediated through reciprocal connections between the insula and the somatosensory cortices (Augustine 1996). More research is required to identify how the insula responds to touch on skin containing CT afferents compared to glaborous skin. It is possible that touch on different skin areas produces different temporal patterns in insula activation, reflecting the different pathways that CT afferents and Aβ fibers take to reach the insula.

The posterior insula is known to play an important role in generally processing both pleasant touch (Löken et al. 2009) and pain (Afif et al. 2008; Ostrowsky et al. 2002) and has also recently been demonstrated to contain separate somatotopic maps for pain (Mazzola et al. 2009) and pleasant touch (Björnsdotter et al. 2009). This suggests that the posterior insula plays a role in representing the affective dimension of cutaneous stimulation. More broadly, the insula is also involved in the regulation of social emotions and empathy (Lamm and Singer 2010), in interoceptive awareness (Craig 2009; Critchley et al. 2004), and in body representation (Tsakiris 2010). These various functions, combined with the direct input the insula receives from CT afferents points to this area playing a key role in the emotional and social aspects of intersubjective touch.

Another key brain area involved in the processing of affective touch is the orbitofrontal cortex (OFC). An fMRI study by Francis et al. (1999) found that, while neutral touch activates the somatosensory cortices, pleasant touch leads to the additional activation of the OFC. In another fMRI study, McCabe et al. (2008) found that, compared with touch to the palm, touch to skin containing CT afferents produces activation in the mid-OFC. It has also been found that separate parts of the OFC are involved in the processing of pleasant and painful tactile stimuli and that painful and pleasurable touch leads to increased activation in the posterior and inferior cingulate respectively (Rolls et al. 2003).

The OFC is known to be activated by large number of rewarding and punishing stimuli, including: taste (Rolls et al. 1989), odor (Zald and Pardo 1997; O'Doherty et al. 2000), monetary incentives (Thut et al. 1997; Knutson et al. 2001), and erotic stimuli (Karama et al. 2002). This suggests that the OFC is involved in the processing of rewarding and punishing stimuli regardless of modality (Rolls 2010; Kringelbach 2005).

5.3   Neural Processing of the Context of Intersubjective Touch

We have already noted the importance of culture, gender, situation, and personal beliefs on influencing intersubjective touch. Given this evidence, it is clear that the understanding how higher-level cognitive factors can modulate the neural representation of intersubjective touch is vital for developing a complete neural explanation of intersubjective touch.

In the only study to date to investigate this issue, McCabe et al. (2008) found that participants gave higher ratings for the pleasantness of touch when a cream was rubbed on their arm while they viewed a label that said "rich" than when they viewed a label that said "thin." fMRI analysis indicated that activation in the insula, ventral striatum and parietal area 7 was modulated by touch when accompanied by the rich rather than the thin label. The pleasantness ratings were negatively correlated with activation in the lateral OFC and positively correlated with activation in the pregenual cingulate. These findings provide further evidence that the hedonic experience of touch involves activity in the OFC and demonstrate that brain areas involved in processing the bodily sensation of touch are affected by high-level cognitive processes such as language. At present, however, it is unclear whether other cognitive influences, such as culture or personal beliefs, modulate the neural basis of intersubjective touch in a similar manner.

5.4   The Relationship between the Insula and Orbitofrontal Cortex in the Processing of Intersubjective Touch

The relationship between the processing of pleasant touch in the insula and OFC is currently unclear. However, considering current knowledge of their functional roles, it seems likely that the insula is responsible for the integration of the specific sensory characteristics of intersubjective touch into a dynamic model of the state of the body (Craig 2009; Tsakiris 2010) and for the localized experience of pleasant touch on an area of the body (Blakemore et al. 2005). Activation in the OFC, by contrast, appears to reflect the activation of a modality-independent neural system for reward and punishment that processes the general hedonic experience of touch (Rolls et al. 2003; Rolls 2010) and can be modulated by high-level contextual factors (McCabe et al. 2008). The two areas are, however, linked by strong reciprocal connections (Cavada et al. 2000; Kringelbach 2005) and it is clear that, while the insula feeds information on the somatic effects of touch forward to the OFC, the OFC also affects processing in the insula and somatosensory areas. Further investigation of the functional links between these two areas is another promising avenue for future research.

6   Touch and Empathy: Mirroring Touch

The discovery of "mirror neurons" in the Macaque brain, and homologous areas of the human brain, which fire both when the monkey executes and when it observes an action, has led to increased interest in the role that shared neural representations play in allowing us to understand others (for recent reviews, see Rizzolatti and Fabbri-Destro 2008; Decety 2010). Recent studies have found evidence that the use of shared representations occurs not only in brain areas involved in processing actions but also in areas processing emotional and somatosensory stimuli (for a review, see Keysers and Gazzola 2009). Most research into somatosensory empathy has focused on empathy for pain (Avenanti et al. 2005; Singer and Frith 2005), but there is growing evidence linking the perception and experience of touch on one's own body and the observation of touch on another 's body.

6.1   Seeing Enhances Feeling of Touch

Observing touch on other people influences the perception of touch on one's self, suggesting that, rather than merely involving the same brain areas, observation and perception of touch also involve shared cognitive representations (Singer and Lamm 2009; Serino and Haggard 2010).

One example of this interaction between observation of another's body and perception of touch on one's own body is that judgments about tactile stimuli are affected by congruent or incongruent observation of others' bodies (Maravita et al. 2002; Thomas et al. 2006). Another is the finding that observation of others being touched leads to visual enhancement of touch (VET). Haggard (2006) showed that judgments of the orientation of a grating felt by the fingertips was improved when participants observed a hand, but not a nonbody object, at the same time.

In investigating VET further, Longo et al. (2008) used the rubber hand illusion (Botvinick and Cohen 1998) to induce a sense of body ownership over a prosthetic hand. They found that VET was stronger when the rubber hand was felt to be part of participants' bodies than when it was perceived as an external object. This indicates that, while some of the effects of VET can be related to intersubjective body representation, self-specific representations of body ownership also contribute to VET. Comparable findings, reported by Serino, Pizzoferrato, and Làdavas (2008), showed that self-specific representations play a role in VET. Participants received near-threshold tactical stimulation while viewing images of either their face, the face of another, or an object being touched. It was found that participants' judgments of whether they were being touch were most accurate when observing their own face and least accurate when observing a nonface object. A second study (Serino, Giovagnoli, and Làdavas 2009) showed that, in addition to being modulated by distinctions between self and other, VET is also modulated by participants' in-group identifications. Participants were more accurate in detecting touch when they observed fingers touching a face from the same ethnic or political group as themselves. The finding of an effect of political affiliation suggests that, in addition to being modified by the *physical* similarity of the observed person, VET can be modulated by more abstract identifiers of in-group/ out-group status.

6.2   Shared Neural Networks for Experience and Observation of Touch

Keysers et al. (2004) were the first to explicitly investigate the neural link between experience of touch on one's body and observation of touch on others. Using fMRI, they found that the secondary somatosensory cortex (SII), but not the primary somatosensory cortex (SI), showed increased bold activation during both the observation of touch and perception of touch, and that this activation was the same if the touched object was a body part or an inanimate object. The strongest evidence for the existence of shared brain areas for both experience and observation of touch comes from a form of synesthesia known as "mirror-touch synesthesia" in which observing others being touched cause tactile sensation on the corresponding part of one's own body. Interestingly, mirror touch synesthetes score high on empathy questionnaires compared to both nonsynesthetes and synesthetes with other forms of synesthesia (Banissy and Ward 2007), which suggests a link between somatosensory empathy and other more cognitive forms of empathy.

Blakemore et al. (2005) used fMRI to discover the brain areas activated by observation of touch to the face or neck in C, a subject with mirror-touch synesthesia. Compared with nonsynesthetes C showed greater activation in areas of the somatosensory cortex. The bilateral insula was also activated during the observation of touch, an area that was not activated by nonsynesthetic participants. As well as being involved in the processing of tactile stimuli (Olausson et al. 2002), the insula is also known to have a role in the attribution of actions to one's self as opposed to others (Farrer and Frith 2002). It seems likely therefore that this insula activation, along with the hyperactivation of C's somatosensory cortices, is responsible for C's tactile experience of touch to others.

In addition to finding evidence of abnormal activation patterns in the somatosensory cortex and insula of a mirror-touch synesthete, Blakemore et al. (2005) observed that even for nonsynesthetes observation of touch led to increased BOLD activation in somatosensory areas. However, in contrast to Keysers et al., they found a significant increase in somatosensory activation when observing humans as opposed to objects being touched, and they also found activation in SI that was organized in a somatotopic manner, with touch to the face causing activation of the face area in SI. Blakemore et al. (2005) suggest that these differences may be due to the fact that their study used faces because the social importance of faces might make them particularly salient for somatosensory mirroring. The finding of SI activity by Blakemore et al., but not by Keysers et al., may also relate to whether the observed touch was perceived as intersubjective or not. In Keysers et al.'s study the touch that participants observed was made by a rod, whereas in Blakemore et al.'s study the observed touch was made by a human hand. The relevance of the touching object being a hand rather than an inanimate object is also suggested by Ebisch et al.'s (2008) finding that activation in the BA2 region of SI is greater when observing touch from a human hand than when observing touch from an inanimate object. Furthermore, this activation was modulated by participants' perception of the intentionality of the touch, suggesting that this activation was related to mirroring of the touching, rather than touched, hand.

As shown in both Keysers et al.'s and Blakemore et al.'s studies, SII is also involved in touch mirroring. Several other studies have found SII activation in response to the observation of touch on the hands (Ebisch et al. 2008; Schaefer et al. 2009) and to the erogenous zones of both men and women (Ferretti et al. 2005). Activation in SII has been a more consistent finding than SI activation and seems to be less dependent on the type of touching object than SI activation (Ebisch et al. 2008). Given the large receptive fields found in SII (Krubitzer et al. 1995), it seems likely that SII activation is involved in representing the experience of being touched in a similar way to the observed touch rather than in specifying the location on the body where touch occurred (Keysers, Kaas, and Gazzola 2010).

At present, what we know of the neural basis of empathy for touch suggests that both the primary and secondary somatosensory cortices possess "mirror" properties, with activation of the BA2 region of SI being largely, although not entirely, involved in the mirroring of active touch, while mirror activation in SII relates to the mirroring of felt touch on another's body. However, it is also clear from the findings of Blakemore et al. (2005) that this functional separation in sensorimotor contagion for touch is not wholly complete. An interesting question for future investigation is whether the amount of activation in SI and SII can be modulated by the perspective people put themselves in while watching intersubjective touch. Based on previous findings, one might expect to find greater SI activation when participants identify with the initiator of touch and greater SII activation when they identify with the recipient.

Finally, a recent study by Ebisch et al. (2011) provides evidence that the insula is particularly tuned to the observation of affective intersubjective touch. Participants observed four videos of types of touch: touch from a neutral object, neutral touch from a hand, positive touch from a hand (caress), and negative touch from a hand (hit). Analysis of BOLD activation found that activation in the insula was modulated when positive or negative touch was observed. This activation was in contrast to activation in SII, which was activated by the observation of both animate and inanimate touch.

6.3 Representing Touch to Self and Other

How does the somatosensory system distinguish between experiencing a touch on one's self and observing touch on another? As mentioned above,

Blakemore et al. (2005) found that observed touch led to activation of the insula in their mirror-touch synesthetic participant but not in controls, suggesting that insula activation is necessary for observed touch to be consciously experienced on one's body. An alternative explanation is offered by Keysers, Kaas, and Gazzola (2010), who note that only in the case of an actual touch on the participants body is the BA3 region of SI, the primary area for processing tactile stimuli, activated. They draw a parallel between this primary somatosensory processing and the case of blindsight patients, who have damage to the primary visual cortex but process visual information in higher-level visual areas without experiencing conscious vision (Stoerig and Cowey 1997). This suggests that lack of activation in BA3 may be responsible for the lack of conscious experience of touch in the case of sensorimotor empathy.

6.4 Intersubjective Multisensory Stimulation: Touch on the Self and the Other

Another key theme regarding the observation of touch in others emerges from recent work on self–other distinctions and the mapping of other people's sensory experiences onto one's own central nervous system. These effects seem to be modulated by the perceived physical similarity between self and other, at least as evidenced by the behavioral evidence to date. Several recent studies have looked at how the sensorimotor system processes bodies that are either similar or dissimilar to one's own. In addition to the findings of Serino et al. (2009; see above) it has been found that motor cortex excitability, as measured by single-pulse TMS over the primary motor cortex, which presumably reflects activity within the human mirror-neuron system, is greater when participants observe gestures performed by an actor of the same ethnic background compared to an out-group member (Molnar-Szakacs et al. 2007). More recently, it has been shown that racial bias reduces empathetic sensorimotor resonance in response to the observed pain of out-group hands (Avenanti, Sirigu, and Aglioti 2010).

However, studies using multisensory integration to study the malleability of self-representations (see Tsakiris 2010 for a review) suggest that sensory processing might alter the perceived physical (Longo et al. 2009) and psychological similarity between self and other, and possibly between in-group and out-group members. For example, seeing a face being touched at the same time as your own face might affect not only your self-face recognition (Tsakiris 2008) but also the felt closeness to the other (Paladino et al. 2010). To the extent that perceived physical similarity is a strong modulator of social cognition and its underlying neural processes, understanding if and how multisensory processing can alter self-representations might be important for probing the sensorimotor basis of social cognition.

7 Conclusion

In this review, we have attempted to give a picture of current research into the social effects of touch across cognitive science. In doing so, we have noted both the powerful effects that touch can have on social interactions and how the interpretation of intersubjective touch can be affected by contextual and sociocultural factors. At the same time, we have highlighted recent advances in neuroscience that show both that intersubjective touch has an anatomically distinct basis in the central nervous system and the role of our own brains' representations of the body when perceiving touch on others. However, to date, the operationalization of intersubjective touch in social cognitive neuroscience has lacked the fine-grained distinctions and manipulations implemented in social psychology literature. Future research should aim to utilize these distinctions to, for example, investigate how neural representations of touch are modulated by the individual performing the touch or by the situation that the touch occurs in. The role of touch on interpersonal multisensory experience is also ripe for future research. It is, for example, currently unknown how far the in-group/out-group identity of the other person being stimulated affects how much they are included in one's self-representation or whether multisensory stimulation can modulate recently observed in-group/out-group biases in sensorimotor empathy.

Finally, it is clear that a full understanding of the social functions of touch will require an interdisciplinary approach. Developing a mature science of intersubjective touch will require researchers from across the cognitive sciences to join hands and work together.

Acknowledgments

Harry Farmer and Manos Tsakiris were supported by the Volkswagen Stiftung, European Platform for Life Sciences, Mind Sciences, and Humanities.

References

Afif, A., D. Hoffmann, L. Minotti, A. L. Benabid, and P. Kahane. 2008. Middle short gyrus of the insula implicated in pain processing. *Pain* 138 (3):546–555.

Alagna, E., S. Whitcher, J. Fisher, and E. Wicas. 1979. Evaluative reaction to inter-personal touch in a counseling interview. *Journal of Counseling Psychology* 26:465–472.

Anisfeld, E., V. Casper, M. Nozyce, and N. Cunningham. 1990. Does infant carrying promote attachment? An experimental study of the effects of increased physical contact on the development of attachment. *Child Development* 61 (5):1617–1627.

Atkinson, J., and O. Braddick. 1982. Sensory and perceptual capacities of the neonate. In *Psychobiology of the Human Newborn*, 191–220, ed. P. Stratton. London: John Wiley.

Augustine, J. R. 1996. Circuitry and functional aspects of the insular lobe in primates including humans. *Brain Research Reviews* 22:229–244.

Avenanti, A., D. Bueti, G. Galati, and S. M. Aglioti. 2005. Transcranial magnetic stimulation highlights the sensorimotor side of empathy for pain. *Nature Neuroscience* 8:955–960.

Avenanti, A., A. Sirigu, and S. M. Aglioti. 2010. Racial bias reduces empathic sensorimotor resonance with other-race pain. *Current Biology* 20 (11):1018–1022.

Banissy, M. J., and J. Ward. 2007. Mirror-touch synesthesia is linked with empathy. *Nature Neuroscience* 10 (7):815–816.

Baranek, G. T. 1999. Autism during infancy: A retrospective video analysis of sensory-motor and social behaviors at 9–12 months of age. *Journal of Autism and Developmental Disorders* 29:213–224.

Barnett, K. 1972. A theoretical construct of the concepts of touch as they relate to nursing. *Nursing Research* 21:102–110.

Björnsdotter, M., L. Löken, H. Olausson, A. Vallbo, and J. Wessberg. 2009. Somatotopic organization of gentle touch processing in the posterior insular cortex. *Journal of Neuroscience* 29 (29):9314–9320.

Blakemore, S.-J., D. Bristow, G. Bird, C. Frith, and J. Ward. 2005. Somatosensory activations during the observation of touch and a case of vision-touch synaesthesia. *Brain* 128 (7):1571–1583.

Blakemore, S.-J., T. Tavassoli, S. Calo, R. M. Thomas, C. Catmur, U. Frith, and P. Haggard. 2006. Tactile sensitivity in Asperger syndrome. *Brain and Cognition* 61:5–13.

Botvinick, M., and J. Cohen. 1998. Rubber hands "feel" touch that eyes see. *Nature* 391 (6669):756.

Brockner, J., B. Pressman, J. Cabbitt, and P. Moran. 1982. Nonverbal intimacy sex and compliance: A field study. *Journal of Nonverbal Behavior* 6:253–258.

Burgoon, J. K. and A. S. Ebesu Hubbard. 2005. Cross-cultural and intercultural applications of expectancy violations theory and interaction adaptation theory. In *Theorizing about Intercultural Communication*, 149–171, ed. W. B. Gudykunst. Thousand Oaks, CA: Sage Publications.

Cavada, C., T. Company, J. Tejedor, R. J. Cruz-Rizzolo, and F. Reinoso-Suárez. 2000. The anatomical connections of the macaque monkey orbitofrontal cortex: A review. *Cerebral Cortex* 10:220–242.

Cohn, J. F., and E. Z. Tronick. 1989. Specificity to infants' response to mothers' affective behavior. *Journal of the American Academy of Child and Adolescent Psychiatry* 5 (7):663–667.

Craig, A. D. 2009. How do you feel—now? The anterior insula and human awareness. *Nature Reviews: Neuroscience* 10 (1):59–70.

Craig, A. D., K. Chen, D. Bandy, and E. M. Reiman. 2000. Thermosensory activation of insular cortex. *Nature Neuroscience* 3:184–190.

Critchley, H. D., S. Wiens, P. Rotshtein, A. Ohman, and R. J. Dolan. 2004. Neural systems supporting interoceptive awareness. *Nature Reviews: Neuroscience* 7 (2):189–195.

Crusco, A. H., and C. G. Wetzel. 1984. The Midas touch: The effects of interpersonal touch on restaurant tipping. *Personality and Social Psychology Bulletin* 10:512–517.

De Thomas, M. T. 1971. Touch power and the screen of loneliness. *Perspectives in Psychiatric Care* 9:112–118.

Decety, J. 2010. To what extent is the experience of empathy mediated by shared neural circuits? *Emotion Review* 2:204–207.

de Waal, F. 1989. Food sharing and reciprocal obligations among chimpanzees. *Journal of Human Evolution* 18 (5):433–459.

DiBiase, R. and J. Gunnoe. 2004. Gender and culture differences in touching behaviour. *Journal of Social Psychology* 144 (1):49–62.

Disbrow, E., D. Slutsky, and L. Krubitzer. 1998. Cortical and thalamic connections of the parietal ventral area (PV) in macaque monkeys. *Society for Neuroscience Abstracts* 24:130.

Duhn, L. 2010. The importance of touch in the development of attachment. *Advances in Neonatal Care* 10 (6):294–300.

Dunbar, R. I. M. 1980. Determinants and evolutionary consequences of dominance among female gelada baboons. *Behavioral Ecology and Sociobiology* 7:253–265.

Dunbar, R. 1993. Coevolution of neocortical size, group size, and language in humans. *Behavioral and Brain Sciences* 16 (4):681–693.

Dunbar, R. 2010. The social role of touch in humans and primates: Behavioral function and neurobiological mechanisms. *Neuroscience and Biobehavioral Reviews* 34 (2):260–268.

Ebisch, S. J. H., F. Ferri, A. Salone, M. G. Perrucci, L. D'Amico, F. M. Ferro, G. L. Romani, and V. Gallese. 2011. Differential involvement of somatosensory and interoceptive cortices during the observation of affective touch. *Journal of Cognitive Neuroscience* 23 (7):1808–1822.

Ebisch, S. J. H., M. G. Perrucci, A. Ferretti, C. Del Gratta, G. L. Romani, and V. Gallese. 2008. The sense of touch: Embodied simulation in a visuotactile mirroring mechanism for observed animate or inanimate touch. *Journal of Cognitive Neuroscience* 20 (9):1611–1623.

Elfenbein, H. A., and N. Ambady. 2002. On the universality and cultural specificity of emotion recognition: A meta-analysis. *Psychological Bulletin* 128:203–235.

Ellsworth, C. P., D. W. Muir, and S. M. J. Hains. 1993. Social competence and person-object differentiation: An analysis of the still-face effect. *Developmental Psychology* 29:63–73.

Farrer, C., and C. D. Frith. 2002. Experiencing oneself *vs.* another person as being the cause of an action: The neural correlates of the experience of agency. *NeuroImage* 15:596–603.

Feldman, R., M. Keren, O. Gross-Rozval, and S. Tyano. 2004. Mother and child's touch patterns in infant feeding disorders: Relation to maternal, child, and environmental factors. *Journal of the American Academy of Child and Adolescent Psychiatry* 43:1089–1097.

Fergus, E. L., J. Schmidth, and J. Pickens. 1998. Touch during mother-infant interactions: The effects of parenting stress, depression and anxiety. Poster presented at the meeting of the International Society of Infant Studies, Atlanta, Georgia, April 1998.

Ferretti, A., M. Caulo, C. Del Gratta, R. Di Matteo, A. Merla, F. Montorsi, V. Pizzella, et al. 2005. Dynamics of male sexual arousal: Distinct components of brain activation revealed by fMRI. *NeuroImage* 26:1086–1096.

Field, T. 2001. *Touch.* Cambridge, MA: MIT Press.

Field, T. 2011. Touch for socioemotional and physical well-being: A review. *Developmental Review* 30 (4):367–383.

Fischer, J. D., M. Rytting, and R. Heslin. 1976. Hands touching hands: Affective and evaluative effects of interpersonal touch. *Sociometry* 39:416–421.

Forden, C. 1981. The influence of sex-role expectations on the perception of touch. *Sex Roles* 7:889–894.

Francis, S., E. T. Rolls, R. Bowtell, F. McGlone, J. O'Doherty, A. Browning, S. Clare, and E. Smith. 1999. The representation of pleasant touch in the brain and its relationship with taste and olfactory areas. *Neuroreport* 10:453–459.

Frank, L. K. 1957. Tactile communication. *Genetic Psychology Monographs* 56: 209–225.

Friedman, D. P., E. A. Murray, J. B. O'Neill, and M. Mishkin. 1986. Cortical connections of the somatosensory fields of the lateral sulcus of macaques: Evidence for a corticolimbic pathway for touch. *Journal of Comparative Neurology* 252:323–347.

Gallace, A., and C. Spence. 2010. The science of interpersonal touch: An overview. *Neuroscience and Biobehavioral Reviews* 34 (2):246–259.

Goffman, E. 1956. The nature of deference and demeanor. *American Anthropologist* 58:473–502.

Goldman, A., and F. de Vignemont. 2009. Is social cognition embodied? *Trends in Cognitive Sciences* 13 (4):154–159.

Goldstein, A. G., and J. Jeffords. 1981. Status and touching behavior. *Bulletin of the Psychonomic Society* 17:79–81.

Grafton, S. T. 2009. Embodied cognition and the simulation of action to understand others. *Annals of the New York Academy of Sciences* 1156:97–117.

Guerrero, L. K. and P. A. Andersen. 1994. Patterns of matching and initiation: Touch behavior and touch avoidance across romantic relationship stages. *Journal of Nonverbal Behavior* 18 (2):137–153.

Guéguen, N. 2002. Touch, awareness of touch, and compliance with a request. *Perceptual and Motor Skills* 95:355–360.

Guéguen, N., and C. Jacob. 2005. The effect of touch on tipping: An evaluation in a French bar. *International Journal of Hospitality Management* 24 (2):295–299.

Gusella, J. L., D. Muir, and E. Tronick. 1988. The effect of manipulating maternal behavior during an interaction on three- and six-month-olds' affect and attention. *Child Development* 59:1111–1124.

Haggard, P. 2006. Just seeing you makes me feel better: Interpersonal enhancement of touch. *Social Neuroscience* 1 (2):104–110.

Hall, E. T. 1966. *The Hidden Dimension*. New York: Doubleday.

Hall, J. A. 1996. Touch, status, and gender at professional meetings. *Journal of Nonverbal Behavior* 20 (1):23–44.

Hall, J. A., and E. M. Veccia. 1990. More "touching" observations: New insights on men, women, and interpersonal touch. *Journal of Personality and Social Psychology* 59 (6):1155.

Harlow, H. F., and R. R. Zimmerman. 1959. Affectional responses in the infant monkey. *Science* 130:421–432.

Helders, P. J., B. P. Cats, and S. Debast. 1989. Effects of a tactile stimulation/range-finding programme on the development of VLBW-neonates during the first year of life. *Child: Care, Health, and Development* 15:369–380.

Henley, N. M. 1973. Status and sex: Some touching observations. *Bulletin of the Psychonomic Society* 2:91–93.

Hertenstein, M. J. 2002. Touch: Its communicative functions in infancy. *Human Development* 45 (2):70–94.

Hertenstein, M. J., R. Holmes, M. McCullough, and D. Keltner. 2009. The communication of emotion via touch. *Emotion* 9 (4):566–573.

Hertenstein, M. J., and D. Keltner. 2011. Gender and the communication of emotion via touch. *Sex Roles* 64:70–80.

Hertenstein, M. J., D. Keltner, B. App, B. Bulleit, and A. R. Jaskolka. 2006a. Touch communicates distinct emotions. *Emotion* 6 (3):528–533.

Hertenstein, M. J., J. M. Verkamp, A. M. Kerestes, and R. M. Holmes. 2006b. The communicative functions of touch in humans, nonhuman primates, and rats: A review and synthesis of the empirical research. *Genetic, Social, and General Psychology Monographs* 132 (1):5–94.

Heslin, R., and D. Boss. 1980. Nonverbal intimacy in airport arrival and departure. *Personality and Social Psychology Bulletin* 6 (2):248–252.

Hollins, M. 2010. Somesthetic senses. *Annual Review of Psychology* 61:243–271.

Hornik, J. 1992. Effects of physical contact on customers shopping time and behaviour. *Marketing Letters* 3:49–55.

Hubble, M., E. Nobel, and S. Robinson. 1981. The effect of counselor touch in an initial counseling session. *Journal of Counseling Psychology* 13:218–223.

Johansson, R. S., M. Trulsson, K. A. Olsson, and K. G. Westberg. 1988. Mechanoreceptor activity from the human face and oral mucosa. *Experimental Brain Research* 72:204–208.

Johansson, R. S., and A. B. Vallbo. 1979. Tactile sensibility in the human hand: Relative and absolute densities of four types of mechanoreceptive units in glabrous skin. *Journal of Physiology* 286:283–300.

Jones, S. E. 1986. Sex differences in touch communication. *Western Journal of Communication* 50 (3):227–241.

Jourard, S. M. 1966. An exploratory study of body accessibility. *British Journal of Social and Clinical Psychology* 5:221–231.

Karama, S., A. R. Lecours, J. M. Leroux, P. Bourgouin, G. Beaudoin, S. Joubert, and M. Beauregard. 2002. Areas of brain activation in males and females during viewing of erotic film excerpts. *Human Brain Mapping* 16 (1):1–13.

Kaufman, D., and J. M. Mahoney. 1999. The effect of waitresses' touch on alcohol consumption in dyads. *Journal of Social Psychology* 139:261–267.

Keysers, C., B. Wicker, V. Gazzola, J.-L. Anton, L. Fogassi, and V. Gallese. 2004. A touching sight: SII/PV activation during the observation and experience of touch. *Neuron* 42:335–346.

Keysers, C., and V. Gazzola. 2009. Expanding the mirror: Vicarious activity for actions, emotions, and sensations. *Current Opinion in Neurobiology* 19 (6):666–671.

Keysers, C., J. H. Kaas, and V. Gazzola. 2010. Somatosensation in social perception. *Nature Reviews. Neuroscience* 11:417–428.

Kleinke, C. 1977. Compliance to requests made by gazing and touching experimenters in field settings. *Journal of Experimental Social Psychology* 13:218–223.

Knutson, B., C. M. Adams, G. W. Fong, and D. Hommer. 2001. Anticipation of increasing monetary reward selectively recruits nucleus accumbens. *Journal of Neuroscience* 21:159.

Kringelbach, M. L. 2005. The human orbitofrontal cortex: Linking reward to hedonic experience. *Nature Reviews: Neuroscience* 6:691–702.

Krubitzer, L., J. Clarey, R. Tweedale, G. Elston, and M. Calford. 1995. A redefinition of somatosensory areas in the lateral sulcus of macaque monkeys. *Journal of Neuroscience* 15:3821–3839.

Lamm, C., and T. Singer. 2010. The role of anterior insular cortex in social emotions. *Brain Structure & Function* 214:579–591.

Lee, J. W., and L. K. Guerrero. 2001. Types of touch in cross-sex relationships between coworkers: Perceptions of relational and emotional messages, inappropriateness, and sexual harassment. *Journal of Applied Communication Research* 29 (3):197–220.

Lehmann, J., A. H. Korstjens, and R. I. M. Dunbar. 2007. Group size, grooming, and social cohesion in primates. *Animal Behaviour* 74:1617–1629.

Löken, L. S., J. Wessberg, I. Morrison, F. McGlone, and H. Olausson. 2009. Coding of pleasant touch by unmyelinated afferents in humans. *Nature Neuroscience* 12:547–548.

Longo, M. R., F. Schüür, M. P. M. Kammers, M. Tsakiris, and P. Haggard. 2009. Self-awareness and the body image. *Acta Psychologica* 132 (2):166–172.

Longo, M. R., S. Cardozo, and P. Haggard. 2008. Visual enhancement of touch and the bodily self. *Consciousness and Cognition* 17 (4):1181–1191.

Major, B., A. M. Schmidlin, and L. Williams. 1990. Gender patterns in social touch: The impact of setting and age. *Journal of Personality and Social Psychology* 58 (4):634–643.

Maclean, K. 2003. The impact of institutionalization on child development. *Development and Psychopathology* 15:853–884.

Main, M., and J. Stadtman. 1981. Infant response to rejection of physical contact by the mother: Aggression, avoidance, and conflict. *Journal of the American Academy of Child Psychiatry* 20:292–307.

Maravita, A., C. Spence, C. Sergent, and J. Driver. 2002. Seeing your own touched hands in a mirror modulates cross-modal interactions. *Psychological Science* 13:350–355.

Mazzola, L., J. Isnard, R. Peyron, M. Guénot, and F. Mauguière. 2009. Somatotopic organization of pain responses to direct electrical stimulation of the human insular cortex. *Pain* 146:99–104.

McCabe, C., E. T. Rolls, A. Bilderbeck, and F. McGlone. 2008. Cognitive influences on the affective representation of touch and the sight of touch in the human brain. *Social Cognitive and Affective Neuroscience* 3 (2):97–108.

McGlone, F., and D. Reilly. 2010. The cutaneous sensory system. *Neuroscience and Biobehavioral Reviews* 34 (2):148–159.

McGlone, F., Å. Vallbo, H. Olausson, L. Löken, and J. Wessberg. 2007. Discriminative touch and emotional touch. *Canadian Journal of Experimental Psychology* 61 (3):173–183.

Molnar-Szakacs, I., A. D. Wu, F. J. Robles, and M. Iacoboni. 2007. Do you see what I mean? Corticospinal excitability during observation of culture-specific gestures. *PLoS ONE* 2 (7):e626.

Morrison, I., L. S. Löken, and H. Olausson. 2010. The skin as a social organ. *Experimental Brain Research* 204 (3):305–314.

Morrison, I., M. Bjornsdotter, and H. Olausson. 2011. Vicarious responses to social touch in posterior insular cortex are tuned to pleasant caressing speed. *Journal of Neuroscience* 31 (26):9554–9562.

Montagu, A. 1971. *Touching: The Human Significance of the Skin.* New York: Columbia University Press.

Nelson, C. A. 2007. A neurobiological perspective in early human deprivation. *Child Development Perspectives* 1:13–18.

Nordin, M. 1990. Low-threshold mechanoreceptive and nociceptive units with unmyelinated (C) fibres in the human supraorbital nerve. *Journal of Physiology* 426:229–240.

O'Doherty, J., E. T. Rolls, S. Francis, R. Bowtell, F. McGlone, G. Kobal, B. Renner, and G. Ahne. 2000. Sensory-specific satiety-related olfactory activation of the human orbitofrontal cortex. *Neuroreport* 11:399–403.

Olausson, H., Y. Lamarre, H. Backlund, C. Morin, B. G. Wallin, G. Starck, S. Ekholm, I. Strigo, K. Worsley, and Å. B. Valbo. 2002. Unmyelinated tactile afferents signal touch and project to insular cortex. *Nature Neuroscience* 5 (9):900–904.

Olausson, H., J. Cole, K. Rylander, F. McGlone, Y. Lamarre, B. G. Wallin, H. Krämer, et al. 2008. Functional role of unmyelinated tactile afferents in human hairy skin: Sympathetic response and perceptual localization. *Experimental Brain Research* 184 (1):135–140.

Ostrowsky, K., M. Magnin, P. Ryvlin, J. Isnard, M. Gueno, and F. Mauguière. 2002. Representation of pain and somatic sensation in the human insula: A study of responses to direct electrical cortical stimulation. *Cerebral Cortex* 12:376–385.

Paladino, M.-P., M. Mazzurega, F. Pavani, and T. W. Schubert. 2010. Synchronous multisensory stimulation blurs self-other boundaries. *Psychological Science* 21 (9):1202–1207.

Remland, M. S., T. S. Jones, and H. Brinkman. 1995. Interpersonal distance, body orientation and touch: Effects of culture, gender, and age. *Journal of Social Psychology* 135 (3):281–297.

Reite, M. 1990. Touch, attachment, and health: Is there a relationship? In *Touch: The Foundation of Experience*, 195–228, ed. K. Barnard and T. Brazelton. Madison, CT: International Universities Press.

Rizzolatti, G., and M. Fabbri-Destro. 2008. The mirror system and its role in social cognition. *Current Opinion in Neurobiology* 18:179–184.

Rochat, P. 2009. *Others in Mind: Social Origins of Self-Consciousness*. Cambridge: Cambridge University Press.

Rolls, E. T. 2010. The affective and cognitive processing of touch, oral texture, and temperature in the brain. *Neuroscience and Biobehavioral Reviews* 34 (2):237–245.

Rolls E. T., Z. J. Sienkiewicz, and S. Yaxley. 1989. Hunger modulates the responses to gustatory stimuli of single neurons in the caudolateral orbitofrontal cortex of the macaque monkey. *European Journal of Neuroscience* 1 (1):53–60.

Rolls, E. T., J. O'Doherty, M. L. Kringelbach, S. Francis, R. Bowtell, and F. McGlone. 2003. Representations of pleasant and painful touch in the human orbitofrontal and cingulate cortices. *Cerebral Cortex* 13 (3):308–317.

Rose, S. A. 1990. Perception and cognition in preterm infants: The senses of touch. In *Touch: The Foundation of Experience*, 299–323, ed. K. Barnard and T. Brazelton. Madison, CT: International Universities Press.

Scafidi F., T. Field, S. Schanberg, C. Bauer, K. Tucci, J. Roberts, and C. Kuhn. 1990. Massage stimulates growth in preterm infants: Replication, A. *Infant Behavior and Development* 13:167–188.

Schaefer, M., B. Xu, H. Flor, and L. G. Cohen. 2009. Effects of different viewing perspectives on somatosensory activations during observation of touch. *Human Brain Mapping* 30:2722–2730.

Scherer, K. R., T. Johnstone, and G. Klasmeyer. 2003. Vocal expression of emotion. In *Handbook of Affective Sciences*, 433–456, ed. R. J. Davidson and H. H. Goldsmith. London: Oxford University Press.

Serino, A., G. Giovagnoli, and E. Làdavas. 2009. I feel what you feel if you are similar to me. *PLoS ONE* 4 (3):e4930.

Serino, A., and P. Haggard. 2010. Touch and the body. *Neuroscience and Biobehavioral Reviews* 34: 224–236.

Serino, A., F. Pizzoferrato, and E. Làdavas. 2008. Viewing a face (especially one's own face) being touched enhances tactile perception on the face. *Psychological Science* 19 (5):434–438.

Shapiro, L. 2010. *Embodied Cognition*. London: Routledge.

Silk, J. B., S. C. Alberts, and J. Altmann. 2003. Social bonds of female baboons enhance infant survival. *Science* 302:1231–1234.

Singer, T., and C. Frith. 2005. The painful side of empathy. *Nature Neuroscience* 8:845–846.

Singer, T., and C. Lamm. 2009. The social neuroscience of empathy. *Annals of the New York Academy of Science* 1156:81–96.

Smith, G., J. Gier, and F. Willis. 1982. Interpersonal touch and compliance with a marketing request. *Basic and Applied Social Psychology* 3:35–38.

Stier, D. S., and J. A. Hall. 1984. Gender differences in touch: An empirical and theoretical review. *Journal of Personality and Social Psychology* 47 (2):440–459.

Stoerig, P., and A. Cowey. 1997. Blindsight in man and monkey. *Brain* 120:535–559.

Thomas, R., C. Press, and P. Haggard. 2006. Shared representations in body perception. *Acta Psychologica* 121 (3):317–330.

Thut, G., W. Schultz, U. Roelcke, M. Nienhusmeier, J. Missimer, R. P. Maguire, and K. L. Leenders. 1997. Activation of the human brain by monetary reward. *Neuroreport* 8:1225–1228.

Trevarthen, C., and P. Hubley. 1978. Secondary intersubjectivity: Confidence, confiding, and acts of meaning in the first year. In *Action, Gesture, and Symbol*, ed. A. Lock. London: Academic Press.

Tronick, E. Z., H. Als, L. Adamson, S. Wise, and T. B. Brazelton. 1978. The infant's response to entrapment between contradictory messages in face-to-face interaction. *Journal of the American Academy of Child Psychiatry* 17:1–13.

Tsakiris, M. 2008. Looking for myself: Current multisensory input alters self-face recognition. *PLoS ONE* 3 (12):e4040.

Tsakiris, M. 2010. My body in the brain: A neurocognitive model of body-ownership. *Neuropsychologia* 48 (3):703–712.

Vaidis, D. C. F., and S. G. M. Halimi-Falkowicz. 2008. Increasing compliance with a request: Two touches are more effective than one. *Psychological Reports* 103:88–92.

Vallbo, A. B., H. Olausson, and J. Wessberg. 1999. Unmyelinated afferents constitute a second system coding tactile stimuli of the human hairy skin. *Journal of Neurophysiology* 81:2753–2763.

Vogt, B. A. 2005. Pain and emotion interactions in subregions of the cingulate gyrus. *Nature Reviews: Neuroscience* 6:533–544.

Weiss, W. J., P. W. Wilson, and D. Morrison. 2004. Maternal tactile stimulation and the neurodevelopment of low birth weight infants. *Infancy* 5:85–107.

Willis, F. N., and H. K. Hamm. 1980. The use of interpersonal touch in securing compliance. *Journal of Nonverbal Behavior* 5:49–55.

Willis, F. N., Jr., and R. Dodds. 1998. Age, relationship, and touch initiation. *Journal of Social Psychology* 138 (1):115–123.

Zald, D. H., and J. V. Pardo. 1997. Emotion, olfaction, and the human amygdala: Amygdala activation during aversive olfactory stimulation. *Proceedings of the National Academy of Sciences of the United States of America* 94:4119–4124.

Zwaigenbaum, L., A. Thurm, W. Stone, G. Baranek, S. E. Bryson, J. Iverson, A. Kau, et al. 2007. Studying the emergence of autism spectrum disorders in high risk infants: Methodological and practical issues. *Journal of Autism and Developmental Disorders* 37:466–480.

6 Touch and the Sense of Reality

Matthew Ratcliffe

1 Introduction

Most philosophical discussions of perception concentrate on vision. There is also a tendency in some areas of philosophy to think of our more general perceptual and cognitive relationship with the world as sight-like: subjects gain access to constituents of an "external" world through a cognitive process somehow akin to seeing, the ultimate aim of which is to achieve a maximally detached or objective "view" of the world.[1] The association between vision and objectivity is at least partly attributable to the fact that vision reveals a world *seemingly* uncorrupted by any relationship with the viewer, a world that appears as "out there," "independent of me." Touch (or "tactual perception"), in contrast, is either closely associated with or partly constituted by perception of one's body. Whichever the case, one cannot perceive the world tactually without perceiving oneself in the process. As Merleau-Ponty (1962, 316) remarks, vision "presents us with a spectacle spread out before us at a distance," whereas, when I perceive something through touch, "I cannot forget in this case that it is through my body that I go to the world" (ibid.). Merleau-Ponty is critical of the tendency to construe our relationship with the world in a voyeuristic way. He maintains that even a seemingly detached "view" quietly presupposes a background sense of participation that is integral to all perception and also underlies all our cognitive achievements. Many others similarly suggest that an understanding of our relationship with the world is somehow obfuscated by the tendency to model it on vision (and, more specifically, on an overly spectatorial and alienating conception of vision).[2]

I think there is something right about such claims, but what exactly is wrong with an emphasis on sight—what is it that gets left out? A full answer to this question might well involve a long inventory of separate complaints. But, in this chapter, I will address just one of them: the failure to adequately characterize our sense of reality and belonging. Whenever we perceive something, visually or otherwise, the experience incorporates a sense that we and it are both "there," situated in a common world. This sense of being part of the world is elusive and difficult to characterize. In fact, it is easy to overlook altogether, if one thinks of perceptual experience in terms of something being presented to a voyeuristic subject. I will argue here that touch is partly constitutive of the sense of reality and belonging, whereas other kinds of sensory experience presuppose it. Hence, touch has a kind of phenomenological primacy over the other senses. Without vision or hearing, one would inhabit a very different experiential world, whereas one would not have a world at all without touch.

My discussion relates to the topic of "the hand" in two principal ways. First of all, I think that an appreciation of the phenomenology of touch has been marred not only by an overemphasis on vision but also by a tendency to take touch with the hands as the paradigmatic example of touch, from which generalizations are then made about all tactual experience. For instance, in *The World of Touch*, David Katz places the emphasis on active touch with the hands, stresses how the hand is a "wondrous tool" with "astonishing versatility," and adds that "one could speak of the world of touch as under the dominion and representation of the hand" (Katz 1989, 28). I certainly do not want to detract from the importance of the hand, an importance that other contributions to this volume make readily apparent. What I do maintain, though, is that exclusive focus on *certain kinds* of tactual experience involving the hands serves to eclipse another form of touch, a "background touch" that is integral to our sense of reality. A second way in which my chapter takes the hand as its theme is in its focus on the experience of one's two hands touching each other, an experience that takes on considerable significance in Merleau-Ponty's work. I revisit and adapt his example, in order to draw a slightly different lesson from it. Contrary to Merleau-Ponty, I show that not all self-touch with the hands can be construed in terms of two poles, perceiver and perceived. Some instances involve a lack of differentiation between perceiver and perceived. Taking these latter cases as a starting point, I suggest that a more pervasive lack of differentiation is integral to our sense of reality and belonging. I also develop a case for the phenomenological importance and, indeed, the primacy of touch.

The importance of touch is a recurrent theme in the phenomenological tradition. In *Ideas II*, Husserl maintains that it is the only sense without which one could not have an "appearing Body [*Leib*]" (1989, 158). His discussion serves as inspiration for Merleau-Ponty (1962, 1964, 1968), who likewise ascribes an important phenomenological role to touch. And an explicit case for the *primacy* of touch over the other senses is sketched by Jonas (1954). As his title "The Nobility of Sight" might suggest, Jonas acknowledges that vision has a number of important attributes that the other senses lack. It is, he says, distinguished from them by its "image performance," which consists of (a) simultaneous perception of many different perceptual contents, (b) "neutralization" of causality (we can perceive something visually without perceiving its effect on us in the process), and (c) perception across distances. In addition, vision facilitates much finer-grained discriminations than the other senses (Jonas 1954, 507).

According to Jonas, vision is implicated in our capacity for objectivity, as it is the only sense that allows a "standing back from the aggressiveness of the world" and presents us with a realm of enduring, spatiotemporally organized entities that exist independent of their impact on us. Even so, he goes on to claim that touch has a kind of primacy over vision. Central to our sense of reality is the experience of affecting and at the same time being affected by things. Vision offers us "a calmed abstract of reality denuded of its raw power," whereas touch supplies what it lacks and is therefore the "true test of reality." For Jonas, vision and touch are not simply distinct senses, with one but not the other supplying a sense of reality. Instead, visual experience *presupposes* a relationship with the world that is partly constituted by touch; sight "rests on the understructure of more elementary functions in which the commerce with the world is carried on on far more elementary terms." Although vision might be the most phenomenologically salient and "sublime" sense, touch is the "primary" sense, in virtue of this presupposed phenomenological role (Jonas 1954, 516–517).[3]

In what follows, I will maintain that touch does indeed have this kind of phenomenological primacy. We already experience ourselves as part of the world when we see, smell, hear, touch, or taste something, part of the same world as what we encounter. But touch is the only one of those senses that is indispensable to this aspect of experience; it is presupposed by the possibility of perceiving anything as "there" or "present." Of course, much hinges on what is meant by "touch," and the bounds of tactual perception are far from clear. It is also debatable whether the "sense of touch" is to be individuated via phenomenological criteria, nonphenomenological criteria, or both, or indeed whether any combination of criteria will suffice.[4] I am concerned here with certain kinds of experience that we *do* refer to as "touch," rather than with whether a principled distinction can be made between a circumscribed "sense of touch" and however many other senses.

In fact, I think that such a distinction cannot be made, though I will not address this here (Ratcliffe 2012).

My strategy will be to reinterpret the phenomenology of touch by starting from what are—according to both everyday and scientific classifications—fairly uncontroversial instances of tactual perception, rather than to import an unfamiliar or controversial conception of touch from the outset. In the process, I will draw on themes in the work of Husserl, Jonas, and especially Merleau-Ponty, but I will do so critically. In my view, all three phenomenologists overstate the importance of certain kinds of tactual experience. Husserl and, more so, Merleau-Ponty overgeneralize from the experience of touching one hand with the other, whereas Jonas errs in emphasizing the perception of "force." All three succeed, to some extent, in conveying the importance of a sense of relatedness, commerce, communion, or reciprocity that is integral to touch. However, they do not provide an adequate account of (a) those tactual experiences that are characterized by a lack of differentiation between perceiver and perceived, and (b) the diversity of tactual experience. I will suggest that both (a) and (b) are central to the contribution made by touch to the sense of reality and belonging.

2 Reality and Belonging

Before addressing the phenomenology of touch, it is important to make clear what is meant by a sense of "reality" and "belonging." Consider the experience of seeing something, such as a cat. In attempting to determine what is integral to the perceptual experience, one might ask whether there is a visual perception with some content that is *followed by* categorization of the perceived entity as a member of the kind "cat," or, alternatively, whether one simply *sees that it is a cat*. In other words, there is an issue over what the contents of perception are and, more specifically, whether they include conceptual contents.[5] For current purposes, I adopt a fairly permissive conception of perceptual experience, according to which we have at least some perceptual grasp of the kinds to which entities belong and the kinds of properties they possess. We do—in at least some cases— perceive a cat, rather than something else that is afterward assigned to the kind "cat" by a nonperceptual process. So long as this much is conceded (and, if one's account aspires to respect the relevant phenomenology, then it should be), a further aspect of perceptual experience also needs to be acknowledged: seeing an entity involves a sense of its being "there" or "present." We do not usually see something and only *afterward* judge it to be really there. The same goes for what we touch—when we make contact with a hard surface, we do not first have a perception of the surface and then affirm its existence.[6] As Husserl puts it, a perceived entity is "naturally and simply there for us as an existing reality as we live naively in perception" (2001, 35); there is a "believing inherent in perceiving" (ibid., 66). This "believing" does not concern what kind a perceived entity belongs to or what properties it might have. It is the belief *that* the entity *is*. And, by "believing," Husserl does not mean in this instance a propositional attitude of the form "I believe that *p*," where *p* is "the entity I see is actually there." It is not a matter of assigning the status "present" to each perceived entity in turn. For Husserl, it consists of a practical, bodily, felt sense of participation in a world, which at the same time amounts to a default acceptance of perceived entities as "there."

An appreciation of something as *there* cannot be fully accounted for in terms of specific perceptual "contents" (regardless of how permissive we are about content), as it is dissociable from them. A perception is sometimes characterized by a sense of uncertainty or doubt—is there really something there, and is it really what it appears to be? One might insist that this involves making a judgment on the basis of some prior perceptual content, rather than there being an additional ingredient of perceptual experience. But I think this is unlikely: doubt or uncertainty often takes the form of a *feeling* that things somehow look wrong, rather than a *judgment* that what is perceived is lacking in some specific respect (see Husserl 1973). We are often unable to attribute this feeling to any perceived property—it just "doesn't look right somehow." Furthermore, there are many other kinds of experience where a sense of something's *being there* is altered or diminished, in a way that is perceptual and yet inexplicable in terms of changes in perceived properties. For instance, at times, things may look strangely unfamiliar, detached, uncanny, wrong, or not quite there. The change sometimes encompasses experience as a whole, rather than specific objects of perception, as often occurs in anomalous experiences associated with psychiatric illness. For instance, what Karl Jaspers calls "delusional atmosphere" presents us with the seeming paradox of everything looking exactly as it did before and yet somehow different or wrong, in a way that cannot be accounted for by any inventory of perceptual contents: "perception is unaltered in itself but there is some change which envelops everything with a subtle, pervasive and strangely uncertain light. A living room which formerly was felt as neutral or friendly becomes dominated by some indefinable atmosphere" (Jaspers 1962, 98).[7]

Of course, "there" is not the same as "real." We not only perceive things to be real; we also adopt nonperceptual beliefs to the effect that something is the case, is not the case, or might be the case. And we can appreciate that an entity is real, even if it is not currently experienced as "there"; a perception or memory of something real can be distinguished from something we imagine or remember imagining. So a sense of something as real is not specific to perceptual experience, and it is not just a matter of sensed presence.[8] But, by the "sense of reality," I do not simply mean "the taking of some entity or other to be real." Rather, I refer to the ability to discriminate between possibilities such as "there," "not there," "possibly there," "perceived," "remembered," "imagined," "real," and "unreal." In the absence of an ability to distinguish such possibilities, we would not be able to encounter anything as "real." In other words, the sense of reality is a modal space that is presupposed by the possibility of perceiving or thinking of anything as real or otherwise. Now, although "real" is not to be identified with "there," the possibility of encountering things as "there" is surely indispensable to a more encompassing sense of reality. Not all of the things we take to be real are perceived by us. However, the possibility of things appearing to us as "there"—as part of the same realm as ourselves, able to affect us and be affected by us—is still an essential aspect of the sense of reality.

According to Husserl, at least in his later work, what I call the sense of reality consists of a kind of habitual, practical sense of belonging to the world that structures all our experience and thought. We do not ordinarily encounter the world as an object of perception or belief. Instead, it is presupposed by perceiving and believing, a context in which things are able to show up as "there," as "real," or otherwise: "It belongs to what is taken for granted, prior to all scientific thought and all philosophical questioning, that the world is—always is in advance—and that every correction of an opinion, whether an experiential or other opinion, presupposes the already existing world" (Husserl 1970, 110).[9]

We can only encounter things in the various ways we do insofar as we already belong to a world. To put it simply, without a sense of being part of a world, we could not perceive or think of things as residing in the same world as ourselves. Hence the sense of reality is not a matter of voyeuristic access to some external realm, from which the voyeur remains separate. Instead, it is phenomenologically inseparable from the sense of belonging. But what does this sense of reality and belonging actually consist of; how might we further characterize the relevant phenomenology? In what follows, I aim to contribute to that task through an examination of tactual experience, starting with the example of one's two hands touching.

Before proceeding, I will defend an assumption about touch that I make throughout, which is similarly made by Husserl, Jonas, Merleau-Ponty, and most others who have written about it. Although one might claim that the "phenomenology of touch" is restricted to what we experience through cutaneous sensation, most of the experiences that I will address also involve perception of bodily position (proprioception) and bodily movement (kinesthesis). This is because what we can perceive through cutaneous sensation alone bears little resemblance to the rich phenomenology routinely associated with our sense of touch. Even passive point contact with an entity incorporates some sense of where on one's body the event is occurring. Most tactual experiences are also inextricable from a sense of whether and how one's body is moving. Perceiving properties such as texture does not involve a series of static snapshots, which are then somehow pieced together: "smoothness is not a collection of similar pressures, but the way in which a surface utilizes the time occupied by our tactile exploration or modulates the movement of my hand" (Merleau-Ponty 1962, 315). Most of the properties that we are said to perceive through touch would be imperceptible without a sense of bodily movement, and it is not simply "movement" that counts. The specific style of movement is important; different kinds of movement are associated with the perception of different properties. For instance, smoothness is perceived through lateral movements, and hardness through up-down movements (Krueger 1982, 9). So touch is phenomenologically intertwined with a sense of bodily position and movement. Separating it from them would leave us with an impoverished abstraction from tactual experience, consisting of little more than bare, nonlocalized sensation, perhaps with some degree of valence.[10]

3 Touching Hands

Husserl and Merleau-Ponty both focus on the experience of tactually perceiving our own bodies and, more specifically, that of touching one hand with the other. For Husserl, this experience is phenomenologically significant because it incorporates a "double sensation," where each hand is experienced as both perceiver and perceived: "the sensation is *doubled* in the two parts of the Body [*Leib*], since each is precisely for the other an external thing that is touching and acting upon it, and each is at the same time Body" (Husserl 1989, 153). This duality of perceiver and perceived facilitates a sense of the body as both that *through* which we experience the world and an entity situated in the perceived world. According to Husserl, touch is the only sense that facilitates perception of what does the perceiving. As he points out, "an eye does not appear to one's own vision," and the experience of seeing one's body does not involve the same level of perceiver-perceived intimacy as self-touch (1989, 155).[11] The same example is later adopted by Merleau-Ponty, who refers to it in several places. In *Phenomenology of Perception*, he remarks:

> if I can, with my left hand, feel my right hand as it touches an object, the right hand as an object is not the right hand as it touches: the first is a system of bones, muscles and flesh brought down at a point of space, the second shoots through space like a rocket to reveal the external object in its place. In so far as it sees or touches the world, my body can therefore be neither seen nor touched. What prevents its ever being an object, ever being "completely constituted" is that it is that by which there are objects. It is neither tangible nor visible in so far as it is that which sees and touches. (1962, 92)

Merleau-Ponty emphasizes the reversibility of the roles of perceiver and perceived more so than Husserl does. Insofar as the body perceives, it is not itself perceived. This is exemplified by the fact that, when the touching hand becomes an object of perception, it ceases to perceive in the way that it previously did. When we try to attend to the touching hand in the act of perceiving, the roles of the hands reverse: perceiver becomes perceived and vice versa. It is this reversibility that facilitates a sense of the body as both perceptual subject and worldly object:

> Thus I touch myself touching; my body accomplishes "a sort of reflection." In it, through it, there is not just the unidirectional relationship of the one who perceives to what he perceives. The relationship is reversed, the touched hand becomes the touching hand, and I am obliged to say that the sense of touch here is diffused into the body—that the body is a "perceiving thing," a "subject-object." (Merleau-Ponty 1964, 166)[12]

According to both Merleau-Ponty and Husserl, the importance of this kind of experience extends beyond perceiving our own bodies: it is only insofar as I can be an object to myself that I can acquire a more general sense of being an entity among others, part of the world. For Merleau-Ponty, the perceiver–perceived duality of touch is also more specifically implicated in the possibility of intersubjectivity. The reversibility that I experience in one of my hands when it touches or is touched by the other also comes into play when I hold another person's hand. This constitutes the sense that I am in contact with another subject, rather than with a mere object of perception: "his hand is substituted for my left hand . . . he and I are like organs of one single intercorporeality" (ibid., 168). He also maintains that there is an analogous reversibility involved in seeing another person and being simultaneously seen by him or her. However, this presumably depends in some way on the reversibility of touch. As noted by Husserl, seeing one's own body does not incorporate the same kind of duality of perceiver and perceived, a duality that is, according to Merleau-Ponty, inextricable from the capacity for intersubjectivity.[13]

Merleau-Ponty's last use of the touching hands example is in his unfinished work *The Visible and the Invisible*, where again the duality and reciprocity that characterizes self-touch is taken to be a source of important phenomenological insight: "It cannot be by incomprehensible accident that the body has this double reference; it teaches us that each calls for the other" (1968, 137). He indicates here that reversibility serves to reveal a philosophically elusive commonality between the perceiver and the perceived, which is presupposed by the phenomenological distinction between them. And he uses the term "flesh" to refer to this "formative medium of the object and the subject" that has no established name (ibid., 147). I will suggest that Merleau-Ponty is right to emphasize the reciprocity between perceiver and perceived, and also to posit a kind of commonality that this reciprocity presupposes. However, he overstates the significance of reversibility in touch and more generally. According to Merleau-Ponty, a hand cannot at the same time occupy the positions of perceiver and perceived: "the two hands are never simultaneously in the relationship of touched and touching to each other" (1962, 93). Hence the hands respect a phenomenological distinction between perceiver and perceived; they can switch between the two poles, but the existence of those poles remains a consistent feature of the experience. However, not all self-touching—or touching more generally—has this kind of structure. Although the example does serve to illustrate a sense of relatedness and reciprocity between perceiver and perceived, I will argue that a different kind of touch, which the touching-hands experience presupposes, is better able to illuminate the sense of commonality underlying the perceiver–perceived relation, a sense of commonality that is presupposed by the possibility of encountering anything as "there."

4   Touch without Differentiation

When I touch my own body, and when I touch things more generally, the part of my body that touches need not oscillate between being a transparent medium of perception and an object of perception that surrenders completely its role as perceiver. When I touch and when I am touched, parts of my body can become perceptually salient to differing degrees and in various ways. Consider what happens when you are writing and your hand starts to tire. It becomes gradually more conspicuous and uncomfortable. As this happens, it becomes progressively less effective as an organ of perception and more something that is perceived. There is a continuum between tactual experiences that are primarily a matter of perceiving one's own body and others that are directed through the body at something else.[14] In addition, it is important to recognize that bodily conspicuousness takes many different forms. The body may well feel conspicuous while being massaged but not in the same way as when holding the pen. And different kinds of bodily conspicuousness may interfere with the body's perceiving to different extents. For instance, although one's skin might feel very conspicuous when touched by another person, perception of what that person is doing often seems undiminished, perhaps even heightened.

In this respect, touch differs from sight. When we see something, the body operates as perceiver rather than as perceived; seeing is not a matter of perceiving one's eyes.[15] Some have claimed that vision is much more touchlike than is generally acknowledged (e.g., Gibson 1979; Noë 2004); it involves a dynamic process of active exploration, rather than a static, detached gaze upon a picturelike world. Even if this is broadly right, there is an important difference between the two senses. Whereas vision occupies one point on a continuum between perceiving body and perceived body, touch—in its many varieties—embraces the whole continuum, and it is the only sense that does.[16] Through touch, we perceive our bodies to varying extents and in a range of very different ways. And, importantly, the phenomenology of touch is relational; the way in which we perceive something through our bodies is inextricable from how we experience our bodies. When grasped by a tired, sweaty hand, a pen *cannot* feel comfortable to the touch. How the hand feels is at the same time a way of perceiving the pen; the two aspects of the experience are inseparable.

However, this does not exhaust the variety of tactual experiences. Essential to a sense of belonging to the same world as what we perceive is, I propose, a kind of touch that does not ordinarily incorporate clearly defined subject and object positions. Instead, at least a partial lack of differentiation exists between the perceiver and her surroundings. I will develop an account of it by first turning, as Merleau-Ponty does, to touching hands. Put your palms together and hold them still, as if to pray. However hard you try to distinguish the perceiving from the perceived hand, you experience only the one unified touch. Physically speaking, two separate parts of the body come into contact, but the tactual experience does not respect the visually perceivable boundaries between their surfaces. This example is problematic insofar as it involves passive touch. One might respond to it by maintaining that perceiver and perceived are undifferentiated here only because passive touch is comparatively impoverished. For the most part, we tactually perceive through movement. Hence, lack of differentiation can be attributed to lack of perception. So let us instead consider an instance of active touch: running one's hand along a surface and perceiving its texture. Here, it is principally the texture that is perceived, while the bodily aspect disappears into the background. But what if the touched surface were the palm of one's other hand? Close your eyes and put your palms together again but, this time, rub them slowly back and forth against each other repeatedly, so that as one hand moves away from the body, the other moves toward it. Again, the experience involves not an alternation between perceiving and perceived hands but a unitary perception of the rubbing.

Two features of the hands-rubbing experience, when taken together, serve to illuminate a pervasive aspect of tactual perception that is eclipsed by an emphasis on reversible poles. First of all, physical contact involving different parts of the body need not amount to experience of distinct touches. What applies to self-touching also applies more generally: two or more different points of contact can be implicated in the same touch. As J. J. Gibson (1962, 481) notes, when we pick up an object with our fingers, what we generally feel is the unitary touch of that object, rather a combination of separate "local signs."[17] Try picking up a cup from above, with all the fingers of one hand spread evenly around the rim. Although you *see* that your fingers are physically separate from each other and from the cup, and that they make contact with the object in different places, the touch is unitary. This not only applies to the fingers of one hand. If you pick up the cup and roll it back and forth between your two hands, again there is (in the absence of interfering factors, at least) a unitary tactual perception of a cylindrical object, rather than two phenomenologically separable touches, from which the presence of a cylindrical object is inferred. The point also applies to passive touch. I am currently resting my lower right arm on the top of a desk. My elbow is touching the surface of the desk, as is most of my forearm. There is a gap between my wrist and the surface, but the surface comes back into contact with the outer side of my right hand, up to the base of my little finger. Despite the fact that a large area of my body makes contact with the surface and that there is a gap between two areas of contact, what I feel is the single, unitary touch of the desk.

What is it that touches something or is touched by it, if not a localized part of the body? Merleau-Ponty maintains that tactual experience involves more than just the part of one's body that comes into contact with something. The perceiving body is a unified whole, rather than a system of parts each of which perceives in isolation from the others: "Each contact of an object with part of our objective body is, therefore, in reality a contact with the whole of the present or possible phenomenal body. . . . The body is borne towards tactile experience by all its surfaces and all its organs simultaneously, and carries with it a certain typical structure of the tactile world" (1962, 317).

Nevertheless, this does not imply that the touching body is phenomenologically undifferentiated. Even if my body as a whole is somehow implicated in every touch, it is still the case that I generally experience myself as touching with my hands or with some other specific part of my body. But is the body as a whole also unified in a single, undifferentiated touch? I think that it is. However, to characterize the form that this touch usually takes, we also need to draw upon a second characteristic of the rubbing-hands experience: the lack of a phenomenological distinction between perceiver and perceived. Perhaps this applies to very few instances of active touch with the hands, and so we should be cautious in generalizing from it.[18] Even so, although it is important not to understate the tactual significance of the hands (which are sometimes referred to as the "fovea" of touch), touch is not solely an achievement of the hands. Lack of perceiver–perceived differentiation is ubiquitous in other kinds of tactual experience, and is not exclusive to those that involve contact between parts of the body. Gibson recognizes the existence of something that is neither active nor passive touch in the familiar sense, but a kind of orientation that both take for granted: "The 'sense of touch' is not ordinarily thought to include the feeling of cutaneous contact with the earth. Nevertheless, the upward pressure of the surface of support on some part of the body provides, for every terrestrial animal, a constant background of stimulation" (1962, 480).

This "background touch" is not something that is phenomenologically localized in the skin of the feet, though, along with a number of other concurrent locations. Don Ihde describes sitting on a comfortable couch: "I find that the cloud-like couch-me experience is so vague that not even any clear distinction between me and where I end and couch is capable of being made. Inner and outer, subject and object are here not at all clear and distinct" (1983, 97).

It is not simply a *part of my body* that I experience as touching the couch. Rather, there is a diffuse tactual perception encompassing the whole body. Of course, some parts of my body are not in physical contact with the couch. But it would be wrong to say that only the bit of me touching the couch feels comfortable. It is my body as a whole that feels comfortable, and the feeling of comfort is inextricable from the feeling of touch. The touch is not localized anywhere. Instead it pervades bodily experience. In addition, this comfort is characterized by a partial lack of differentiation, not just between parts of my body but also between my body and its surroundings. Could this really be so? The experience of rubbing one's hands together clearly illustrates that (a) physically separate touches need not be phenomenologically separate, and (b) touch can be characterized by a lack of distinction between what touches and what is touched. If this much is accepted, there is nothing a priori objectionable to the claim that certain touches take the form of a unitary, allover feeling, which is often characterized by a lack of differentiation between body and surroundings.[19] And the claim is also phenomenologically plausible: although the tactual background is seldom at the forefront of awareness, it is phenomenologically conspicuous on occasion. Ihde (1983, 98) offers the example of jumping into cold water and then feeling an allover sense of bodily exhilaration as you emerge into the warm air. When suddenly immersed into icy water, it is one's body as a whole that feels the unified "touch" of the water. The experience involves a heightened distinction between one's body as a whole and the medium that surrounds it. However, this is not an objection to the view that background touch more usually involves a lack of perceiver–perceived differentiation. On the contrary, the fact that this sense of distinctness from one's surrounding is unusual suggests that background touch ordinarily has a more diffuse structure.[20]

Background touch is not an experience of coming into *physical contact* with something. We are always in physical contact with some medium, whether it be the air, a comfortable couch, or a hard object. Take the experience of swimming in the sea, in peaceful, warm water. The water is not encountered as a number of localized touches or even as something coming into contact with the whole body. Rather, there is a diffuse sense of inhabiting some medium: the body as a whole is "in touch" with it, not differentiated from it in the way it would be from a hard, rough object that is felt by the hands. It is within this context that localized touches are felt—the movement of the water between one's fingers as one's hands sweep through it, or the brushing of seaweed against one's toes.

In order to appreciate the phenomenology of background touch, we need to escape from a conception of touch that is tacitly constrained by the surreptitious imposition of sight upon it. When we reflect on tactual experience, there is—I suspect—a tendency to construe that experience in terms of what is seen. This can happen in two ways. First of all, visual perception of touching no doubt affects, to some degree, the actual tactual experience (something I will say more about in section 6). Second, what one sees—physical convergence between surfaces—will also influence how one interprets the experience. The examples of touch that are typically addressed involve localized contact of body parts, usually the hands, with solid entities—precisely the kinds of touch that are most easily amenable to visual inspection. Consequently, it is easy to assume that the phenomenology of touch conforms to visually accessible boundaries, with the perceiving on one side and the perceived on the other.[21] But one cannot "see" the unified touch of the couch or the sea in the same way. Touch is not constrained by what is visually available; touch as a form of perception needs to be clearly distinguished from touch as perceived physical contact.

Once it is recognized that not all tactual experiences conform to (a) visually perceivable differences and (b) a perceiver–perceived structure, the structure of background touch can be better appreciated. Central to it, I have proposed, is a lack of complete differentiation between perceiver and perceived. When we perceive an entity as distinct from us, by means of touch or some other sense, we do so against a backdrop of tactual immersion in the world. The phenomenological differentiation between us and the realm in which objects of perception reside is never complete. Hence, we are never fully separate from what we perceive. This contributes to a sense of "being there," of sharing a world with what we perceive. Nevertheless, it is not sufficient. In the remainder of this chapter, I will argue that touch also contributes to the sense of reality and belonging in virtue of the diverse range of significant relations between perceiver and perceived that it encompasses.

5   Significance, Possibility, and the Primacy of Touch

Like background touch, tactual experience more generally is not simply a matter of *perceiving physical contact*. Touch is of course frequently associated with physical contact and also causally enabled by it, but physical contact between perceiver and perceived is not essential to tactual experience. Merleau-Ponty offers the example of a blind person using a cane to perceive, and notes that what we perceive through touch need not be something in contact with the skin. The person perceives *through* the cane, rather than perceiving the point of contact between hand and cane, and then indirectly perceiving or inferring the properties of surfaces that come into contact with the cane: "Once the stick has become a familiar instrument, the world of feelable things recedes and now begins, not at the outer skin of the hand, but at the end of the stick" (Merleau-Ponty 1962, 152). Such examples do not, however, serve to dissociate tactual experience from physical contact. Physical contact is still involved; the most we can say is that the boundary between perceiver and perceived has shifted outward.

However, other kinds of tactual experience involve neither physical contact nor the experience of physical contact. Consider the tactual perception of absence. Drawing on the work of Katz, Merleau-Ponty also discusses running one's hand between threads of linen or between the bristles of a brush. Between the threads and the bristles, he says, is not a total absence of tactual experience but, rather, a "tactile background" (1962, 316). One perceives the space, and perceives it *as* a space, as an absence of something. The sculptor Rosalyn Driscoll (unpublished) describes how this kind of experience can play an important role in the tactual appreciation of sculpture. When touching a sculpture, space can, she says, be "charged with the unexpected absence of an expected presence." Driscoll maintains that empty space is not tactual nothingness; it has a distinctive kind of perceptual presence: "*When touching, space is felt as presence. What you find with your hands are the spaces, while your eyes find the objects.*" Many different kinds of tactual experience involve perception of absence. For example, in contrast to touching a sculpture, imagine trying to find your way out of a dark, narrow cave, moving your hands around only to discover smooth, cold walls, with occasional indentations. Eventually, you come across a space and, as your arm enters, you feel a cool draft against it. There is an experience of its meeting with no resistance, of its not being touched.

Why is tactual perception of absence not just an absence of tactual perception? In the cases of the brush, the sculpture, and the cave, the experience is shaped by expectation, which might be conscious and articulate in nature and/or bodily and habitual. Often associated with this is a sense of the significance of things. Indeed, tactual experiences of much the same stimulus can differ radically, depending on its significance and the nature of one's expectations. For example, a caress from one's lover in a familiar setting, and to some degree anticipated, feels very different from the unexpected and unwelcome caress of a stranger. Significance is not a matter of some conceptual content that is *separable* from a coexisting tactual perception; it is inextricable from how something feels.[22] This is not to suggest that conceptual content has no role to play in tactual perception—inseparability is not absence. However, it is doubtful that all kinds of tactually felt significance require conceptualization; some things *feel* consistently pleasant, unpleasant, comforting, repulsive, threatening, enticing, and so forth, in a way that can conflict with a conceptual understanding of their significance. Even so, it is clear that conceptually grasped significance and explicit expectation do *at least* inform how things feel to the touch. In so doing, they no doubt interact in intricate ways with a bodily, felt sense of significance. For example, in a scene from the film *Flash Gordon*, characters are required to endure the trial of inserting an arm into one of many holes in the stump of a tree, where a deadly "tree beast" lies in wait. How might that feel? As the hand slowly enters, its surface is unusually conspicuous and vulnerable; any slight touch—a leaf or some moss—is at the forefront of awareness, and the draft of air against the sweating palm is also pronounced. The hand is primed for touch: the expectation of touching, the tactual perception of absence, and the sense of anticipated presence themselves amount to a kind of tactual experience. And this profoundly affects the experience of actual contact. It is also worth noting that the place a tactual experience occupies on the body–world continuum is partly a matter of the kind of significance attached to it. In this example, the hand is phenomenologically conspicuous and vulnerable, and the experience is very much a bodily one. In contrast, when I habitually reach out to sip some tea from the mug in front of me, I perceive the mug more so than my hand. What one perceives and how one perceives it are both bound up with how things matter.[23]

Of course, physical contact or force often is implicated in touch. According to Jonas, it is for this reason that touch plays a privileged role in constituting the sense of reality. Touch is the only sense that involves perception of force and thus of interaction between perceiver and perceived. A sense of reality is essentially a matter of affecting and at the same time being affected by things, rather than of viewing them in a seemingly detached fashion: "Reality is primarily evidenced in resistance which is an ingredient in touch-experience. . . . Touch is the sense, and the only sense, in which the perception of quality is normally blended with the experience of force, which being reciprocal does not let the subject be passive; thus it is the sense in which the original encounter with reality as reality takes place" (Jonas 1954, 516).

However, it is misleading to suggest that touch incorporates a discrete experience of force. It is important to appreciate the diversity of tactual experiences, a diversity that is largely attributable to the many different kinds of significance that things can have for us. Touch incorporates a sense of significant contact (or significant lack of contact), rather than mere force. And it is not a case of "force plus significance," either. Gently squeezing part of a lover's body in the act of lovemaking and squeezing a rubber ball that has a similar consistency do not incorporate a common experience of force. Force is an abstraction from what is perceived, rather than an isolable component of the perception.

As indicated by examples such as tactually perceived absences, there is more to touch than perceiving the actual. Tactual perception also incorporates a sense of salient possibilities. Merleau-Ponty draws attention to this in his discussion of Schneider, where he claims that the patient's experience of touch no longer incorporates a full range of tactual possibilities; it is afflicted by a "confinement to the actual" and there is a "collapse of the sense of potential touch" (Merleau-Ponty 1962, 117). With this, the phenomenology of actual touching is also diminished. Without the experience of being surrounded by significant possibilities that could be actualized by one's activities, the temporal dynamic that ordinarily structures tactual experience is lost, and the object that is touched loses its "carnal presence and facticity" (ibid., 109). In addition, a perceived world that is no longer characterized by tangibility is no longer configured in terms of goals and potential practical projects, as their intelligibility presupposes the possibility of tactual contact. Consequently, the overall structure of experience is transformed; the patient's world has somehow "contracted."

Although Merleau-Ponty's interpretation is—as he acknowledges—debatable, a "loss of potential touch" (possibly taking several different forms) does seem an apt way to describe something that features in many reports of anomalous experience. People suffering from certain kinds of psychiatric illness often complain of something like this. For example: "It's almost like I am there but I can't touch anything or I can't connect. Everything requires massive effort and I'm not really able to do anything. Like if I notice something needs cleaning or moving, it's like it's out of reach, or the act of doing that thing isn't in my world at that time . . . like I can see so much detail but I cannot be a part of it. I suppose feeling disconnected is the best way to describe it" (patient quoted by Horne and Csipke 2009, 663).

There is an intimate association here between perception of things as somehow intangible, an inability to pursue one's projects, and a sense of being disconnected from everything, not part of the same "world."[24] As the example indicates, it is not only tactual perception that incorporates

tactual possibilities. What we perceive visually can sometimes appear curiously intangible, whereas—in the normal case—the world we see offers up
a host of tactual possibilities. As Husserl, Merleau-Ponty, and others emphasize, perceptual experience of something by means of one sensory modality
incorporates a structured system of possible perceptions by means of that
modality and others, as well as possible actions. These systems comprise
what Husserl and Merleau-Ponty call the perceived object's "horizon." As
Merleau-Ponty puts it, "any object presented to one sense calls upon itself
the concordant operation of all the others" (1962, 318). Husserl (1989, sec.
I, ch. 2) similarly claims that visual perception incorporates tactual possibilities and vice versa. And both insist that the senses are united by their
common dependence upon bodily experience and activity. To quote
Husserl: "The Body [*Leib*] is, in the first place, the *medium of all perception*;
it is the *organ of perception* and is *necessarily* involved in all perception"
(1989, 61).[25]

   I want to maintain that tactual possibilities, along with the tactual
background, are indispensable to a sense of reality and belonging—they
connect us to things. The sense of reality presupposed by sight depends
on them; without our experience of potential touch, what we see would
not appear as "there." When looking at the blade of a sharp knife that
glimmers in the sun, you *see* its sharpness, its potential to cut—among
other things—you. Sometimes you can almost *feel* it sliding against your
hand; you might pull back your hand or clench your fist. Perception of
sharpness is, in some instances at least, phenomenologically inseparable
from a sense of the significance of the sharpness. That significance is felt
as a tactual potentiality, rather than contemplated in an abstract manner.
Without potential touch, one could not experience a seen entity in this
way. Although not all potential touch is quite so salient, it is still a ubiquitous feature of experience—things around us appear tangible, as exemplified by anomalous forms of experience where everything looks curiously
intangible.

   Of course, the dependence between vision (and other senses) and touch
is mutual; touch is permeated by visual possibilities too. When you tactually explore an object without seeing it, you have some a sense of how it
will look from various angles. And the influence of vision on touch is not
restricted to mere possibilities; what we see can also affect the experience
of actually touching. For example, Jonathan Cole (pers. comm.) reports a
patient with a spinal cord injury and no feeling below the waist, who felt
her leg being touched whenever she saw it being touched. Given that
vision can induce tactual experience in the absence of any tactual stimula-

tion, it seems plausible to assume that it also has some effect on those tactual experiences where there is tactual stimulation.

However, let us contemplate the comparative effects of losing vision and touch. Loomis and Lederman (1986, 2), among others, make the point that you can close your eyes or ears and thus cease to see or hear, but you cannot similarly switch off all touch, and that consequently it is harder to fully appreciate the role that touch plays in our lives. It is important not to understate the experiential changes associated with loss of vision or hearing. Closing one's eyes and no longer seeing anything is phenomenologically different from losing one's sight. John Hull explains how, when he became blind, he initially retained—to some extent—the world of the sighted person; he lived as a sighted person without sight. However, he gradually came to inhabit a world of process and change, rather than a realm of enduring objects that sighted people take for granted: "Mine is not a world of being; it is a world of becoming. The world of being, the silent, still world where things simply are, that does not exist" (Hull 1990, 82).[26] On the basis of his account, the difference between not seeing and loss of sight seems to be the difference between a world that continues to offer visual possibilities, albeit possibilities that one cannot actualize, and a world from which the possibility of seeing has been removed.[27]

Even in the absence of both hearing and vision, an experiential world of some kind remains, a world that can still be imbued with many different kinds of significance (see, e.g., Helen Keller's *The World I Live In* [2003]). However, I do not think that the same applies to the absence of touch. Granted, one's existential predicament would be transformed by a loss of vision or hearing. But without touch, one could not have any kind of existential predicament.[28] This is not to say that touching and the having of an existential predicament are one and the same; they are not. But, regardless of what else the sense of belonging to a world might involve, it is inextricable from the possibility of touching.[29] I cannot appeal to empirical evidence in support of this claim, as there are no recorded instances of the total loss of tactual perception. There are well-documented cases of partial loss, such as that of Ian Waterman (Cole 1991). However, Waterman has a fully intact sense of touch above the neck, and others with similar injuries all retain touch in some part of the body. To further complicate matters, even if someone did lose all experience of *actual* touch and retained a sense of being part of the world, potential touch would also have to be absent in order to refute my position. And it would be very difficult to ascertain whether or not this were so.

So why is touch uniquely indispensable? The reason, I suggest, is that what we call "touch" incorporates a remarkable diversity of kinds of significant relation between perceiver and perceived. Merleau-Ponty suggests that perception is a kind of "communion" between perceiver and perceived (1962, 320); a "vital communication with the world" (ibid., 52–53), as exemplified by the to-and-fro that characterizes touching hands. I agree that a sense of reciprocation—experienced against the backdrop of the partial lack of differentiation that I described earlier—is important. But it needs to be supplemented by the appreciation that touch embraces a wide range of ways of communing with the perceived, which are masked by the use of a single term "touch." Take the claim "I see x." Although x might be any number of different entities, all instances of "seeing" involve much the same way of encountering something. Of course, there are many different variants, such as "I gaze upon x," "I glance at x," and "I visually explore x." However, they are all variants of a form of perceiving in which the body and world poles remain fixed; the body is perceiver and is not itself perceived in the act of perceiving. Now compare this to "I touch x." To begin with, we need to add "I am touched by x," so as to acknowledge the broad distinction between active and passive touch. But we can then go on to further distinguish many different forms of touching and being touched, such as exploring, caressing, prodding, squeezing, stroking, being stroked, caressed, prodded, and so forth. These can be further differentiated: timid tactual exploration, involving a sense of vulnerability, is very different from eager, confident exploration. In all these cases, there are considerable phenomenological differences with respect to the conspicuousness of one's body, the way it is felt, the extent to which something other than one's body is perceived, and the character and extent of the experienced relationship between one's body and what it touches. Consequently, touch is implicated in numerous very different kinds of significant connection with our surroundings, which are phenomenologically diverse to an extent that visual and auditory experiences are not. The experience of being there is, in part, a matter of having the potential to affect and be affected by things in these various significant ways. The possibility of these kinds of contact is inextricable from the possibility of meaningful activity, from having any kind of practical project, and thus from any sense of participation in the world. Together they comprise a sense of connectedness and communion that is far richer and more diverse than what might be achieved through abstract perception of force.

    In the absence of any sense other than touch, one can retain a potential to encounter things as there, as real. However, given the phenomenological breadth and diversity of what established practices classify as touch, all or almost all kinds of significant contact involve or presuppose tactual contact or its possibility. A complete loss of potential touch would involve an inability to appreciate what it would even be to touch or to be touched (comparable to the loss of visual possibilities reported in some experiences of blindness, such as Hull's). Were this to occur, the sense of reality and belonging would be absent not just from perception but also from thought. One might be able to utter "the object is there." But, as one could summon no sense of what it would be to relate to it, to connect with it, to affect it and be affected by it, the utterance would be empty. Of course, we also have experiences of affecting and being affected by things and people in numerous nontactual ways. However, when we are hurt or moved by the words of another person, for example, the experience presupposes (ordinarily, at least) a sense of both parties as participants in a shared world. And it is this presupposed sense of participation in the world that incorporates the possibility of tactual forms of contact.

6 Conclusion

According to Jonas (1954, 509–510), a phenomenological analysis of touch is more difficult than an analysis of any other sense because touch is "the least specialized." My view is that it is so wide ranging as to be inseparable from the having of a world. This is obscured when it is misinterpreted voyeuristically and physically, in terms of surfaces coming into physical contact with each other, and the registration of pressure or force. To make contact with the world is to be "in touch" with it, but it is about significant contact rather than just physical contact; the latter is an abstraction from the richness of tactual experience.[30] Also implicated is a tactual background, which involves a certain lack of differentiation between the perceiver and her surroundings. In emphasizing the sense of commonality, significance, and communion that is integral to all perception, my conclusion complements much of what Merleau-Ponty says, particularly in his later work. However, through an appeal to tactual experiences that incorporate lack of differentiation, I have suggested that the sense of commonality is phenomenologically accessible, rather than being something quietly presupposed by the phenomenon of reversibility. I have also proposed that we can better appreciate the indispensability of touch to a sense of things as "there" by emphasizing the diversity of tactual experience and, in the process, the extent to which a sense of belonging to a significant world depends on tactual contact.

Acknowledgments

I am very grateful to Jonathan Cole, Rosalyn Driscoll, Zdravko Radman, and the audience at the 2009 International Congress of the German Society for Phenomenological Research in Würzburg, for helpful comments and advice.

Notes

1. This is exemplified by works such as Thomas Nagel's (1986) *The View from Nowhere*, which contrasts a subjective perspective with a detached, objective view of the world and asks how the two might be reconciled. For Nagel, reality is progressively revealed as we approximate a fully objective view of things. The problem, as he points out, is that this view fails to accommodate the subjectivity of the viewer.

2. For example, Heidegger challenges the entrenched philosophical association between truth and "pure beholding" (1962, 214). For more recent discussions of whether and how current Western thought might be vision-centric, see, e.g., Levin 1993.

3. Claims for the primacy of touch are not restricted to phenomenology. The view dates back at least to Aristotle, who claimed that, although hearing, vision, taste, and smell are contingent capacities that animals may or may not have, touch is a necessary condition for the being of an animal body. (See, e.g., Freeland 1995 for a discussion of Aristotle on the primacy of touch.) A contemporary proponent of the same view is Brian O'Shaughnessy, who maintains that touch is the only sense "essential to the animal condition as such"; it is "broad enough to overlap with the sheer capacity for physical action on the part of its owner" and is therefore the most "primordial" of the senses (1989, 37–38).

4. There is more general debate concerning which criteria are used and should be used to individuate senses. See, e.g., Nudds 2003 and Keeley 2002 for two contrasting approaches.

5. See, e.g., Siegel 2006 for a good discussion of this issue.

6. This applies to hearing too (regardless of whether it involves perception of mere "sounds" or also accommodates richer contents such as "footsteps"), but perhaps not to many instances of smell and taste. One can experience a smell or taste without it being recognized as a smell or taste *of* anything; there is no object of perception to perceive as "there" or otherwise.

7. See Ratcliffe 2008, 2009, for a phenomenological study of the nature and variety of these experiential changes.

8. The word "real" is used in a range of different ways, and it is important not to confuse my restricted use of it with others. For example, "real leather" is not to be contrasted with unreal leather, but with fake leather. See Austin 1962 for a discussion of how the terms "real" and "unreal" operate.

9. We find something similar in Heidegger's *Being and Time*: "Whenever we encounter anything, the world has already been previously discovered, though not thematically" (1962, 114). Merleau-Ponty (1962, 333) similarly stresses that the world is not an object of perception.

10. A distinction is sometimes made between "haptic" and "tactile" perception, where the tactile perception is a product of cutaneous sensation alone and haptic perception is generated by a combination of sensory inputs from the skin, joints, tendons, and muscles. Achievements that fall under the category "touch" are generally taken to be a matter of haptic rather than tactile perception (Loomis and Lederman 1986; Klatsky and Lederman 2002).

11. See Mattens 2009 for a good recent discussion of Husserl on touch.

12. In the essay *The Philosopher and His Shadow*, Merleau-Ponty (1964) also acknowledges at length the extent of his debt to Husserl.

13. See Stawarska 2006 for a discussion of Merleau-Ponty on the reversibility of touch and vision. Seeing one's own body, Stawarska points out, differs in important respects from touching it. One sees one's arm but there is no associated feeling in the arm of being seen.

14. This is something emphasized by David Katz (1989), as well as Krueger (1982) and Loomis and Lederman (1986, 4). However, it is something that Merleau-Ponty does not sufficiently acknowledge. This is not to say that he was unaware of it; his discussion of touch in *Phenomenology of Perception* draws heavily on Katz's work.

15. There are occasional exceptions to this, such as when we see a dazzling light and it "ceases to be light and becomes something painful which invades our eye itself" (Merleau-Ponty, 1962, 315).

16. Thanks to Rosalyn Driscoll for this point.

17. Gibson (1962) blindfolded his subjects so as to investigate the nature of their tactual experiences without interference from vision.

18. Even if the experience is infrequent, it complicates Merleau-Ponty's account of interpersonal touch. When I clasp my hands together (as opposed to actively touching one with the other), I again experience a lack of differentiation, rather than oscillation between perceiver and perceived. It is only when I squeeze one hand with the other and then vice versa that I get reversibility. The same arguably applies to shaking someone else's hand. There may well be alternation between a sense of one's hand as perceiver and as perceived. However, there is also a certain lack of

differentiation and consequently a sense of commonality. Perhaps this kind of tactual diffusion is as central or even more central to the constitution of interpersonal experience.

19. Vision and hearing, in contrast, remain for the most part fixed at one end of the body–world continuum. One perceives the world, rather than the body, and so the experience does not incorporate a lack of differentiation between the two. Taste, like touch, does not respect firm boundaries between the body and its surroundings. Sometimes, there is just "the taste." However, taste is consistently phenomenologically localized in one part of the body. Hence there is no taste equivalent of all-encompassing background touch.

20. I am not suggesting that the "existential background" to perception consists solely of tactual feeling. And it need not be a matter of touch *plus* something else. Just as tactual stimulation of two different parts of the body need not be felt as two separate touches, it could be that tactual and other kinds of bodily feeling (such as emotional feelings) are phenomenologically inseparable. See Ratcliffe 2008 for a discussion of the phenomenological structure of "existential feeling." See Slaby 2008 for the view that some affective feelings have a diffuse phenomenology.

21. It is debatable whether and to what extent Husserl and Merleau-Ponty fall prey to this particular mistake, but others clearly do. For example, Martin (1995) explicitly addresses the *phenomenology* of touch and the body, but construes the experienced body as a bounded object ending at the skin, and claims that touch is primarily a boundary sense.

22. Merleau-Ponty (1962, 319–320) insists that all perception incorporates a sense of the significance of things. I will add to this an emphasis on the diversity of ways in which we find things significant, and the claim that most of these kinds of significance depend on the possibility of touch.

23. Neurobiological evidence supports the view that touch is shaped by expectation. For example, Blakemore, Wolpert, and Frith (2000) claim that self-tickling is ineffective because (in this case, nonconscious) anticipation of being touched in some specific way can regulate sensory input from the relevant area.

24. Alterations in the sense of reality and belonging can occur in various psychiatric conditions, including schizophrenia, depersonalization-derealization and forms of depression. See Ratcliffe 2008, 2009, for a discussion of the range of such changes.

25. See Gallagher 2005 for a detailed defense of the view that the body plays this kind of role in structuring perception. Gallagher suggests that a "pre-noetic body schema" operates as a structuring framework for all kinds of sensory experience. His case for this draws not just on phenomenology but also on a wide range of complementary scientific findings.

26. Merleau-Ponty similarly remarks on how the structure of one's world is transformed by the absence of vision; the "whole significance of our life . . . would be different if we were sightless" (1962, 225).

27. I do not wish to suggest that all experiences of blindness take the form of a complete loss of visual possibilities. See Sacks 2005 for a discussion of the various ways in which a visual phenomenology can persist in the absence of sight.

28. Maxine Sheets-Johnstone (2009, chs. 3 and 4) also emphasizes the diversity of touch and links this to its primacy.

29. I have assumed throughout that smell and taste do not play as important a phenomenological role as vision, hearing, and touch in humans. For many kinds of organism, though, smell is more important. But my claim for the primacy of touch is not human specific. It would apply equally to any kind of organism with a sense of reality and belonging.

30. Although I have proposed that the phenomenological structure of touch is inextricable from the having of any kind of experiential world, I do not wish to imply that a particular physiology is also necessary. By "touch," I am not referring to a contingent physiology, but to experiences that are attributed to touch. Only if the relevant phenomenology were impossible without a certain physiology would that physiology be indispensable too.

References

Austin, J. L. 1962. *Sense and Sensibilia*. Oxford: Clarendon Press.

Blakemore, S. J., D. Wolpert, and C. Frith. 2000. Why can't you tickle yourself? *Neuroreport* 11:R11–R16.

Cole, J. 1991. *Pride and the Daily Marathon*. London: Gerald Duckworth.

Driscoll, R. Unpublished. *By the Light of the Body*.

Freeland, C. 1995. Aristotle on the sense of touch. In *Essays on Aristotle's* De Anima, ed. M. C. Nussbaum and A. O. Rorty, 227–248. Oxford: Oxford University Press.

Gallagher, S. 2005. *How the Body Shapes the Mind*. Oxford: Oxford University Press.

Gibson, J. J. 1962. Observations on active touch. *Psychological Review* 69:477–491.

Gibson, J. J. 1979. *The Ecological Approach to Visual Perception*. Hillsdale, NJ: Erlbaum.

Heidegger, M. 1962. *Being and Time*. Trans. J. Macquarrie and E. Robinson. Oxford: Blackwell.

Horne, O., and E. Csipke. 2009. From feeling too little and too much, to feeling more and less? A non-paradoxical theory of the functions of self-harm. *Qualitative Health Research* 19:655–667.

Hull, J. M. 1990. *Touching the Rock: An Experience of Blindness*. New York: Pantheon Books.

Husserl, E. 1970. *The Crisis of European Sciences and Transcendental Phenomenology*. Trans. D. Carr. Evanston, IL: Northwestern University Press.

Husserl, E. 1973. *Experience and Judgment*. Trans. J. S. Churchill and K. Ameriks. London: Routledge.

Husserl, E. 1989. *Ideas Pertaining to a Pure Phenomenology and to a Phenomenological Philosophy: Second Book*. Trans. R. Rojcwicz and A. Schuwer. Dordrecht: Kluwer.

Husserl, E. 2001. *Analyses Concerning Passive and Active Synthesis*. Trans. A. Steinbock. Dordrecht: Kluwer.

Ihde, D. 1983. *Sense and Significance*. Atlantic Highlands, NJ: Humanities Press.

Jaspers, K. 1962. *General Psychopathology*. Manchester: Manchester University Press.

Jonas, H. 1954. The nobility of sight. *Philosophy and Phenomenological Research* 14:507–519.

Katz, D. 1989. *The World of Touch*. Trans. L. E. Krueger. Hillsdale, NJ: Erlbaum.

Keeley, B. L. 2002. Making sense of the senses: Individuating modalities in other animals. *Journal of Philosophy* 99:5–28.

Keller, H. 2003. *The World I Live In*. New York: New York Review of Books.

Klatsky, R. L., and S. J. Lederman. 2002. Haptic perception. In *Encyclopedia of Cognitive Science*, 508–512, ed. N. Nadel. Basingstoke: Palgrave Macmillan.

Krueger, L. E. 1982. Tactual perception in historical perspective: David Katz's *World of Touch*. In *Tactual Perception: A Sourcebook*, 1–54, ed. W. Schiff and E. Foulke. Cambridge: Cambridge University Press.

Levin, D. M., ed. 1993. *Modernity and the Hegemony of Vision*. Berkeley: University of California Press.

Loomis, J. M., and S. J. Lederman. 1986. Tactual perception. In *Handbook of Perception and Human Performance*, vol. 2, 1–41, ed. K. R. Boff, L. Kaufman, and J. P. Thomas. New York: John Wiley.

Martin, M. 1995. Bodily awareness: A sense of ownership. In *The Body and the Self*, 267–289, ed. J. L. Bermúdez, A. Marcel, and N. Eilan. Cambridge, MA: MIT Press.

Mattens, F. 2009. Perception, body, and the sense of touch: Phenomenology and philosophy of mind. *Husserl Studies* 25:97–120.

Merleau-Ponty, M. 1962. *Phenomenology of Perception*. Trans. C. Smith. London: Routledge.

Merleau-Ponty, M. 1964. The philosopher and his shadow. In *Signs*, 159–181, trans. R. C. McCleary. Evanston, IL: Northwestern University Press.

Merleau-Ponty, M. 1968. *The Visible and the Invisible*. Trans. A. Lingis. Evanston, IL: Northwestern University Press.

Nagel, T. 1986. *The View from Nowhere*. Oxford: Oxford University Press.

Noë, A. 2004. *Action in Perception*. Cambridge, MA: MIT Press.

Nudds, M. 2003. The significance of the senses. *Proceedings of the Aristotelian Society* 102:31–51.

O'Shaughnessy, B. 1989. The sense of touch. *Australasian Journal of Philosophy* 67:37–58.

Ratcliffe, M. 2008. *Feelings of Being: Phenomenology, Psychiatry, and the Sense of Reality*. Oxford: Oxford University Press.

Ratcliffe, M. 2009. Understanding existential changes in psychiatric illness: The indispensability of phenomenology. In *Psychiatry as Cognitive Neuroscience*, 223–244, ed. M. Broome and L. Bortolotti. Oxford: Oxford University Press.

Ratcliffe, M. 2012. What is touch? *Australasian Journal of Philosophy* 90:413–432.

Sacks, O. 2005. The mind's eye: What the blind see. In *Empire of the Senses*, 25–42, ed. D. Howes. Oxford: Berg.

Sheets-Johnstone, M. 2009. *The Corporeal Turn: An Interdisciplinary Reader*. Exeter: Imprint Academic.

Siegel, S. 2006. Which properties are represented in perception? In *Perceptual Experience*, 481–503, ed. T. S. Gendler and J. Hawthorne. Oxford: Clarendon Press.

Slaby, J. 2008. Affective intentionality and the feeling body. *Phenomenology and the Cognitive Sciences* 7:429–444.

Stawarska, B. 2006. Mutual gaze and social cognition. *Phenomenology and the Cognitive Sciences* 5:17–30.

7 Perception and Representation: Mind the Hand!

Filip Mattens

They are the eyes of the skin.
—J. Cole and A. Dawson

From Aristotle and Galen to Charles Bell and Charles Féré, many presume a connection between the hand and our intellectual capacities. In this study, however, I will point out how the hands outsmart the intellect, as they trick us into a central epistemological credo and partake in a strategy that upholds its plausibility.

In the first section, "Sensory Reversibility and Natural Discontinuity," I will point out the consequences of the fact that the hand sets the terms for our take on touch. To begin with, this explains why the distinctive feature of touch has always been regarded as a mere by-product (1.1). It suffices to disimagine the hand to comprehend that the skin is not a perceptual organ but the basis for a fundamentally different kind of sense. By reversing the logic of touch, however, the hand renders it plausible to see touch as an object-sense (1.2), the result of which is that epistemology situates its value in the perceptual overlap with our other object-sense, eyesight. This generates misguided expectations that blind us to the facts of haptic perception (1.3).

In the second section, "Figural Reversibility and Manufactured Conformity," I introduce another sort of reversibility, exclusive to manual perception (2.1). This provides me with an example that challenges the common treatment of shape perception (2.2). The analysis of this example reveals a mechanism that explains why we are so easily prepared to ignore the facts (2.3). The example will also enable me to point out how this mechanism is to be understood; a telling experiment supports the general hypothesis (2.4).

To conclude, I address the inverse question: What is the effect of our epistemological expectancies on our understanding of the hand? The result of the manufactured conformity of sight and touch is that philosophers fail to see the hand as a strategic tool that developed under the skin, finessing the sense of touch from within. Instead of taking into account its unique perceptual powers, they treat the hand as if it were an organ of compensation.

1   Sensory Reversibility and Natural Discontinuity

1.1   An Incidental Surplus
Moving the fingertips of my right hand over the back of my left hand, I feel a soft membrane taut over a bony structure. Repeating this action with the tip of my pen, moving it softly over the back of my hand, a feeling runs, so to speak, through the skin on the back of my left hand, parallel to the pen's movement. Obviously, if the pen moves along the fingertips of my right hand, I sense the movement going through my fingertips in just the same way. Connecting these examples shows that, already in the first case, while feeling over the bony structure of my left hand, I also experienced a flow of sensations in the fingertips of my right hand. The point is not merely that my right hand and left hand perceived each other simultaneously, but rather that each one was itself being felt. Hence we must hold that whenever I touch something, I also sense the part of my body that is touching: when I shift my attention away from the object's features, I feel my hand.

These observations are taken from Edmund Husserl's 1907 lectures on perception (Husserl [1907] 1997, §47). They were introduced in psychology by the work of his then doctoral student David Katz. Katz captures the crux of the matter in Husserlian terms: the bipolarity of the tactile sense is like a shift of attitude; one takes a different stance toward the same sensory input (Katz 1925, 20).[1] Depending on the concrete situation, such a shift of attention requires more or less effort. This particular fact already caught Thomas Reid's attention. One can easily have the sensation of hardness, Reid observed, by pressing one's hand against the table while setting aside all thought of the table, but it is extremely difficult to attend to it as a distinct object of reflection (Reid [1764] 1997, 56). Husserl's point, however, is that there is no distinct object of reflection; rather, the same sensory impression that represents the texture of the table becomes, after a shift of attention, the perception of my hand (Husserl [1913] 1989, §36). More recently, Michael Martin has compared this phenomenon with so-called aspect-shifts, since one can switch one's attention between the object and how one's body feels (Martin 1993, 206).

This reversal of perceptual direction can easily be noticed—for example, by moving one finger first slowly and then swiftly over an uneven surface (like a laptop's keyboard). It is therefore quite puzzling that, more than two thousand years after Aristotle's analyses, Husserl's description of this phenomenon appeared as a phenomenological discovery. A brief look at the manner in which three different authors contend with this issue will explain why.

(a) Let us assume that Reid was the first to notice it. Reid's remark was not an attempt at descriptive accuracy but rather the fruit of a biased approach to perception: in an effort to fend off Berkeley's skepticism by using its own tools, Reid wanted to drive a wedge between subjective sensations and objective qualities for the sole purpose of concluding that their connection is utterly inexplicable. No one will ever be able to give a reason, Reid argues, why these physical stimuli produce those sensory impressions; in just the same way, it is unintelligible why, say, a warm sensation is rigidly linked to heat as a quality in the fire. Therefore, the link must be installed by nature. The sense of touch, then, is of particular importance for Reid's plea against skepticism because it is widely considered to be our access to external reality. At this point, Reid remarks that, even though we tend to overlook the sensations while perceiving the objects' qualities, we can focus on the subjective hard sensation while pushing against the table. Here, the reversibility of touch is little more than a convenient illustration to convince the reader of a radical distinction between objective properties and subjective sensations in general.

(b) It seems fair to say that the notion "sensation," as *equally* applicable to all the senses, is a functional term, rather than a descriptive fact; what it stands for is not obvious from experience and definitely not in accordance with ordinary language. Even though this is only one aspect of the original Latin word *sentire*, today we say that we sense (instead of "perceive") something whenever we notice a bodily feeling; the derivative word "sensation," then, is used to refer to the bodily feeling itself.[2] When Husserl wanted to mark the class of experiences that allow a reversal of attention, he could have used the common word "sensation"; yet because of the philosophical meaning, dating back to the sixteenth century, he was forced to coin a new, variant term: "sensings." Sensings are localized in the body; this is what distinguishes them from so-called visual sensations. No shift of attention can turn the surface colors of objects into an experience of the eye. A color that appears is never the color of the eye; hence, a color experience is never a perception of the eye. Even looking at one's eyes in the mirror does not show eyesight, while touching makes us feel our hands.

Katz argues along the same line, contrasting touch with sight, because a subjective, body-directed component conjoins the object-directed pole.

(c) Katz's talk of a subjective and objective pole echoes Heinrich Weber's 1834 experimental report *De Tactu*. Weber calls a particular sense an "objective sense" when we use it to perceive objects; if we only perceive the effect of an object and not the objects themselves, it is a "subjective sense." However, Weber does not oppose tactile experiences to visual phenomena; he introduces the notion of subjective sense for distinguishing between two methods of discovering the weight of objects: one depending on the subjective sense of muscular kinesthesia and another on the "objective sense of touch" (Weber [1834] 1978, 55). Before discussing heat-induced pain, Weber introduces a similar distinction, saying that all perceptions can be divided in "subjective" and "objective perceptions," of which the latter "arise when we not only sense the change in our organs . . . but also seem to sense the object itself stimulating our organs" (ibid., 91). The general idea is that the five sense organs enable us to have objective perceptions, whereas other parts of the body only admit subjective perceptions (e.g., pain); it happens, however, that, as the effect of disease, a sense organ loses "its ability to communicate objective perceptions to the mind" so that stimulation only causes pain (ibid., 92).

The three preceding theoretical contexts indicate why the reversible structure of touch remained unnoticed in the literature on the senses and perception. The observations of Reid, Weber, Husserl, and Katz have one thing in common: whether it is to the result of a shift of attention, an unnatural stance, or a physiological disturbance, in each case the bodily subjective aspect is presented as something *secondary*. For Reid, Husserl, and Katz, it concerns something one can *also* notice. For Weber, it is what remains after degeneration of the organ. What can explain this fact that the unique and distinctive feature of touch is regarded by both philosophers and physiologists as if it were an incidental surplus?

Weber believes that in vision, "we do not perceive only the change in the eye: we also distinguish between the light and the objects producing the light—or so it seems. . . . The same thing happens in the ear when we hear sounds. We also have the same experience with the tactile organs" ([1834] 1978, 91). He concludes that these are called organs of sense for that very reason. According to this picture, if touch involves a subjective perception, this can only mean that we fail to perceive the object. However, as a description of experience, these observations may seem perverted. I do not "perceive" a change in my ears when I hear birds in the garden or a car in the distance. Neither do I "perceive" a change in my eye when looking at the bookshelf across the room. To the contrary, the books appear away from me; the colors that I see are seen over there, as the colors of the books on the shelf, and it is impossible to perceive them otherwise. It is precisely by pointing out the phenomenal difference between seeing and feeling that Husserl and Katz bring the reversible nature of tactile impressions to the fore. Hence, that the subjective aspect occurs as a surplus follows from the fact that they contrast touch to visual perception. However, the very idea of comparing cutaneous sensitivity with ocular perception is itself the result of a deceptively trivial assumption going back to Aristotle; in Weber's version: "the unique property of the sense-organs is that they assist us to become the partakers of objective perceptions" (ibid., 91).

Whether the bodily aspect of tactile sense really is something secondary comes down to the question of whether perceiving the object (that causes the stimuli) is of primary importance. The term "primary" inevitably hints at something like an original or true biological function. It is a risky enterprise, if not simply a bad practice, to ask "why" a biological function exists, for this suggests that it evolved for a reason. However, the way touch has usually been described in philosophical psychology, and accordingly used in epistemological arguments, seems so implausible from an evolutionary point of view that I will reconsider what seems most vital to tactile sense. To avoid misunderstandings, my attempt to retrieve its "vital function" is not meant to establish something like the essence of touch but rather a first step in bringing out a number of hand-inspired presuppositions that trick us into certain epistemological fictions.

1.2   A Fundamentally Different Sense

Because, from the start, touch is considered on a par with the other senses, its subjective bodily aspect can only occur as an "accompaniment" (Martin 1993) of a most natural object-relation. At best, it is understood as an additional form of object perception, directed at another object: "a second object, the body" (Husserl [1913] 1989, §37). In this way, both poles of tactile sense remain within the paradigm of object perception: its object is either an external corporeal entity or a body part.[3] However, both cases seem to fly in the face of biological "reason": (a) it seems of no use to perceive one's own entire body(-surface) as an object; and (b) it seems of no use to be able to perceive an object with one's entire body(-surface).

To understand that object perception is a rather exceptional aspect of touch, it suffices to think over again what is most basic in touch. We ascribe touch to the vast majority of animals. We do so on the basis of the kind of sensitive integument that humans have in common with other mammals and birds. The most general observation about the skin is that it typically envelops the animal's entire body, and so does skin sensitivity. This indicates that cutaneous sensitivity serves an animal to know where it is touched; that is to say, rather than serving an animal to know what it touches, cutaneous sensitivity serves an animal to know that it is being touched. It suffices to think of a dolphin, horse, or turkey to understand that reaching out toward an object and, in so doing, making skin contact *in order to* perceive the object's properties is extremely rare in animals that are equipped with a tactile system comparable to ours.

Although it may seem counterintuitive at first, it is sufficient to separate touch from the other senses and isolate it from a context, in order to understand that touch is actually not an object-sense at all. Imagine yourself lying on an operating table, waiting for a painful injection; at this point, a wasp stings you. Clearly, you will not feel what kind of object touched you. We should see this situation as the rule rather than an exception. When something contacts my skin, I cannot tell, *by touch alone,* what touched me. The following observations show that *tactile sense is not fundamentally object directed.*

(1) Whereas it is easy to imagine odors as disconnected from any object, we tend to think of touch as essentially related to material things. However, if the link to an external object fell away, olfaction would completely lose its vital function, while tactile sense would not. Imagine a person with unusually high skin sensitivity such that the slightest touch would cause an intense feeling (note that it does not have to be unpleasant or pleasant). This person might no longer be able to identify any object through skin contact alone. The same is possible when eyesight becomes oversensitive to light. The point is that, whereas vision would simply become useless, cutaneous sensitivity would retain its vital function. This person would only overreact to tactile stimuli, withdrawing at the slightest contact—more or less like normal subjects do in an unfamiliar, dark environment.

(2) A smell naturally signals the presence of some edible or noxious object. Phenomenologically speaking, it is the rotten smell itself that is disturbing; yet, biologically, it essentially points beyond itself to something rotten. In touch, however, all that matters is what happens at the surface; who or what causes it is irrelevant. Clearly, it is not irrelevant to me whether the sting comes from a wasp's attack or a doctor's injection. However, the fact that, against my instincts, I do not withdraw from the doctor's injection is due to (nontactual) background knowledge of objects

like needles, doctors, and so on; for an infant or an animal, there simply is no difference.

(3) The wasp's sting shows that the scope of tactile sense itself does not go beyond the body. We sense what happens to our body, but the very same mechanical impact can be caused by an open series of very different objects. We are highly sensitive to what objects do to the skin, but the objects themselves are tactually insignificant. We tend to neglect this because we already see objects in function of what they might do to our body. However, in themselves, objects are neither harmful nor harmless. A wolf might caress a lamb just as the lamb could bite off the wolf's ear. The point is that, strictly speaking, all one *feels* is *how the skin is acted upon*: that one is being scratched, punched, stung, and so on.

To summarize these observations, when something contacts any given spot of my body, I feel *where* I am touched, *how* I am touched, but not *what* touches me. Because impact is nonetheless physical contact, we think of touch as the ultimate access to particular objects. However, the skin itself only registers various forms of impact; we fill in an object.

Touch is often said to be necessary to animality, but our take on it is usually anthropocentric through and through. We rethink tactile sense in terms of its possible contribution to knowing objects (i.e., taking this corporeal entity to be such and such). Several researchers have complained that early experimental psychology focused only on passive touch and neglected haptic exploration. The motivation for such complaints is the worry that passive touch experimentation would not fully bring out the object-perception potential of touch. However, my point is precisely that we test passive touch as if it had the same interest as active touch. We transfer our vision-dominated, multimodal, object-oriented interests onto the skin. Weber's seminal experiments test the skin's discriminatory capacities *qua* differences in weight and temperature of objects. Similarly, Katz's famous experiments test our ability to identify objects based on their texture (e.g., paper of varying grain sizes).[4] However, even though the skin's performances are sometimes remarkable, it still does not follow that our entire body surface is equipped with cutaneous sensitivity for these purposes. If discriminating differences for comparing or (re)identifying objects were among the vital functions of touch, mammals and birds would be equipped with specialized antennae and feelers to check external objects. Also, they would be better off without the subjective aspect of cutaneous impressions. Indeed, arms, hands, and fingers are appendages that function as feelers. However, the development of the hand in primates did not render sensitivity on the back of the hand (and all over the body surface) redundant, and neither did reversible sensitivity in the fingertips decrease.

In sum, the subjective aspect of cutaneous impressions partakes in their vital function; the way touching complements human object-directed interests is a supplementary side effect. This brings us to a dual conclusion:

• The skin's anatomical characteristics as well as its perceptual logic clearly indicate that the *tactile sense serves an organism not for touching, but for sensing that it is being touched*: an organism withdraws from contact as a matter of reflex, not because it presumes the presence of a harmful corporeal entity on that location. If "perception" is object related (implying that a perceiver takes a thing to be a certain way), then, in its most basic form and most vital function, as it occurs in birds and mammals, *tactile-sense is not a perceptual sense, but a vigilant sense.*

• The hand, however, reaches out in order to palpate, and that is an intentional action meant to find out more, beyond what is actually being felt right now—*this* makes it object directed. The hand reverses the logic of touch from a vigilant sense to an object-sense: from "instinctively withdrawing from where one is being touched" to "intentionally reaching out toward something that is (seen) over there with the specific intent to get to know its tangible properties." This highly exceptional condition is responsible for what seems most ordinary to us: touching and looking are two mutually complementing channels for finding out what an object is like, suggesting that, basically, *touch and vision are perceptual modalities in much the same manner.*[5]

1.3   Misguided Expectations

The vital feature of touch is regarded as a by-product, not so much because the hand is in the foreground as the key example of most analyses, but rather because the hand sets the theoretical background for the analysis of touch in general. In its vital role, tactile sense is not exploratory, since the skin's action radius simply coincides with the region it must cover. The empty hand, however, must first find its object before it can put its unique perceptual strategies to work. What remains after imagining away the hand is not just passive tactual perception, but a different sort of sense. The general problem, then, is that, if a reflection on touch implicitly takes manual object-perception into account from the start, it will develop a one-sided picture that induces misguided expectations. These are the three components of such a one-sided outlook:

(1) As we can palpate, pick up, and handle objects, it is beyond question that touch is a means to get to know objects. Hence, touch shares a crucial capacity with vision: we not only feel *and* see the soil we are walking on, but we can also visually *and* tactually "single out" a corporeal entity, a spatially individuated body. This implies that we must have access, through both vision and touch, to the properties that are constitutive of spatial entities. Hence, it seems natural to say that we perceive spatial properties through vision and touch. Without further specifications, we assume that, when a property is spatial, it can be perceived visually and tactually.

(2) In turn, spatial properties acquire a special epistemological status because they are common to vision and touch, that is, not dependent on or reducible to a specific sensory experience. As a consequence, philosophers tend to overestimate the affinity between touch and spatial properties and, hence, to underappreciate the contribution of cutaneous sensitivity to object perception.

(3) Since an object has only one size and one shape, it seems self-evident that vision and touch perceive the same property, only by a different means. At best, perceiving, say, a shape visually might be an experience with a different character than perceiving it tactually. But this can only be a side effect due to specifically tactile or visual sensational properties, not a property of the shape itself *qua* spatial property, since, say, a visual circle and a tactile circle cannot be different in any relevant way.

We are facing a concatenation of facts and suppositions: the hand enables us to single out spatially individuated entities; we tactually perceive spatial properties; insofar they are spatial, we tactually perceive the same as we perceive visually; and, whatever cutaneous sensitivity contributes must be subordinated under the sensational by-product. In a word, as soon as we picture touch as an object-sense, we end up with the idea that "of all the (outer) senses touch is the one sense that necessarily is concerned with spatial properties" (O'Shaughnessy 2000, 671).

This idea has inspired remarkably fictional theories and speculative claims. Motivated by optical illusions and blinded by the inverse projection problem, early modern philosophers came to believe that touch is the only proper means to perceive spatial properties, so that infants first have to learn to interpret visual appearances on the basis of tactual experience. The general idea, usually ascribed to Berkeley, but proclaimed by many, including Reid and Mill, is summarized in the slogan: Touch educates vision. This view, which strikes us today as "perverse" (Epstein 1995, 3), had been the orthodoxy until the twentieth century and continues to have a mysterious cogency (see Smith 2000). Similarly, Kant, among others, stated that, without the sense of touch, it would be impossible for us to have a notion of solid objects' shapes (see Cassam 1997, 79–82). I will set such claims aside and instead distinguish two variant views about tactual spatial perception itself: a strong thesis implicit in such claims and, on the other hand, a less outspoken and milder claim more common today.

(a) The strong claim is the assumption that touch can perceive any spatial property, without restriction or exception. It can be found in the context of the early modern understanding of perception where spatial properties are said to be "tangible properties," because visual appearances are spatially unstable. Vision facilitates perception because it merely indicates the real, tangible properties. Reid ([1764] 1997, 78), for example, explains that for a blind person it would take a lifetime to perceive the shape of St. Peter's in Rome. This example shows that Reid adheres to the strong claim, actually believing that one can tactually perceive the shape of something as complex as a building.

(b) The milder version remains silent about the scope of spatial touch, but states that whenever we tactually perceive spatial properties, we come to know the same as what we know when we see these properties. It is supported by the truism that a visual shape cannot be different from a tactile shape *qua* shape, from which it follows that the difference between seeing and feeling the same shape can only be due to an extrinsic additive of sensational properties. Whereas Reid's example shows that the strong claim leads to an untenable exaggeration of the scope of touch, the weaker version may seem more reasonable.

The aim of the following sections is twofold: first, to demonstrate that even the weaker version goes against the facts, and, second, to explain why it nevertheless seems so reasonable. To achieve this twofold goal, I will first introduce another sort of "reversibility," different from sensory reversibility, but nevertheless unique to haptics. The analysis of this form of reversibility will allow me to show how the hand sustains misguided expectations about spatial perception by taking up an ambivalent position between sight and touch. More importantly, it will provide me with an example that, in the same stroke, challenges the weaker claim and hints at the mechanism that upholds its persuasiveness.

2 Figural Reversibility and Manufactured Conformity

2.1 Showing What I Saw

When our hands meet an object, we get a sense of its surface condition as well as its material character. Prehension naturally implies a multifaceted form of perception. One perceptual facet involved in grasping originates in the motility and flexibility of each hand and of both hands considered together. Namely, as the hand molds itself around an object to enforce its grip on it, we get a sense of its spatial features. As is well known, before and during prehension, the hand, just like the dual hand, *preshapes and reshapes itself.*

Shaping, however, entails a unique feature in our cognitive relation to perceptual objects: the hand can literally "represent" certain properties of an object that it previously explored. This, in turn, enables one to communicate these features to someone else. Within a given conversation, saying that an object is approximately two feet wide can be perfectly equivalent to holding up one's hands about two feet from one another. This means that, in a situation where linguistic communication is impossible, I can still respond to an interrogation about a particular object: I can answer questions such as how big the object was, what shape it had, how it moved, and so on. Because of my hands, I can still be an eyewitness.

One's ability to represent objects with one's hands initially centers on shaping and moving one's hands as if one were in contact with the object (i.e., touching or holding it). This ability is further expanded by acts that imitate manipulation. Eventually, hand movements can even trace out the object's contours. For example, when I am asked which object was used as the murder weapon, it might not be sufficiently unequivocal to imitate hammering, but I can then outline the overall shape of a hammer with my fingers. Thus, the core of actions that mimic grasping and handling a certain object can be complemented with *acts explicitly representing features of the object itself.* For example, when playing as soldiers in a battle, children either pretend to be pulling the trigger by rapidly bending their index finger or they mimic the revolver itself by making a fist (the gun's grip) while stretching out the index finger (the barrel) and bending their thumb (the gun's hammer) up and down.

By performing acts of touching, grasping, palpation, and manipulation in the absence of a particular object, one evokes some of the relevant features to identify it. And, in more purposive circumstances, one can also communicate specific qualities of objects. Thus, our hands enable us not only to perceive but also to reproduce certain properties. At first sight, this faculty establishes a further discontinuity between sight and touch. Whereas vision merely scans external objects, the hand can also act out what it takes in. No doubt, my gaze may display how I feel about what I see, but my eyes cannot retell *what* they saw.[6] Because one acts out properties that one previously took in, I will call this form of communication "manual representation."

Rather than driving them apart, this form of reversibility exclusive to haptics further aligns sight and touch in theoretical reflection. To understand why, it suffices to consider where manual representation outperforms linguistic representation. Words never fully enable me to explain, say, the curving silhouette of Milo's *Venus* or the way a racing driver passed a rival while rounding a bend. However, moving my hand the way I would touch *Venus*'s body or repeating the race car's move with my hand, respectively, bring back these features in a relevant way. There are two points to be made here, which explain why this form of reversibility, related to haptics but impossible in eyesight, nevertheless sustains the tendency to conform our outlook on touch to the role and function of visual perception:

• Whenever I use my hands to communicate the features of external objects, my hands address themselves to someone's eyesight. When I separate my thumb from my index to indicate the size of a spider, when I undulate my hand to imitate a snake, or when I fold my fingers so as to evoke the silhouette of a bird, I do so because I trust one will recognize the respective features *visually*. This brings out a second point:

• My hands cannot represent typically tactile properties; they cannot, of course, imitate the warmth, texture, or other significant material properties of a given object. What enables me to have a nonlinguistic conversation about an object, a scene, or an event is the fact that my hands literally copy its *spatial* features.

Reproducing spatial features, my hands can trace out the silhouette of an object, the composition of a scene, or even the development of an event by sequentially copying the reorganization of its elements' relative positions. Even in its most rudimentary forms, manual representation can be more accurate and less ambiguous with respect to spatial properties than an extensive verbal report. It is important to point out that manual representation occurs in the context of communication—that is to say, in a context that already concerns telling (or lying about) how one takes the world to be. Even though in many cases we are not after a precise spatial reproduction, representing something manually nevertheless involves a content that can be veridical or not.

Even though it might usually intermingle with expressive acts, an attempt to communicate certain properties of an external object by copying them is a fundamentally different action: throwing my hands up in the air to emphasize that something was huge is different from holding my hands in a precise position so as to make sure that someone else gets an accurate idea of the size of the respective object. It does not concern a subspecies of expression. Rather, I seek to avoid expression in order to make clear that I am trying to arrive at a real spatial correspondence between a property of an object and the organization of my hand(s). The actions may seem more basic than expressive gesticulation, but the intention is more sophisticated.

Furthermore, since communication always occurs in some preexisting context, and given the fact that here the addressee is our faculty of visual interpretation, manual representation is not tied to the actual size and the anatomically possible configurations of our hands. Depending on the context, it can be sufficient to reconfigure one's hands in correspondence to the shape of an object or its motion even when its size is beyond the scope of human gestures, or vice versa. That is to say, in manual representation one actually does spatially copy features, but only in a relevant manner.

Since someone's hands can only act out *spatial* properties, but do so in a way that enables someone else to *visually* grasp the relevant content, I will call the modus operandi of manual representation "figural reversibility."

2.2 Feeling What You Show

The flexibility and motility necessary for handling objects entail the possibility of figural reversibility. It occurs "in between" my visual perception and someone else's: what I have seen I can manually retell to someone looking at my hands. In this way, figural reversibility links up visual content and what we are inclined to think we perceive tactually: if my hands can show the spatial determinations of an object, why would they not be able to perceive them?

The apparent affinity between "manual posture" and "visual content" creates a bond between our organ for touching and our visual take on the world, which further evens out the "natural discontinuity" between eyesight and the sense of touch (cf. sec. 1). Even though it adjusts our expectancies about tactual perception to our visual acquaintance with spatial properties, figural reversibility will provide me with an example to demonstrate that these expectations are misguided and, as a consequence, uphold theoretical fictions. Am I not able to perceive tactually all the same spatial properties that my hands can represent to another perceiver? The answer is no. A suitable way to demonstrate the problem is to consider what remains of manual representation in the absence of eyesight, if the message of one person's hands were to be caught by someone else's tactual perception—directly from hand to hand.

Imagine a concrete situation in which linguistic communication is impossible, and then consider what happens when vision is impossible as well so that the interpreter must rely on manual perception. For example, a climber on her way to Mount Lao runs into a local villager. The villager can give directions, but his hands can also act out spatial properties of the path itself (steep, tortuous, bumpy, narrow, bendy, etc.). He can even quite effectively warn the climber for animals along the path ahead. In much the same way as a practitioner of the ancient art of "Chinese shadows," he can simply arrange his hands so as to reproduce the silhouette of, say, a wolf. Now, simply consider what remains of these possibilities if this encounter took place after nightfall. The climber might still understand certain indications. However, if the climber would palpate the hands that represent a wolf, she will feel nothing but a boy's hands and fingers.

What explains this, one might argue, is simply the fact that a "haptic interpreter" does not expect that the other's hands represent something. Let us therefore resume the test in a context where there can be no doubt about the representational intention of the haptic medium. First think of an instruction book on shadow theater; every page shows a line drawing of two hands in a specific posture and, next to it, the same silhouette filled out as a dark patch, showing the resulting shadow. Now consider a classic technique for creating haptic drawings; an item is cut out from a thin sheet of cardboard so that a blind (or blindfolded) perceiver can feel along the inner edge that delineates the sunken part. Combining these two types of reproduction, we could use the sample image of a wolf taken from a handbook on shadow theater to produce such a haptic relief drawing.

When this relief drawing is presented to a visual perceiver and a (blindfolded) tactual perceiver, the results diverge in a revealing manner. A visual perceiver recognizes a wolf in the hands that she sees in the cardboard relief. A tactual perceiver still does not come to think of a wolf, and, more significantly, might not even feel that the relief drawing is of two hands. In other words, if the subject matter is reduced to its shape, then seeing the wolf's silhouette becomes even easier, whereas a touching hand no longer grasps its own likeness.

This example drives sight and touch apart where they are supposed to overlap. In the following sections I first present the facts that we ignore in order to uphold the manufactured conformity between sight and touch, and then I discuss the mechanism that sustains it.

2.3 Tactually Taking In

Unlike, say, dolphins or turkeys, we have bodies that are articulated in a way that allows us to grope over objects and feel their shapes. However, when it comes to tactually perceiving shapes, our abilities are quite restricted. Numerous essays in epistemology and the philosophy of perception point out that shapes can be perceived by sight and by touch, but these essays never seem to consider it of any importance to account for the large inequalities *qua* performance between seeing and feeling shapes. This discrepancy between sight and touch is disregarded, as if it were an insignificant, contingent fact.

Let me first present a series of facts that should not be set aside: (a) Touch (without vision) is extremely efficient at identifying objects we are familiar with in our daily lives. (b) However, the object's shape is of minor importance in this process, as identification is generally completed in a split second on the basis of haptic experience of local features. (c) Blindfolded subjects recognize familiar objects *before* they even had the chance to feel over (a segment of) the object's global shape. (d) Conversely, if a blindfolded subject is asked to identify a familiar object on the basis of its global shape, the task becomes difficult and performance becomes slow; for example, if the shape is reproduced in a relief of the object's relevant contours, success in identification drops significantly. (e) Yet, it seems fair to say that overall shape is highly efficient for visual identification of familiar objects; for example, a comic strip can present an entire scene by means of overall contours while leaving out all other visual qualities and object properties like color, real size, shadow, and so on.

The real difficulty for tactual perception occurs when there is no object to identify. Imagine a complex shape cut out of a piece of cardboard, about the size of a bicycle. What would it mean to say that its "shape" is tactually perceived? Following its irregularly bending contour, I feel its edges, but I do not feel the figure. Of course, one might say that the shape is perceived bit by bit. But this just ignores the problem. The following observation indicates that we know we have a more effective grip on figures through vision than through touch. After feeling an object's contours, I might be surprised upon opening my eyes that the whole is quite different than I had expected. However, the opposite scenario is implausible. Standing in front of a complex figure cut out of cardboard, I will never say that, now that I have closed my eyes and I trace its contours, I finally get a sense of how the figure really is. The reason is that, perceptually, a figure consists not so much in the elements that make up its contour but instead more in the mutual proportions between these elements; contour lines "make" a figure, but the figure is the resulting composition.

One *never really grasps the figure as such,* as was already clear from the fact that tactile identification of familiar objects on the basis of small and simple relief figures is error prone. Identification fails *because* one does not tactually perceive the figure. On the other hand, it is unthinkable that one would see the object's contours but not its shape (which explains why no errors occur in vision). If touch had access to the shape in every respect (including the way it is seen), one would simply recognize the respective object (without exception, as in vision).

One possible response would be to say that "shape" is a "common sensible" only to a limited extent or restricted to certain cases. But this, of course, undermines the epistemological significance of the notion of common sensible. Another possibility is to say that shapes are not common sensibles. But how, then, can we account for the limited overlap, the limited number of cases in which we are undeniably able to perceive a figure by means of vision as well as touch?

The differences in performance and procedure in object identification provide a clue. A brief, slight, and local contact is often sufficient to iden-tify a familiar object, whereas, in vision, it requires specific effort to focus on a local element. This means that, in those cases where brief contact is sufficient, a local tactile experience makes a blindfolded subject think of an object mainly known through vision, which provides, first and fore-most, a global view. As a consequence, in such cases, one is able to deter-mine the overall look of an object without even feeling its shape. Only in those cases where unique tactile cues are absent or unrevealing, a tactile perceiver will have to try to puzzle together the overall spatial structure of the object. Asking a blindfolded subject to perceive a figure in relief is an artificial task; the task is not just highly uncommon, but also requires a shift in mindset. That is to say, the starting point for the task is tactu-ally unusual, while its goal is typically visual. When a blindfolded subject has to tactually perceive a contour relief, the subject is actually trying to imagine what it looks like. When we succeed in figuring out its shape, we draw on visual imagination: *to feel a figure is to imagine it.* Examples of "figural reversibility" in manual representation can show how we should understand this thesis and how not to.

2.4 Visually Figuring Out

Since my hands address themselves to the eye, it seems fair to say that manual representation would not work in the dark. But let us assume that a "haptic interpreter" has been informed about my intention to tell something using my hands. The more that manual representation is figurative, the harder it becomes for a haptic interpreter to get the message. If, for example, I configure my hands so as to imitate a characteristic silhouette of a deer's head, the haptic interpreter would be clueless because it already takes from a visual interpreter an imaginative stance toward the configuration of my hands to "see" them *as* something that they are not, namely a deer's head. A haptic interpreter, on the other hand, precisely relies on imagination in order to grasp the configuration of my hands. Thus, in order to "feel" my hands *as* something that they are not on the basis of their configuration, she would have to take an imaginative stance toward the product of the first form of imagination. This would be like imagining a crack in the wall and, thereupon, to start seeing things in the crack. To consider this as possible is a form of phenomenological self-deception: one can imagine a cloud against the sky; one can also imagine a cloud in the shape of a cauliflower; but one cannot thereupon recognize a friend's face in it, unless one is imagining her face as a cauliflower-shaped cloud. While it is perfectly possible to perform a single imaginative act with a layered object, one simply cannot reiterate the function of imagination itself.

This last example leads to the crucial point. A more effective way to trigger people's imagination is to "use" one's hands to render shadows on a wall. However, as anyone can easily check for oneself, you will not be successful in shadow theater by closing your eyes and fully focusing on your hands, but only by looking at the resulting image: one needs to *see* what the hands produce on the wall. The same is true for any form of manual representation insofar as it involves aspects of shape. This shows that it would be naïve to think that one knows "from within" how one's hands look. Just as the posture of my hands does not manifest itself in the same way as it appears from without, so too does the trajectory of my hands' movements not manifest itself in the manner in which it appears to an observer. As the examples show, the difference is decisive for shape perception. This explains why the fact that I am perfectly able to trace out a specific silhouette is no guarantee that I am able to recognize this silhouette when following a crack that traces out the exact same shape.

The hand has what it takes to be an actor, but in order to be a spectator it must appeal to imagination. The important point, however, is that imagination is nothing like an internal reconstruction; an internal image is not just a fiction, but is even beside the point. To be successful, one must stay focused on the external object itself as one's hand moves over it, instead of compiling a patchwork of local shape characteristics. In a word, tactually perceiving a figure is to succeed in imagining what it would look like if one were to see it. This thesis, following from the previous observations concerning manual representation, is also supported by the next experiment.

Wijntjes et al. (2008) tested blindfolded sighted subjects' abilities to identify objects on the basis of the manual exploration of relief drawings of everyday objects. As is well known, blindfolded subjects as well as blind subjects are rather bad at this, often failing to tactually recognize even half of the items. Wijntjes et al. complemented this test with a further experiment. Whenever participants failed to identify the object, they were given pen and paper and could freely draw the lines they previously explored haptically. In a significant number of cases, subjects were able to correctly identify a previously mis- or unidentified item after *seeing* their own sketch. To assure a correct interpretation of these results, the participants' actions were recorded to register at what point they were able to identify the object, and in a second test group the blindfold was not removed (i.e., they could not look at their drawing hand nor see their sketch). The results indicate that the improvement is not generated by the act of sketching or by the observation of one's own sketching hand, but is instead due to the visual access to their own sketch upon completion.

Even though the blindfolded participants' fingers traced the contours themselves, it is eyesight's take on the whole of these traces that makes the difference. The fact that they could act it out afterward by tracing a sketch, and even recognize the item of their sketch correctly, proves that the information they tactually assembled, recollected, and reproduced was spatially accurate, though not sufficiently relevant as such for shape perception. They already had all the relevant spatial information at their disposal, but, in order to get a grip on the shape, they had to see it. It follows that, when seeing it, they gain a significantly different take on a collection of elements that compose a shape.

3   An Organ of Compensation?

The hand interferes with our understanding of perception in a way that is hard to retrieve. Because "manual shaping" belongs to touch but works hand-in-glove with vision, it acts as the linchpin of a series of intuitions that reinforce one another. As handling things and reaching out in order to feel what something is like are done in the same movement and by the same means, it seems self-evident that we should understand touch as a perceptual tactic that is after a corporeal entity's constitutive properties, thus overlapping with vision in their common ability to assemble the spatial skeleton of one's environment. It may even seem natural to think of the hand as a more apt instrument than the eye for gaining information about the spatial organization of the things around us. In sum, we assume that touch is, or must essentially be, about finding out spatial properties, since spatial properties are epistemically important in that they do not depend on a single sense, and, in turn, this assumption is supported by the intuitive appeal of the idea that the spatial nature of the organs of touch is a god-given precaution against the perspectival deformations that undermine the epistemological status of spatial vision. The importance of the issue is twofold:

(1) The idea that touch is naturally related to spatial properties owing to the spatial nature of its organs and the conviction that this would entail a more reliable access to the spatial nature of other physical bodies are epistemological fictions, because they are built on the silent assumption that the *spatiality of the body is perceptually significant before it is perceived itself.* In order to preclude any spatioperceptual error, and thus entail the desired epistemic advantage, a bodily subject should have to know its body in a relevant manner prior to perceptual experience. Of course, one can use one's thumb or hand to measure an object and thus come to know its length. But the above assumption actually requires that, say, the length of one's hand has a spatial significance prior to any encounter with perceptual objects. This amounts to assuming that, in principle, prior to experience, you could tell that a glove fits your hand upon seeing the glove (without having seen your hand), or that, after grasping and palpating a shoe, you could tell whether it fits (without also palpating your foot). The point is that the dimensions of one's body parts gradually gain perceptual significance through interaction with objects. The statement that a book measures four thumbs is meaningful to me only because perceptual experience and interaction with other objects have given me an idea about the size of my thumb. Using a part of one's body to measure an object is actually an act of comparison: whenever the spatiality of a body part partakes in a perceptual task, this presupposes comparative perceptual experience. Simply stated, physical objects tell one how big one's body is and vice versa.

While the point is quite obvious with respect to size, the aim of my analysis of "figural reversibility" has been to bring out the corresponding situation in tactually detecting objects' shapes. Examples have shown that certain spatial configurations escape tactual perception even though they are within the hand's reach (sec. 2.3). The hand's capacity to shape itself and to trace out shapes does not automatically entail the capacity to perceive the same variety of shapes. The reasons are less obvious than in the case of size because the hand allies itself with our visual take on shapes, which in turn suggests that, when we consider such cases of tactually detecting an object's overall shape by moving one's body over its contours, we set touch upon a task with a typically visual goal.

In vision just as in touch, the variations in a sensuous expanse are motivated by various systems of spontaneous self-movement, which allows a perceiver to get a sense of spatiality and spatially individuated entities. What distinguishes the original sensorimotor principle developed by Husserl ([1907] 1997; Drummond 1979) from contemporary variants is that the phenomenological approach understands bodily movements as constitutive of our perceptual space without ever passing on to the idea that the spatiality *of* one's body and movements would provide the "spatial content" of a particular perceptual experience. For the same reasons as in the above problem with size perception, it would be naïve to assume that the spatial properties of one's movements are what make us perceive the shape of objects. Hence, one should not situate a natural connection between visual and tactile shape perception in the fact that the movements of the respective organs can be spatially congruent. Rather, a theory of sensory perception should explain why vision's take on figures is much more refined than touch's *despite* the fact that the perception of spatiality in both vision and touch originates in the same functional dependence of sensory variations on self-movement.

There can be no doubt that the hand enhances our abilities to get to know shapes in the absence of vision. But to say that the hand enables us to perceive a thing's overall shape is to make a claim that is fundamentally different from the claim that the hand enables us to form an idea of what a thing looks like shape-wise. Either way, however, the claim rests on a preceding decision, to wit that the hand can be seen as a means to find out what we typically know through vision. The hand is implicitly understood as a tool for assembling the same information as eyesight. This amounts to treating the hand as an organ that compensates for touch's natural lack of affinity with spatial configurations.

(2) The more important point is that, in treating the hand as an organ of compensation, one fails to appreciate that touch might already be a primitive access to shape providing its own, unique epistemic profits. The examples of relief figures have shown that shapes can escape tactual perception. Could there be a way in which shape manifests itself to touch that is never really grasped as such in vision? And, would we not equally tend to overlook this because we are usually able to anticipate this when looking at it? The answer to these questions must be yes.

Consider an animal with no hands, say, a dolphin. A dolphin can see the silhouettes of other sea animals, but it is absurd to want to maintain that a dolphin could or would *tactually* perceive the overall shape of a starfish or a whale shark. On the other hand, it would also be absurd to want to maintain that a dolphin would sense no difference whether it bumps into a rounded or a jagged lump of solid matter. To say that the difference in what it senses is epistemically equivalent to the difference between how both objects look, or to say that what this dolphin tactually perceives must be something it could just as well see, only cashed out in skin-related sensational properties, is precisely to exchange what is uniquely tactile for our visual take on the world, instead of acknowledging the primitive manifestation of shape to touch.

How does shape manifest itself most primitively to touch? First consider what touch is. The verb "to touch" stands for the mechanical event in which two material objects make contact. For example, we say that the car touched the parking meter or, when someone intentionally makes contact with the car, we say that she touched it. The difference between these examples is that in the first case the word "to touch" stands for one specific mechanical event among many others (e.g., my car did not scratch, push over, bump into . . . the parking meter but merely touched it) whereas in the latter case we use the word "to touch" for a sentient body's proper manner of perceiving which can involve any form of mechanical contact. A sentient body "feels" because, whenever it is being touched, or touches, the mechanical event taking place at the body surface entails a cutaneous feeling (cf. sec. 1.2). When we seek to find out what an object is like, we exploit our skin's sensitivity to the way it is acted upon to fathom the object's material nature. Touching an object, we play off a repertoire of divergent touch-actions to coax such sensory effects from the object, which makes that we immediately feel how the object responds to how it is acted upon; we probe its material character by testing its causal behavior: for example, tapping and pinching an object, we sense how it will affect, or be affected by, another material object and, thus, whether it can be used to obtain a given result in a given mechanical context.

Since "touch" is the event of two material objects making contact, shape manifests itself as the effect it has in this event, that is, the way in which the shape of one object influences the deformation and displacement of another object. Hence, what a material object does to a sentient body is partly determined by its shape. In a tactile experience, we straightforwardly perceive an object's shape in the way it determines what we feel during a given form of bodily encounter. We feel a difference whether we sit down on a flat surface or on a convex surface and we sense the difference between bumping into a sharp corner or into a rounded corner. Moreover, we immediately experience the shape as the shape of a material object. An edge may look dangerously sharp, but we instantly feel that it is compliant, that it won't hurt my body and that my body can't lean on it. Similarly, a black balloon and a replica made of lead may look identical, but in touch, we immediately sense that their shapes will have a different effect in a causal context. Since the mechanical significance of a shape varies with the material nature of the object, touch provides a direct resource for knowing what it means for a materially specified object to be spatially structured in this or that way.

The human hand is a particularly refined instrument for probing an object's material composition. Because of its flexibility and motility, it functions like a pocket-size mechanical test kit. Its share in the perception of shape is, however, more complicated since, in a certain sense, it has no shape itself. A palpating hand in the dark does not approach a thing as a preset, rigid grip. Rather, in order to grasp, the hand must precisely be loose and deformable. Enclosing a physical thing to enforce one's grip on it is a compound act in which the hand's role undergoes a transition: from a loose and flexible physical object that gives in to the way a thing imposes its shape toward a rigid and forceful object that precisely uses its spatial enclosure to physically master this thing—manipulation is a matter of the causal interaction between two physical objects, of which the hand is the actor, not just because it is most active, but because of its power to adjust its own shape, intending a specific effect on the other, solid-shaped body.

However, what the hand is capable of again interferes with our theoretical expectancies about tactual perception. The same flexibility and motility that evolved for grasping is also what enhances our ability to tactually detect figures to the extent where it seems justified to say that we "feel" them. In this way, the hand's powers link up two otherwise unrelated perceptual capacities of bodily subjects: the primitive tactile experience of an object's shape as a causal factor in the encounter of one's own body and another physical body, on the one hand, and, on the other hand, the manuovisual cooperation technique in which the hand helps us to form an idea of what an object looks like globally. The hand occupies an intermediary position, so to speak, between the uniquely tactile experience of solid objects' shapes and the typically visual manifestation of objects' silhouette shapes. As a consequence, the objects that we most often "handle" perfectly support the theoretical alignment of sight and touch. Consider the example of a glass. Who would deny that grasping it is sufficient to know that it is round? However, before one even had the chance to pick up a glass, one always already *saw* the circle segment that the glass's bottom draws on the tabletop. Hence, one spontaneously relates back what one feels to the figural manifestation of the object in vision. When philosophers point out that we see *and* feel shapes, and state that the shape as it is felt cannot be different in *any* relevant sense from the shape as it is seen, it does not occur to them to even consider that grasping a glass tells me that it rolls in one direction, rather than that it is circular; that, when I grasp a glass or step on a bottle, *what I feel* is the nature of something that imposes itself on my body in the way something does that rolls evenly and optimally over a surface, rather than that each point of its cross-section is equidistant from its center.

Of course, something that rolls evenly has a circular cross-section. The difficulty is rather that it does not make sense to say that a circular figure is a typically visual object, while, on the other hand, it cannot be denied that a slightly more complex or irregular figure remains in the blind angle of touch, yet perfectly accessible visually. Like a glass, the objects that usually figure in philosophical arguments as examples to illustrate the fact that we perceive shapes through vision and through touch suggest confirmation bias, as they are often small, humanmade articles of use with a well-defined simple shape (a coin, a plate, a bottle, a die, a ball, a box). As it is beyond doubt that we can perfectly "tell" the shape of such objects by grasping them, it seems entirely legitimate to say that both the visual and tactile experience represent, say, a circle, distinguishable from one another only by the co-occurring color additive or thermal-texture additive, respectively. In line with such examples, one simply extends this view, taking for granted that tactual exploration is about finding out that something is spherical or rectangular by following its contours.

To conclude, let me briefly repeat my central observations. Unlike in psychology, it is not uncommon in philosophy that touch is only dealt with because of spatial perception, or even that investigations into touch focus exclusively on spatial properties. In this context, philosophers often point out that we can perceive the shape of an object through sight and through touch. It seems fair to say that this common observation is a rough abstraction of a multifaceted situation. There is no need to deny that a perceiver knows that she relates to the same objective feature through touch and vision, just as she knows that she sees and feels one and the same object. Nonetheless, it seems more accurate to say that the eye and the body each have their own peculiar take on the same object so that its shape manifests itself to a perceiver differently in seeing and feeling, in each case facilitating a particular practical relation to the object. In vision the object's overall silhouette appears which facilitates the identification of an object as an object of a certain type or the reidentification of a particular item. In touch, on the other hand, we sense the mechanical encounter of our body and another body so that we straightforwardly experience the mechanical significance of the object's shape. The central theme of this chapter is that the human hand is responsible for a general tendency to neglect these differences between sight and touch. The hand makes that there are at least three different forms of tactual access to shape. Just as any other body part, our hand senses how a material object imposes its shape on it during a physical encounter. But, as it encloses small (parts of) objects, the hand can also shape itself around them so that we have a take on their overall shapes that comes close to the way they are seen. Finally, the refined motility of the arm, hand and fingers makes it possible to trace the contours of larger objects so that we can form an idea of their overall shape which approximates, more or less well depending on the concrete case, the shape as we see it. The latter two forms of tactual access to shape correspond to our visual take on an object's shape so that it becomes plausible to say that vision and touch do not only relate to the same objective feature, but actually gather the same information. However, the latter two forms do not occur in animals without hands. Moreover, it is not only unclear how but also why a sighted animal would try to tactually perceive an object's overall shape by tracing its contours. It seems that philosophers focus on shape when discussing touch because overall shape allows us to quickly recognize objects in vision, which is our most prominent modality in the perceptual identification of everyday objects. In so doing, they not only neglect that touch is rather bad at this, but they also ignore what touch is good at; they fail to appreciate that the combination of the skin's sensitivity and various touch-actions puts us in immediate contact with the material nature of objects. Tactile perceivers spontaneously apply various perceptual strategies to fathom an object's material composition and this enables them to identify everyday objects before they even had the chance to trace their overall contours. To summarize the point, consider the example of a polystyrene foam replica of a hammer. Looking for an object to drive a nail in the wall, a visual perceiver will spot a hammer, even by seeing the replica's shadow, whereas a blindfolded person would never start tracing the replica's overall shape, as she instantly feels that this thing cannot be used for hammering. An account of touch that focuses on shape describes what a bodily subject would have to do in order to figure out what is (already) salient in vision, instead of describing what a tactual perceiver usually does to complement vision. Moreover, if tactually detecting figures is a cooperation between the hand and visual imagination, then feeling figures is actually an attempt at seeing rather than a tactile experience, namely a strategy to get the kind of grip we have on shapes visually. Paradoxically, the spatial proficiency of the hand covers up the unique take we have on solid shapes through bodily contact, because from the start philosophical reflections situate touch's epistemic value in the hand's ability to find out what we naturally know through vision, already treating the hand as an organ of compensation.

Acknowledgments

This study was made possible by a 2010 grant from ASL—Academische Stichting Leuven to the author as a Fellow in Philosophy at Harvard University.

Notes

1. By the time Katz published his book on touch, Husserl no longer characterized phenomenology as descriptive psychology, but had incorporated his philosophical psychology into a more ambitious philosophical project. Consequently, the young Katz repeatedly opposes his phenomenology to Husserl's. When introducing the bipolarity of touch, he quotes instead *Contributions to Phenomenology of Perception*, the dissertation of one of his fellow students under Husserl, Wilhelm Schapp, which provided an important source for Maurice Merleau-Ponty's *Phenomenology of Perception*, as did Katz's own work, and Husserl's unpublished manuscripts on this issue, which Merleau-Ponty studied extensively at the Husserl-Archives in Leuven.

2. This usage extends to emotional states, even where the bodily aspect is minimal or even absent (e.g., "I sensed no regret").

3. Reversibility does not characterize all bodily feelings. I cannot shift my attention between, say, a feeling of hunger and my stomach. Hunger signals the state of my stomach, but it does not reveal my stomach itself. In a similar way, one feels effort or tiredness in one's muscles, but not, say, their length. Such bodily feelings are felt *somewhere*, but they do not reveal anything *as a spatial object*. However, if someone embraces your upper arm with both hands and then slides them down along your elbow all the way to your wrist, the spatial object that is your arm somehow appears to you.

4. We tend to make the opposite mistake in color vision. Sensitivity to color contrasts allows frugovores and folivores to locate and identify edible objects; therefore, color vision is indeed most basically a discriminatory sense, which explains a child's difficulties in categorization.

5. Starting from this kinship, philosophical analyses then point out differences (e.g., tactile reversibility), as if they were peculiarities in the periphery of object perception.

6. My gaze is expressive, but it cannot adopt the features of external things. At best, I can copy the facial expression of someone else in much the same way as I can imitate someone else's gestures. But the look on someone's face, just like his or her gesticulation, is precisely *that person's* expression; it is not a reproduction of (the properties of) something else.

References

Cassam, Q. 1997. *Self and World*. Oxford: Clarendon Press.

Drummond, J. 1979. On seeing a material thing in space. *Philosophy and Phenomenological Research* 40:19–32.

Epstein, W. 1995. The metatheoretical context. In *Perception of Space and Motion*, 1–22, ed. W. Epstein and S. Rogers. San Diego: Academic Press.

Husserl, E. [1907] 1997. *Thing and Space*. Trans. R. Rojcewicz. Dordrecht: Kluwer.

Husserl, E. [1913] 1989. *Ideas Pertaining to a Pure Phenomenology and to a Phenomenological Philosophy, Second Book*. Trans. R. Rojcewicz and A. Schuwer. Dordrecht: Kluwer.

Katz, D. 1925. *Der Aufbau der Tastwelt*. Leipzig: Barth.

Martin, M. 1993. Sense modalities and spatial properties. In *Spatial Representation: Problems in Philosophy and Psychology*, 207–218, ed. N. Eilan et al. Oxford: Blackwell.

Mattens, F. 2011. Silhouette and manipulation. In *Life, Subjectivity, and Art*, 303–323, ed. R. Breeur and U. Melle. Dordrecht: Springer.

Mattens, F. Unpublished. What touch is really about.

O'Shaughnessy, B. 2000. *Consciousness and the World*. Oxford: Clarendon Press.

Reid, T. [1764] 1997. *An Inquiry into the Human Mind on the Principles of Common Sense*. Ed. D. Brookes. Edinburgh: Edinburgh University Press.

Smith, A. D. 2000. Space and sight. *Mind* 109 (435):481–518.

Weber, E. H. [1834] 1978. *The Sense of Touch*. Trans. H. E. Ross. London: Academic Press.

Wijntjes, M., T. van Lienen, I. Verstijnen, and A. Kappers. 2008. Look what I have felt. *Acta Psychologica* 128:255–263.

8 Phenomenology of the Hand

Natalie Depraz

I am undertaking this eulogy on the hand in the same way one would fulfill a duty
of friendship. As I am writing, I can see mine arousing my mind, carrying it away.
There they are, those tireless companions who, for so many years, have fulfilled
their task, one holding the paper in place, the other increasing the number of small,
hurried, gloomy and active signs on the blank page. With them, man establishes
contact with the toughness of thought. They bring out the block. They give shape
to it, a contour and, to the writing itself, a style. . . . The human face is made up
mostly of receptive organs. The hand is action: it takes, it creates, and sometimes it
even gives the impression that it is thinking. At rest, it isn't a soulless tool, aban-
doned on the table or hanging alongside the body: habit, instinct, and willingness
to act all reside in it. And it does not take much practice to guess the gesture it is
about to make.

—Henri Focillon

1 The Hands at the Crossroads between Metaphysics and Anthropology

As is beautifully attested in the quotation above, it is quite common in
anthropology and philosophy to consider the hand as a remarkable signa-
ture of our humanity and, in a broader sense, to link the presence of the
hand with the faculty of intelligence and the ability to make decisions.

1.1 From Leroi-Gourhan Back to the Greeks: The "Intelligence" of the Hand

The hand is not the docile servile of the mind: it strives for it, it searches, it goes
through all sorts of adventures, it takes its chances.

—Henri Focillon (2010), 124

To begin, let us mention André Leroi-Gourhan's great work, in which he as a paleontologist rigorously demonstrates the part played by the hand in the evolutionary process of our humanization. Because, at one step of the evolution process, the anterior members of big apes were no longer used for moving and walking, they became useful for other functions, for example to grasp objects, to eat (instead of using claws), and eventually to build tools.[1] The hypothesis of the French anthropologist amounts to linking the bodily standing posture with the development of the language area of the brain, while noting the striking proximity in the brain of the hand and the language areas.[2]

Much earlier, famous Greek philosophers had already stressed the remarkable function of the hand and its essential link with the cognitive abilities peculiar to human beings: Anaxagoras from Clazomenae (500–428 BC) and Aristotle a century later both acknowledged the crucial role the hands play in human intelligence, although they had logically opposite contentions. Anaxagoras states in his *Fragments* that human beings possess "intelligence" (*nous*) *because* they have hands, and he states in his book *Peri phuseos* (*On Nature*) that human beings think because they have hands,[3] whereas Aristotle in his work on biology entitled *On the Parts of Animals* inverted the logical causality: humans possess hands because they are intelligent.[4] By suggesting that intelligence is the rational a priori cause of the bodily presence of hands, Aristotle reacts against Anaxagoras's naturalistic stance, according to which the development of the grasping function of the hand would allow the process of intelligence to emerge. Furthermore, the Ionian natural philosopher gives a primacy to the organ over its function: for him, the organic existence of the hand creates the very function of grasping. On the contrary, following Aristotle, the Latin philosopher Lucrece would also stress the primacy of the function over the organ: the creative dynamics inherent in human intelligence gradually transforms the organ and its initial abilities. But in point of fact, such causal logical argumentation leaves each philosopher with only one side of the experiential truth, which remains unsatisfactory for both. Of course, it is interesting to stress the contrast between these two logical stances, but it is merely a preliminary framework that needs to be further developed and adjusted to what experience shows us in its whole complexity.

1.2   The Thinking Hand: A Dialectical Process?

Man made the hand, I mean that little by little he set it apart from the animal world, he freed it from a natural and antique servitude, but the hand made the man.
—Henri Focillon (2010), 107

If we move forward in time and consider Friedrich Engels's contention on the one side and Henri Bergson's on the other, we discover the same logic-metaphysical contrast, but with more subtleties.

In Friedrich Engels's *Dialectics of Nature*, the hands are introduced as the result of the evolutionary process of humanization[5] (as Leroi-Gourhan would empirically demonstrate a century later), but they also express the working material "manual" activity of the human beings attuned to the immanent dialectics of nature. Although Engels found himself close to Anaxagoras's naturalistic stance, he would never advocate the idea that an organ might create a function: the hands are here to adjust to human activities that simultaneously require a thinking process.[6]

A century later, Henri Bergson would not accept Engels's natural dialectical adaptation of the hand, but instead would present the hand as the "working faculty" (*faculté fabricatrice*) meant to create tools (*outils à faire des outils*). Here the philosopher of *L'évolution créatrice* (*Creative Evolution*) echoed Aristotle's definition of the hand as an *organon pro organon*, a tool for making tools, but nonetheless he would never (like Aristote) consider the hands a derived effect of human intelligence: hands are (in his words) the "*organe-princeps de l'intelligence humaine*," which means that they play the role of a principle of human intelligence, but an "organic" one, that is, not a formal, Kantian one. With such a nice integrated expression (*organe-princeps*), Bergson attempts to understand the hand as a kind of "material a priori" of human intelligence.[7]

In short, there is a significant, circular relationship between the hand and the specifically human thinking process. Does this mean that hands and thoughts are two sides (body and mind) of a single human reality? The hand is commonly associated with the human experience of grasping (Leroi-Gourhan), working (Engels), and building (Bergson), that is, of controlling (monitoring) an object, using it (e.g., grasping a glass in order to drink), or again, transforming it (e.g., using wood to make fire, molding clay to create plates). In all these cases, with various consequences, of course, grasping, working, and modeling amounts to a kind of knowing that derives from "mastering," ultimately as a sign of absolute authority: the Hand of God. In this respect, we may also think of Kant's anthropological rational and voluntary contention of dominating Nature (following Descartes's willful reason), with Nature's peculiar feedback effect, which endowed the human being with very few abilities:

For nature does nothing superfluous and is not wasteful in the use of its means to attain its ends. The mere fact that it gave human beings the faculty of reason and the freedom of will based on this faculty is a clear indication of its intent with regard to their endowments. They were intended neither to be led by instinct, nor to be supplied and instructed with innate knowledge; they were intended to produce everything themselves. The invention of their means of sustenance, their clothing, their outward security and defense (for which it gave them neither the bull's horns, nor the lion's claws, nor the dog's teeth, but only hands), all the joys that can make life pleasant, their insights and prudence, and even the goodness of their will were intended to be entirely the products of their own efforts. Nature seems to have taken pleasure in its own extreme economy in this regard, and to have provided for their animal features so sparingly, so tailored as to meet only the most vital needs of a primitive existence, as if it had intended that human beings, after working themselves out of a condition of the greatest brutishness to a condition of the greatest skill, of inner perfection in their manner of thought, and hence (to the extent possible on earth) to a state of happiness, should take the full credit for this themselves and have only themselves to thank for it. It thus seems as if nature has been concerned more with their rational *self-esteem* than with their well-being. (Kant [1784] 2006; third proposition)

Contrary to other animals, the human being has only her hands and is compelled to work and to develop her will and her reasoning faculty. The direct consequence of this constraint is that she is freed from the constraints of Nature; but another consequence is the risk that she will apply to persons what meant for only Nature and objects. I may grasp the hand of somebody, for example, in order to make her feel my presence, which may lead to possession or the manipulation of a human body, transforming and instrumentalizing it (see, e.g., Ihde 1979). Furthermore, such a tendency may lead us to neglect the importance of aesthetic and ethical contemplation, either of Nature or of persons.

2   Phenomenology I: The Hand in the Light of Receptive Intersubjectivity

The gesture that does not create, the impulsive gesture, causes and defines the state of consciousness. The gesture creates continuous action on the inner life. The hand tears away the touch from its receptive passivity; it organizes it for the experience and for the action.
—Henri Focillon (2010), 128

In the Kantian context, little room is left for a positive experience of passivity, that is, of receptivity, which goes "hand in hand" with freedom and openness. In an intersubjective and receptive framework, however, we experience many ways of using our hands: we may shake hands in order to "say hello" or to congratulate, or simply to wave (goodbye) at a distance (in French, more explicitly: *faire un signe de la main, saluer de la main*); we may gently graze the hand of another person without grasping it; we may practice a "light touch," eventually leading to stroking or tickling, also giving pleasure to the other (or to oneself); we may bring our hands together as a sign of despair, in order to implore or to pray; or we may rub our hands together to warm ourselves or as a sign of joy.

In short, the hand brings with it inner sensations of continuity with emotions (pleasure or displeasure): it may be used alone or in contexts where both hands are engaged (e.g., in praying or playing instruments), and it also reveals various forms of intersubjectivity (e.g., shaking hands, waving, raising one's hand in order to address somebody). In the end, when I hold out my hand to somebody in order to help her or beg her for something, I am passively submitted to the other, no longer grasping or controlling anything.

Moreover, the experience of the hand and of touching are close but not equivalent: I can touch with my foot, with my lips, with my knee, or with my shoulder, and conversely, as mentioned just above, there are many contexts where I use my hand at a distance, without making any direct contact with others. Such an indicative sketch of experiential features belonging to a phenomenology of the hand remains open to multifarious other concrete and particular contexts and examples: I in no way mean here to be exhaustive; I just intend to begin by undoing some prejudices we may have about the meaning of the hand. My hypothesis is that both experiential and linguistic contexts (without priority) are useful for such a practical phenomenology and can help us overcome the often limited distinction between the physical and the symbolic to find a more integrative and encompassing lived experience.

I would now like to review the various contributions to the phenomenological tradition (Husserl, Heidegger, Merleau-Ponty, Sartre, Levinas, Henry, Ricoeur, Derrida) and use them as resources to explore further the various aspects of a more concrete phenomenology of the hand. To what extent are such historical third-person analyses sufficient? In what sense do they anticipate our immediate first-person experience of the hand and contribute to it? How may they be enriched by our own first-person experience and open the way for new descriptive aspects of our experience of the hand?

The hand being a distinctive feature of humanity (nonhuman mammals have legs, feet, paws, and claws, but no hands), it is well known that, within the gradual evolution of humankind, it contributed to the development of a full-blown language as it allowed us to become bipedal. In this way, the hand strongly participates in a high-level understanding of human beings as socially, intersubjectively embedded agents. I therefore will use passive receptivity and intersubjectivity as complementary threads in both my historical and experiential exploration of the hand and suggest a variation between different receptive and intersubjective stances, from the weakest to the strongest.

2.1   A Social Yet Still Individual Phenomenology of the Hand: Practice and Know-How

Given our natural-cultural anthropological starting point, the first immediate intersubjective stance at work related to our experience of the hand is its *educational* practice in the context of its uses, its growing *know-how*, in short, its cultivation.

We have to understand intersubjectivity broadly here, as a socially fabricated practice: the subject is a worker, working with her hands, and thus has cultivated manual abilities (from the Latin *manus*): typing, playing music, writing, hammering, and so on. In this context, the hand is considered alone, each single hand being used as such although both are acting together; furthermore, the hand cultivates its know-how while using tools that develop it further and thus ends up developing into an integrated prosthetic tool, a mixture of a technical and a bodily instrument.

The philosophy of such a social-practical "handling" (in French: *maniement*) was first initiated by Martin Heidegger 1927 in *Sein und Zeit* (*Being and Time*), with his famous conceptual distinction: *Zuhandenheit* versus *Vorhandenheit*, which indicates a distinctive relationship of the hand to the objects of the world, taking them either as tools/instruments to be used as means, or as goals in themselves to be freely contemplated, with the hand at the intersection of the two. Whereas the latter is given ontological primacy because it stresses the free acting power of the subject, the former tends to make the subject an instrument at the service of the used objects (leading to "instrumentalization"; in the French, *manipulation*). Nevertheless, pragmatists like Hubert Dreyfus or Bernard Stiegler with his phenomenology of technics tended to put aside Heidegger's distinctive axiology of the hand as instrument versus the hand as acting goal in order to show the continuity of both dimensions: the handling or even "handing" activity may enter into an expert process because it is not solely "instrumental": the path may become the goal (and pragmatics receives here its full-blown meaning).

Strikingly enough, within such an understanding of the "handing/ handling" subject, the scope of intersubjectivity nevertheless remains too broad to be clearly identified: it is socially and pragmatically determined but never refers to contexts of genuine relationships with the other or others: I use my hands, but I remain alone in such a use, although both hands may be together intensely cultivated. Now, I have two hands, which polarizes my bodily spatial situation in the world, both symmetric and asymmetric: I am more clever with one hand (often the right one) and more clumsy with the other. Such an asymmetry individuates me as a uniquely situated bodily subject and furthermore creates within myself an inner fracture or splitting (Ricoeur speaks in a more general way of "inner alterity"). My contention is that such a primal alterity within me as a bodily self is the very origin of intersubjectivity as a strong, interbodily stance (see Depraz 1995, ch. 5). Given such a core bodily intersubjectivity within myself, it seems difficult to consider further my hands as two identical hands (two right hands or two left hands). Their individualizing asymmetry opens the way for a primal intrasubjectivity and intersubjectivity of the hand.

2.2 Bodily Intersubjective Phenomenology of the Hand: Erotica and Its Pathological *qua* Therapeutic Reversals

Given this fact of the individualizing asymmetry of the hand, Heidegger's analysis (even pragmatically and bodily anchored) strikes me as quite an isolated—solipsistic—contention within the entire phenomenological tradition of the hand. In contrast with this solipsistic view, a whole trend of investigation focuses on the inner intersubjective dimension of the hand experience: with various emphases, nearly every phenomenologist from Husserl onward suggests a specific organic articulation between the hand experience and the eye experience as a grounding intersubjective framework. In this sense, "touching" becomes the common activity of the hands, as opposed to grasping, an activity that objectifies the quality of the relationship with the other.

Touching opens the way for a possible noninstrumentalized and non-active intersubjective experience. As a matter of fact, every phenomenological analysis (Husserl, Merleau-Ponty, Sartre, Levinas, Derrida, Henry, Marion)—each time in a specific way—broaches this issue of the ambivalence of the touching experience of the hands: between intrusive possession (Sartre) and the gentle touch of the light stroke (Levinas).

Within this common stance, we find numerous settings: my hand may touch the hand of the other; another may touch my hand; my hand may touch some other bodily organ of another; the hand of another may touch one of my bodily organs; my hands may be used to call attention to another at a distance; and so on.

a. My Hand and the Other('s) Hand: Various Forms of Touching Contact (Husserl and Merleau-Ponty)

The word "contact" literally names the originally intersubjective dimension of touching, the basic experience being the touched/touching experience. Most of the time when one of two persons is touching the hand of another, the latter is being touched, even if the situation may change over time and become reversed: where I was actor, I may become the receiver, and vice versa. But such a dynamic crossing is not reciprocal insofar as the interactive experience involves a rhythm of alternating polarities: I am touching the other and the latter receives my touching; a fraction of a second later, she may be touching me and I will be touched. The point is: the cosubjectivation of both lived bodies is never complete, but each is inhabited by a recurrent tendency of objectivation. What is still more interesting is that such a structural asymmetry in the very intricacy of activity and passivity also operates within myself (as a self-altered/fractured subject) in my first-person experience of touching myself: my right hand is touching my left hand and experiences a similar (even if not identical) subjectivized quasi objectivation.

In *Ideas II* (1912–1915; §41), Husserl first described such an experiential context of hands intertouching, but also situated it in the broader and complementary framework of the visual perception of the other and the bodily experience of automaticity and resistance: in his view, optic and haptic experiences codetermine my experience of the other. In this section of *Ideas* II (*Hua* IV: 158, 160), Husserl appeals to the hand experience at three main points within a more encompassing description of the oscillating movement of passive mechanical processes of the bodily experience and its free "I can." Here the hand is an example of the body functioning in an individual way: how my hand becomes an object while being touched by my other hand, how it feels the pressure of an object and is able to control it, how felt sensations vary as a function of my own inner disposition. In short, the description of the hand is individual: it is always my hand as a part of my body, or my hands relating to each other within myself, whereas the description found in the intersubjectivity textual material (in *Hua* XIII–XV) reveals the intrinsic intersubjectivity of the hand experience.

How is such a change in perspective possible? In fact, Husserl begins with an ontological distinction between self and other with respect (here) to the hands: my hands and the hands of the other are structurally given to me in an antinomic contrast. Whereas my hands are given to me as being proper to me, the hands of the other will always be given to me as being alien to me. This may seem trivial, but it determines the way I experience and understand the intertouching experience. For Husserl, since I can never feel how the other feels his own hand touching, I will also have a limited access to my "other" hand, which appears, to a certain extent, as an alien hand to my touching hand. In short, what is stated at the intersubjective level (the irreducibility of feeling the touching experience of the other) is partly reverberated at the intrasubjective level: the hand of the other remains alien to me, and so my other hand may in some respects appear to me as not being mine, as Husserl states it in some limit-experiences of imaginary dissociation, where I see my "other" hand as being an "alien" hand (not belonging to myself, but to someone else).[8] In this respect, Merleau-Ponty radicalizes the touching component of inter-subjectivity with the standard contention that vision creates a distance that generates idealization and abstraction, whereas the tactile experience goes hand-in-hand with a concrete proximity. But Husserl is also right to stress the tendency to obscure blindness brought about by touching and the lightness of a tactile vision, which I have called elsewhere its warm lucidity or luminosity.[9] Let us consider the experience of myself stroking my beloved while gazing at him or her. Such a tactile power of the gaze is not a mere image nor simply symbolic; it corresponds to a genuine experience, in the same manner as the voice may also touch us: such synesthetic experiences are at the core of Merleau-Ponty's analysis in *Phenomenology of Perception* and of *The Visible and the Invisible*. Further, touching is also for Merleau-Ponty a "light touch," insofar as it never ends up as a fully objectifying contact, but remains what he calls a "quasi contact," which frees the other to take his or her own lead.

Many daily experiences attest to such an oscillating mixture of passive receptivity and monitoring activity between myself and the other within intertouching experiences: when we shake hands or embrace each other, a heterogeneous dynamics generates the singularity of our relationships; when we (adults and children, or beloved persons) hold hands, we experience a polarized rhythm that gradually strengthens our bond and is based on faithfulness and tenderness. In short, for Merleau-Ponty, the intertouching experience is characterized by the continually replayed "imminence" of its achievement. In short, the experience is never "complete," which would mean objectivation.

In Merleau-Ponty's *The Visible and the Invisible*, what happens between my two hands in terms of reversibility of activity and passivity is mentioned as a possibility between my hand and the hand of the other: "if my left hand can touch my right hand while it palpates the tangibles, can touch it touching, can turn its palpation back upon it, why, when touching the hand of another, would I not touch in it that same power to espouse the things that I have touched on my own?" (Merleau-Ponty 1968, 141). Merleau-Ponty's experiential intuition is to incorporate the hand of the other into a continuity of experience with my experience of my other hand according to the general contention of the experience of a global undifferentiated flesh of the world. Furthermore, even earlier, in *Phenomenology of Perception*, Merleau-Ponty describes in a very striking way the intricate lived continuity of the subjective bodily activity of an organist and his instrument. At one point in development of the organist's cultivated skill, the organist is one with his or her instrument: there is only one "lived" experience, no separation, no distinction any longer between subject and object: "so direct a relation is established that the organist's body and his instrument are merely the medium of this relationship" (Merleau-Ponty [1945] 2012: 168). In short, we have one global hand and handing *qua* touching experience, which takes place as an expressive and receptive sensible field and no longer belongs to a subject, an other, or even to an object: there is simply handing and touching. From Husserl to Merleau-Ponty, we find a change in the hand's experience: individualized, localized, and contrasted with the hands of the other in Husserl; belonging to a global body, understood as a general power and reception of the touching process of things themselves in Merleau-Ponty. Beyond such a contrast, both phenomenologists focus on the kinesthetic and synesthetic experience of the hand and do not bring to the fore its affective-emotional component.

b.   The Affective Intertouching Experience: Stroking (Levinas and Sartre)

Receptive intersubjectivity also unavoidably reveals the emotional component of intertouching hands, which is not brought about solely by the bodily pragmatics of the hand: touching thus intrinsically generates affects, and not only physical or lived sensations. In this respect, Levinas and Sartre are more acute, given their common stress on the "stroking" experience as a particular way of touching. This sort of stroking experience of our hands is remarkably striking, but it also appears local, a bit narrow. We need to have in view the more global and ecological experience of inter-touching: I may touch the other with my hand or vice versa (touching his or her shoulder, I am touching the whole of the person); I may "touch" the other while begging something of her or him; I may touch the other by crying in front of him, which also shows the relevance of a "distant" touch, that is, conversely, of a possible "touching" vision. Such a global approach to the hand is crucially at work in Sartre, with the "localiza-tion" of such an experience leading, according to him, to a pathology of dissociation:

I never see an arm raised alongside a motionless body. I perceive Pierre-who-raises-his-hand. . . . we need only recall the horror we feel if we happen to see an arm which looks "as if it did not belong to any body," or we may recall any one of those rapid perceptions in which we see, for example, a hand (the arm of which is hidden) crawl like a spider up the length of a doorway. . . . this disintegration is apprehended as extraordinary. . . . It comes as a shock when a photograph registers an enormous enlargement of Pierre's hands as he holds them forward (because the camera grasps them in their own dimension and without synthetic connection with the corporal totality). (Sartre [1945] 2012, 346–347)

Here touching is not reduced to the physical tactile experience of contact, but includes feeling and moving, which the French language is better able to name with the same word: *toucher* means both touching and "moving" (resonating with "emotion").

c. From Lightly Stroking to Holding the Hand to Me (Levinas)

With Levinas's analysis in *Totality and Infinity* (1980), a continuity between tactile sensations, psychic emotions, and ethical feelings provides an aston-ishing extension, that is, an extended experiential meaning to the bodily intersubjectivity at work in the experience of our hands.

Here, more than *my* hands, the hands of *the other* are given primacy: the experience of "contact" is best described through the stroking experi-ence, through which I am the one who is subjected to such a "light touch," which creates within me a deepened sensibility and also generates for me a form of fragility. In short, stroking is not a superficial experience; it involves a renewed understanding of my bodily sensations as being in inner continuity with ethical feelings:

Animation can be understood as an exposure to the other, the passivity of the "for-the-other" in vulnerability, which refers to maternity, which sensibility signifies. . . . A sinking in that never goes far enough, the impatience of being sated by which it has to be defined, can be discerned in the fusion of the sensing and the sensed. (Levinas [1976] 1981, 71, 72)

Such a hypersensitivity (which may lead to "erotization") toward the other unavoidably generates ambivalent feelings, which express themselves through hand gestures: praying, supplicating the other, being exposed to the other and his or her possible violence while he or she holds his or her hands up to me.

d.   The Intrusive Pressure Generated by the Hand (Sartre)

"I can feel his hand pressing on mine": such a first-person testimony exemplifies the intersubjective touching-touched hand experience Husserl and Merleau-Ponty first described as a fully subjectified experience. What both refused to discuss is better revealed by the realistic account provided by Sartre: my hand experience of the other brings about possession and control of his or her freedom.

Levinas's example of a stroking hand understood as a light touch thus becomes in Sartre's analysis of the relationship with the other in *Being and Nothingness* a "pressure" that goes hand in hand with possession and intrusion: "the caress is an appropriation of the Other's body . . . the caress is not a simple stroking; it is a *shaping*. In caressing the Other, I cause her flesh to be born beneath my caress, under my fingers . . . it seems that I lift my own arm as an inanimate object and that I *place* it against the flank of the desired woman, that my fingers which I run over her arm are inert at the end of my hand" (Sartre [1945] 1956, 476–477). Sartre here directly "answers" Levinas, although it might be historically the reverse: stroke is not an *affleurement*, but a *façonnement*. Furthermore, contrary to Husserl's and Merleau-Ponty's sensitive neutral descriptions of the hands, Levinas's and Sartre's are clearly "eroticized" and therefore "situated": it is not an "I" who touches by stroking an "other," but a "man" who lightly touches or presses on the "desired woman," as Sartre expressively writes. Starting from a third-person description (impersonal, infinitive) of the "other" (*Autrui*) being touched, Sartre switches to an incredibly detailed and precise organic first-person writing: "I cause her flesh to be born . . . under my fingers . . . I lift my own arm . . . I *place* it against the flank . . . my fingers which I run over her arm are inert at the end of my hand." As for Levinas, his erotic description of stroking is as situated as Sartre's, but the figure of the woman, which is put in the forefront, is different: not the woman as object of desire, the "whore," but the maternal woman. It therefore produces the opposite figure of the "masculine": the seducer (the sadist, the pimp) for Sartre; for Levinas, the small boy, the child. Besides, Levinas's description is in the first person in a different way than Sartre's grammatical "I"-laden account: although the style of the former is broadly "impersonal," it generates through its general nominalization process an immediate experiential contact that Sartre's predicative verbal propositional discourse lacks (see Depraz forthcoming). Levinas and Sartre thus suggest exactly mirror-experiences of the other, where the stroke of the hand is crucial but endowed with fully opposed meanings and experiential consequences: Sartre describes the intersubjective relationships in the light of the pathology of sadomasochist perversion, where manipulation and objectification is the rule; Levinas furnishes the stroking contact of myself by the other with a sensitive absolute ethical (but hardly pragmatic) dimension. In contrast with Husserl and Merleau-Ponty, both engage in affective and thus liminal pathological experiences of hand-intersubjectivity. They therefore exceed the still local view of the hand and its touch into more a view of more global/ecological lived bodies, touching where the hands are at work but are not alone.

e. Hand-Therapy and Inner Move (Bois)
We have seen how the way we use our hand(s) is never neutral: it affects the other in his or her flesh (as a global personal unity of lived body, emotions, and thoughts); it is not possible to use our hands without also engaging their potentially intruding or freeing component. As we gradually become aware of such pathological or normative effects, it also becomes necessary to develop our own hands' know-how and thus become expert in using our hands toward ourselves for the sake of the other.

Such an ability brings back the general "pragmatic" orientation mentioned earlier of Heidegger, Dreyfus, and Stiegler, but with a more operative approach of an affective bodily intersubjective experience of the hands. Such a "culture" of the hand may be developed and educated spontaneously through our own daily experiences and intersubjective maturing.

Strikingly enough, however, it is worth noticing that such a "handskill" was developed in a very much detailed and disciplined way by kinesitherapists and motor-psychologists. I would like to mention here the hand-practice first initiated by Danis Bois[10] and his work on what he calls the *toucher interne* ("inner touch"): going through such therapy helps one to become aware in an incredible way of such objectifying tendencies of hand-touching; furthermore, being able to guide such therapies enables one even more to go into very subtle touching of the *masse corporelle*, of the "tonus," of the "orientations" and "directions" of one's flesh, in short of the thin materiality of one's lived body.[11] The hands of these manual practitioners are able to reveal one's whole self in its full intimacy, and not only one's body as a set of organs that can be independently resensitized.

In this respect, the therapeutical hands are not mere instruments; they participate in the entire process of my enactive knowing myself while I embody myself, and, at the same time, the hand-expert also embodies him- or herself further through such a kinetic hand practice.

3   Phenomenology II: The Hands in the Depths of Felt Intrasubjectivity

Given the above account of intersubjective hand-embodiment through therapeutic skill, I would like to come to my final contention of a self-cultivated intrasubjective phenomenology of the hand.

I have already noted, while discussing Husserl's and Merleau-Ponty's bodily intersubjective hand-intertouching, how such an experience is both and co-occurrently inter- and intrasubjective: I mentioned this point just above, leaving aside the experiential singularity of my hand touching and being touched by my other hand. The common ground between hands from two different people touching each other and hands from the same person touching him- or herself lies in such ambivalence between subjectivation and objectivation (explored just above with Sartre and Levinas at an extremely affective level). In this respect, Husserl and Merleau-Ponty both conclude in the following way: objectivation is never complete, but always leaves a space for the freedom of the other and for my own freedom toward myself.

I would like now to go beyond this standard view (which is not false, but limited) and take into account the pathological aspect revealed by both Sartre (perversion) and Levinas (unconditioned ethics). While doing so, I will offer another phenomenological account built on a practical, educated ethics. In this exploration, various resources (mainly Henry and Ricoeur) will prove helpful.

3.1   The Self-Affecting-Affected Inner Hand (Henry)

Thanks to Levinas's analysis of the stroke as a "light touch," I have already mentioned the continuity between sensation and affection, that is, between feeling and moving.

For Michel Henry, the subtle sensitivity of my hand is in no way "object"-directed, be it in a pragmatic context of using a tool, even in a prosthetic way, or be it in an intersubjective context of touching the other, whatever touch it may be, freeing or alienating.

In writing this, I apply Henry's view of the self-affecting-affected flesh to the "hands," insofar as he does not provide a systematic *positive* thematic description of the hand. Indeed, his entry into the "hand"-experience is mainly critical: both when he deals with Maine de Biran's analysis of the example of my touching a piece of wood and feeling its resistance in myself as an inner effort (*Philosophie et phénoménologie du corps*), and also when he tackles in *Incarnation* (2001) the Merleau-Pontian issue of the touching-touched hand experience and rejects it because of its primarily hetero-affective stance.

Regarding the latter, let's consider his main criticism: "It is truly incorrect to say along with Merleau-Ponty that while my right hand which was touching my left hand lets itself be touched, it gives up its self-control at the same time, its touching condition, only to find itself absorbed in the touch understood as a tangible something, as a sensitive something similar to all the material bodies of the universe" (Henry 2001, 240; my translation). What Henry rejects in the analysis in *The Visible and the Invisible* is the subjective bodily organ necessarily becoming an objective thing, that is, what is experientially grounding in Merleau-Ponty (but also Husserl), the objectivation-subjectivation processes inherent in the passive-active bodily experience. On the contrary, he radically claims the "unity" of one unique originary flesh, at once touched and touching, touched *where* the subject is touching.[12] Hence the vocabulary of "self-impressionality" and of "self-movement" as a unique power of the flesh. How is this possible? What is the "inner place" of such a biprocessual bodily move? Let's get back to Henry again, who expresses such an "inner move" in quite an amazing way: "whereas the resisting continuous gives in until its material immobilization under the thrust of primal flesh, it is this practical organic continuum that presently stops, maintains, or rejects the instinct, this instinct thus becoming the pathos of a suffered constraint" (ibid.). What is remarkably described as a bodily flow, slowing down at one point and unfolding again into another flow, interestingly echoes D. Bois's "inner move" and requires further investigation.

Henry's description of the hands as self-affecting inner organs and as leading to an inner experience of various sensible flows "underneath the skin" is found in crucial passages in *Incarnation*:

It is the anguish of the one . . . who can caress the skin of it in such a way that the one who caresses will not only feel, on his own hand as it is moving, the impression of smoothness, of freshness or of warmth that the other person's skin will communicate to him. While moving his hand on the hand, he will also provoke a series of impressions—of freshness, of warmth, of pleasure or of terror. On this Other's skin, that is, under it, at this moving limit of the organic body of the Other, whereas, breathing more slowly, the body will either carry it, immobilize it, or will hold it in the "I can" of its native flesh . . . is he indeed going to stretch his hand out to the magic object, put it on this living flesh, standing there next to him, and seeming open to his grip, trying to feel it where it feels itself, at the very place where its sensuality is the strongest, in its sexual difference—"take" it and hold it in its very power? . . . Where touching this body, this sex, would mean touching the spirit itself where it is indeed spirit, touch life where it feels itself, in its own self, irreducible to any other. (Ibid., 288–289)

Interestingly enough, Henry arrives here at the same "touching" experience as Levinas and Sartre—the erotization power of stroking, but with a different conclusion: it is neither a light touch as in Levinas nor an intrusive pressure as in Sartre; it is rather a global "feeling" full of anxious and distressed emotion of not being able to feel the other, but also risking to feel the other where it feels itself.

3.2 The Caring Handling Self (Ricoeur)

On a more ethical level, Paul Ricoeur opens the way in *Oneself as Another* (1992) for a "pluralistic" intrasubjective phenomenology of the hand. Whereas Henry provides his description on the basis of the self-affected individual subject, both hands contributing to her radical self-feeling and individuation, Ricoeur's thematic of "care" for the other gives way to hands as given and offered hands. Helping, caring for the other is originally associating with mutual hand-activities: I hold out my hand(s) and the other equally holds out her hand(s). While Henry's similar "holding out my hand" (*tendre la main vers*) is each time seen as being at risk of "possessing" the other (hence the anxious and distressed feeling), Ricoeur's pragmatic-ethical statement relies on an embodied ethics of (care/handling) "*soin*" and of appeal or request (*sollicitation*) to the other.

In this respect, many daily expressions attest in French to our "handing" interactions with others: *prêter main forte, mettre la main à la pâte*, and also relying on helping gestures: holding a child's hand in the street, asking the latter to hold one's hands while crossing the street, and so on.

4 Conclusion: The Inner Intersubjective Expressivity of the Hand?

Such "intimate" hand experiences, be they erotic or ethical, all suggest a "diffusive" inner feeling that embraces the whole person: in this respect, the hand experience cannot be reduced to a local and localized grasping or even touching experience. It takes place in a more global "hyperaesthetic" expressive feeling (for more on this term, see Depraz 2001) that the French historian of art Henri Focillon nicely accounted for in his elegant essay *Eloge de la main*:

The hands' eloquence is extraordinary. And it is by them that language was modeled, first of all experienced by the whole body and mimed in dances. For an everyday usage, the hand gestural gave it its thrush, helped to articulate it, to separate its elements, to isolate them from a wide sound syncretism, to give it rhythm and even to color it with subtle inflexions. There remains something in what the ancients called the art of oratory in this speech mimic, the exchanges between the voice and the hands. The physiological differentiation specified the organs and the functions. They hardly ever collaborate anymore. Speaking with our mouth, we keep quiet with our hands, and, in some lands, it has become inappropriate to express oneself with both the voice and the gestural; by contrast, other lands have eagerly kept this double poetic. (Focillon 2010, 108–109)

Such an account of the aesthetic power of the hand compels us to take fully into account the hands' original expressivity, which definitively leaves aside its manipulatory dimension and lets us enter into its free communicative dynamics.

As mentioned above, the hands are in themselves a language, which is made of gestures. Though nonlinguistic, hand gestures may follow speech, assist it, sometimes clarify it; in the case of deaf and mute people, it offers a substitute to articulated verbal expressions. Sometimes, our hand gestures betray us, express something different from what we say, or may operate alone without any need of speech: praying is a silent communication, where the hands joined together "speak" alone; applauding has its own meaning as a sign of joy and gratefulness.

But hands are also the only way human beings know how to write and, more generally, to create art: far from operating as a "tool for another tool" (e.g., a pencil or computer), the instrumental dimension recedes and becomes transparent. The hands are our very spirit embodied and our thoughts made words and sentences: "I am going to begin to write what I do not know myself, trying, as much as possible, to let my spirit and my pen be guided by their movement, not making any other than that of the hand." Here are the first words of a book by Jeanne Guyon, a French mystic from the seventeenth century, called *Les torrents*: my hand is here the only guide of my spirit, a sentiment that echoes Anaxagoras's initial words: man is intelligent because he has hands. Did not Derrida himself stress that writing is handwriting, and that thinking is acting with our hands? (Derrida 1990, 2005). Beyond the rationalist Aristotelian tradition, the thinker of "touching" with our hands will again recognize both Anaxagoras's and Guyon's insight.

Acknowledgments

I would like to thank my first-year students at the Department of Philosophy of the University of Rouen, who were patient and open minded enough to welcome the course I offered during the first semester (fall 2011) entitled "The Hand(s): Phenomenology between Anthropology and Metaphysics."

Notes

1. "Within a perspective which starts with fish in the Paleozoic era and ends with the human in the Quaternary period, it is as though we were witnessing a series of successive liberations: that of the whole body from the liquid element, that of the head from the ground, that of the hand from the requirements of loco-motion, and finally that of the brain from the facial mask" (Leroi-Gourhan [1964] 1993, 25).

2. "In the development of the brain the relationship between face and hand remained as close as ever: Tools for the hand, language for the face, are twin poles of the same apparatus" (Leroi-Gourhan [1964] 1993, 20). Amazingly, we find this statement already in Saint Gregory of Nyssa's *Treatise on the Creation of Man* (379 AD): "So it was thanks to the manner in which our bodies are organized that our mind, like a musician struck the note of language within us and we became capable of speech. This privilege would surely never have been ours if our lips had been required to perform the onerous and difficult task of procuring nourishment for our bodies. But our hands took over that task, releasing our mouths for the service of speech" (quoted in Leroi-Gourhan [1964] 1993, 25).

3. Anaxagoras, *Fragments, Peri Phuseos, On Nature* (in Tannery 1930, ch. 12, 295–303); *Atlas de la Philosophie, Présocratiques* (1999, I, 31).

4. According to Aristotle:

Now it is the opinion of Anaxagoras that the possession of these hands is the cause of man being of all animals the most intelligent. But it is more rational to suppose that his endowment with hands is the consequence rather than the cause of his superior intelligence. For the hands are instruments or organs, and the invariable plan of nature in distributing the organs is to give each to such animal as can make use of it; nature acting in this matter as any prudent man would do. For it is a better plan to take a person who is already a flute-player and give him a flute, than to take one who possesses a flute and teach him the art of flute-playing. For nature adds that which is less to that which is greater and more important, and not that which is more valuable and greater to that which is less. Seeing then that such is the better course, and seeing also that of what is possible nature invariably brings about the best, we must conclude that man does not owe his superior intelligence to his hands, but his hands to his superior intelligence. For the most intelligent of animals is the one who would put the most organs to use; and the hand is not to be looked on as one organ but as many; for it is, as it were, an instrument for further instruments. This instrument, therefore,-the hand-of all instruments the most variously serviceable, has been given by nature to man, the animal of all animals the most capable of acquiring the most varied handicrafts.

Much in error, then, are they who say that the construction of man is not only faulty, but inferior to that of all other animals; seeing that he is, as they point out, bare-footed, naked, and without weapon of which to avail himself. For other animals have each but one mode of defence, and this they can never change; so that they must perform all the offices of life and even, so to speak, sleep with sandals on, never laying aside whatever serves as a protection to their bodies, nor changing such single weapon as they may chance to possess. But to man numerous modes of defence are open, and these, moreover, he may change at will; as also he may adopt such weapon as he pleases, and at such times as suit him. For the hand is talon, hoof, and horn, at will. So too it is spear, and sword, and whatsoever other weapon or instrument you please; for all these can it be from its power of grasping and holding them all. In harmony with this varied office is the form which nature has contrived for it. For it is split into several divisions, and these are capable of divergence. Such capacity of divergence does not prevent their again converging so as to form a single compact body, whereas had the hand been an undivided mass, divergence would have been impossible. The divisions also may be used singly or two together and in various combinations. (Aristotle, *On the Parts of Animals*, Book IV, 687a–b)

5. As Engels writes:

For erect gait among our hairy ancestors to have become first the rule and in time a necessity presupposes that in the meantime the hands became more and more devoted to other functions. Even among the apes there already prevails a certain separation in the employment of the hands and feet. As already mentioned, in climbing the hands are used differently from the feet. The former serve primarily for collecting and holding food, as already occurs in the use of the fore paws among lower mammals. Many monkeys use their hands to build nests for themselves in the trees or even, like the chimpanzee, to construct roofs between the branches for protection against the weather. With their hands they seize hold of clubs to defend themselves against enemies, or bombard the latter with fruits and stones. In captivity, they carry out with their hands a number of simple operations copied from human beings. But it is just here that one sees how great is the gulf between the undeveloped hand of even the most anthropoid of apes and the human hand that has been highly perfected by the labour of hundreds of thousands of years. The number and general arrangement of the bones and muscles are the same in both; but the hand of the lowest savage can perform hundreds of operations that no monkey's hand can imitate. No simian hand has ever fashioned even the crudest stone knife.

At first, therefore, the operations, for which our ancestors gradually learned to adapt their hands during the many thousands of years of transition from ape to man, could only have been very simple. The lowest savages, even those in whom a regression to a more animal-like condition, with a simultaneous physical degeneration, can be assumed to have occurred, are nevertheless far superior to these transitional beings. Before the first flint could be fashioned into a knife by human hands, a period of time must probably have elapsed in comparison with which the historical period known to us appears insignificant. But the decisive step was taken: the hand became free and could henceforth attain ever greater dexterity and skill, and the greater flexibility thus acquired was inherited and increased from generation to generation. (Engels 1960, ch. 9 [see http://www.marxists.org/archive/marx/works/1883/don/ch09.htm])

6. Engels, again:

At first, therefore, the operations, for which our ancestors gradually learned to adapt their hands during the many thousands of years of transition from ape to man, could only have been very simple. The lowest savages, even those in whom a regression to a more animal-like condition, with a simultaneous physical degeneration, can be assumed to have occurred, are nevertheless far superior to these transitional beings. Before the first flint could be fashioned into a knife by human hands, a period of time must probably have elapsed in comparison with which the historical period known to us appears insignificant. But the decisive step was taken: the hand became free and could henceforth attain ever greater dexterity and skill, and the greater flexibility thus acquired was inherited and increased from generation to generation.

Thus the hand is not only the organ of labour, it is also the product of labour. Only by labour, by adaptation to ever new operations, by inheritance of the resulting special development of muscles, ligaments, and, over longer periods of time, bones as well, and by the ever-renewed employment of these inherited improvements in new, more and more complicated operations, has the human hand attained the high degree of perfection that has enabled it to conjure into being the pictures of Raphael, the statues of Thorwaldsen, the music of Paganini. (Engels [1873–1882] 1960, ch. 9 [see http://www.marxists.org/archive/marx/works/1883/don/ch09.htm])

7. "Thus, intelligence incarnates the power to create 'tools to make tools,' the hand being the first one" (Bergson [1911] 1998; translation modified).

8. Husserl, *Ideas II, Hua* XIII, 46–49. (In French: *De l'intersubjecitivité*, trans. I. N. Depraz, 52–57 [Paris: PUF, 2001].)

9. *Hua* XV, 298–299, 302; Depraz (2001,21–22).

10. See on this matter Bois and Berger 1990; Berger 2006.

11. About the importance of practice and of a cultivated discipline, see Depraz, Varela, and Vermersch 2003.

12. As Henry puts it: "It is the same primal flesh that is at the same time both touched and touching. . . . It is 'touched' where it is 'touching' and in the same way . . . a passivity succeeds to our primal action of 'touching' flesh powers, the 'being touched' whose pure phenomenological matter is the same as that of the action. Activity and passivity are two different and opposite phenomenological modalities, but they are two modalities from the same flesh, their phenomenological phenomena is the same, that same flesh one" (Henry 2001, 230, my translation).

References

Aristotle. 2007. *On the Parts of Animal*. Trans. W. Ogle. Adelaide: eBooks@Adelaide. http://ebooks.adelaide.edu.au/a/aristotle/parts/.

Atlas de la Philosophie, Présocratiques. 1999. Paris: Livre de Poche.

Berger, E. 2006. *La somato-psychopédagogie: Ou comment se former à l'intelligence du corps*. Paris: Point d'Appui.

Bergson, H. [1911] 1998. *Creative Evolution*. Trans. Arthur Mitchell. Mineola: Dover.

Bois, D., and E. Berger. 1990. *Une thérapie manuelle de la profondeur. Méthode Danis Bois, fasciathérapie, pulsologie*. Paris: Guy Trédaniel.

Depraz, N. 1995. *Transcendance et incarnation: L'intersubjectivité comme altérité à soi chez Edmund Husserl*. Paris: Vrin.

Depraz, N. 2001. *Lucidité du corps: De l'empirisme transcendantal en phénoménologie*. Den Haag: Kluwer.

Depraz, N. 2011. L'éloquence de la première personne. *Alter* 19: *Le langage.*

Depraz, N. Forthcoming. Phénoménologie de l'erôs féminin. In *Femme, erôs, philosophie.* Louvain-la-Neuve: UTC (forthcoming).

Depraz, N., F. J. Varela, and P. Vermersch. 2003. *On Becoming Aware: A Pragmatics of Experiencing.* Amsterdam: Benjamins Press.

Derrida, J. 1987. *Geschlecht II: Heidegger's Hand.* Trans. J. P. Leavey, Jr. In J. Derrida, *Deconstruction and Philosophy.* Ed. J. Sallis. Chicago: University of Chicago Press.

Derrida, J. 2005. *On Touching: Jean-Luc Nancy.* Trans. C. Irizarry. Stanford: Stanford University Press. (Originally published as *Le toucher: Jean-Luc Nancy.* Paris: Galilée, 2000.)

Engels, F. [1873–1882] 1960. *Dialectics of Nature.* Trans. J. B. S. Haldane. New York: International Publishers. http://www.marxists.org/archive/marx/works/1883/don/.

Focillon, H. 2010. Eloge de la main. In *Vie des formes.* Paris: P.U.F.

Gregory of Nyssa. *Treatise on the Creation of Man.* (*Traite de la creation de l'homme,* 1944 edition, Paris.)

Heidegger, M. 2010. *Being and Time.* Trans. J. Stambaugh, revd. D. J. Schmidt. Albany: SUNY Press.

Henry, M. 2001. *Incarnation.* Paris: Seuil.

Husserl, E. 1989. *Ideas Pertaining to a Pure Phenomenology and to a Phenomenological Philosophy,* Book II. Dordrecht: Kluwer. *Hua IV . . . Hua XV.*

Husserl, E. 2001. *Hua XIII * De l'intersubjecitivité I.* Trans. N. Depraz. Paris: P.U.F.

Ihde, D. 1979. *Technics and Praxis: A Philosophy of Technology.* Boston: D. Reidel.

Kant, E. [1784] 2006. *Idea for a Universal History from a Cosmopolitan Point of View.* In *Toward a Perpetual Peace and Other Writings on Politics.* Edited and with an introduction by P. Kleingeld. Trans. D. L. Colclasure. New Haven: Yale University Press.

Leroi-Gourhan, A. [1964] 1993. *Gesture and Speech.* Trans. A. B. Berger. Cambridge, MA: MIT Press.

Levinas, E. [1976] 1981. *Otherwise Than Being or Beyond Essence.* Trans. A. Lingis. The Hague: Nijhoff.

Levinas, E. 1980. *Totality and Infinity: An Essay on Exteriority.* Trans. A. Lingis. Dordrecht: Springer.

Merleau-Ponty, M. [1966] 1968. *The Visible and the Invisible.* Trans. C. Lefort. Evanston: Northwestern University Press.

Merleau-Ponty, M. [1945] 2012. *Phenomenology of Perception*. Trans. D. A. Landes. London: Routledge.

Ricoeur, P. 1992. *Oneself as Another*. Trans. K. Blamey. Chicago: Chicago University Press. (Originally published as *Soi-méme comme un autre*, 1990.)

Sartre, J.-P. [1945] 1956. *Being and Nothingness: An Essay in Phenomenological Ontology*. Trans. H. Barnes. New York: Philosophical Library.

Tannery, P. [1930] 1999. *Anaxagore de Clazomene: Pour l'histoire de la science hellène. Atlas de la Philosophie, Présocratiques*. Paris: Livre de Poche.

III MANUAL ENACTION

9 The Enactive Hand

Shaun Gallagher

The hand is the cutting edge of the mind.
—Jacob Bronowski

The enactive view of human cognition starts with the idea that we are action oriented. Our ability to make sense of the world comes from an active and pragmatic engagement with the world, along with our capacities to interact with other people. In this regard, Anaxagoras's observation that we humans are the wisest of all beings because we have hands better reflects an enactive view than Aristotle's claim that "Man has hands because he is the wisest of all beings."[1] In the Aristotelian tradition, the hand is raised to the level of the rational by considering it the *organum organorum*. More generally, in the history of philosophy, hands are inserted here and there to provide a firm grasp on some important philosophical notions. Thus, a statement attributed to Isaac Newton suggests that the thumb is good evidence of God's existence, and Immanuel Kant used his hands (the fact that hands are incongruent counterparts, e.g., a left hand doesn't fit properly into a right-hand glove) to prove that Newton was right about space being absolute.[2] But this is not the main line drawn by the philosophical traditions. With respect to rationality, the eyes have it more than the hands. Both philosophically and scientifically, vision dominates.

1 Sleight of Hand

At least since Plato, rationality has been associated with vision: to understand is to see, and to see the highest of all things is to see that which can be seen—the *eidos*—which is not something you can touch or hold in your hand. Furthermore, to see someone's soul you need to look into her eyes. Hands will tell you only what kind of work the person does, whereas contemplation (gazing theoretically upon the *eide*), even if infrequently

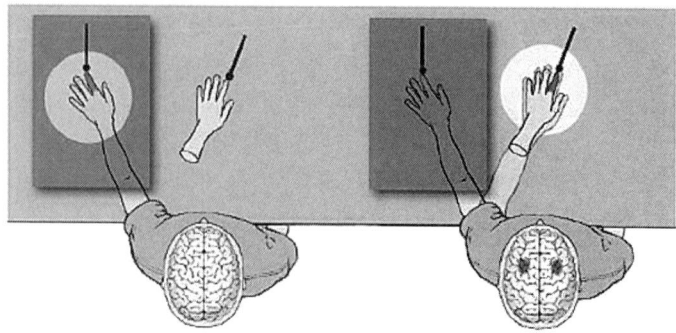

Figure 9.1
Rubber-hand illusion (from Botvinick 2004).

reached, is philosophically more celebrated than the life of action (*vita activa*), and certainly taken to be of a higher nature than the kind of work done with one's hands. The focus on vision leads to high-church cognitivism, idealism, and the main insights of metaphysics and epistemology. Not just the history of philosophy but also contemporary science shows us that vision dominates and almost always trumps the hand. We can find many examples of this. Consider recent findings on the rubber hand illusion (RHI). One of your hands is hidden from your direct view beneath a blind; there is a rubber arm/hand on the table, in front of you, positioned so that it is close to where you would normally see your hand (figure 9.1). The experimenter starts to stroke both your hidden hand and the rubber hand synchronously and you suddenly feel the tactile stimulus in the rubber hand as if it were your hand. You start to feel that the rubber hand is part of your body (your "sense of ownership" for the body part is modulated). On the one hand, the synchrony of the stroking is important for this effect; on the other hand, however, the real force is vision. Close your eyes and the illusion disappears; the sensation shifts back on your real hand (Botovnik and Cohen 1998).

Vision trumps proprioception as well. While experiencing the illusion you are asked to indicate using a ruler on the table the precise position of your hidden hand located beneath the blind. Most subjects indicate that their fingers are closer to the position of the rubber hand than they actually are. This is termed "proprioceptive drift." One might argue, however, that the hands themselves are not fooled. Thus, Kammers et al. (2009) showed that there was proprioceptive drift toward the rubber hand for the ruler-based perceptual judgment about the location of one's hidden hand, but no proprioceptive drift when one is asked to use the other (real) hand to point to the position of the hidden hand. The suggestion is that the motor system does not succumb to the illusion. In a follow-up study, however, Kammers et al. (2010) showed that when grasping action is included, the motor system can also be susceptible to the illusion. If, while experiencing the RHI with all hands shaped in a grasping posture, you are asked to use the hidden hand to reach out to take hold of a visible object, your reach is imperfectly calculated by the degree of proprioceptive drift. So the hand is fooled. Although the claim has been that the illusion disappears when the stroking is asynchronous, certain effects can be had even with vision alone and without the synchronous stroking. Rohde et al. (2011) have shown a dissociation between proprioceptive drift in RHI and the sense of ownership for the rubber hand. While the feeling of ownership for the rubber hand requires synchronous stroking, proprioceptive drift occurs not only in the synchronous stroking condition but also in the two control conditions: asynchronous stroking and vision alone. If you simply stare at the rubber hand long enough, your real hand is subject to proprioceptive drift. Perhaps one could show that this effect is even stronger if you are looking at the real hand of an attractive person sitting across the table from you. Does your hand feel like it's drifting a bit in that direction?

What about the hand being quicker than the eye? Even if the effects of vision are quite powerful, every now and then one finds that the hands take the lead—and a corresponding line of thought (some version of which one can find in the Aristotelian tradition) leads in the direction of embodied cognition and practical (not just contemplative) wisdom—even if it does not fully reach an enactive view.

In a book entitled *Hands*, advertised as "intended for all readers— including magicians, detectives, musicians, orthopedic surgeons, and anthropologists," John Russell Napier states: "The hand is the mirror of the brain; there can be no such combination as dextrous hands and clumsy brains" (1980, 25). One can confirm this in experiments where the hands seem to "outsmart" vision, and indeed where the hands seem to know something that vision doesn't. Thus, if you ask someone to reach out and grab an object he sees in front of him, and then, after he has started reaching you move the object slightly and quickly to the right or left, his hand will adjust its trajectory and precision grip to the new position of the target although the subject is unaware that the target moved (Pélisson et al. 1986). The movement doesn't register in conscious vision (the ventral stream), although it must do so in unconscious vision (the dorsal stream).

This is Goodale and Milner's (1992) well-known distinction between two visual pathways in the brain: one, the ventral, serving recognition, and the other, the faster dorsal system, serving motor control. The hand is faster than conscious vision, but not necessarily faster than unconscious vision. As an agent reaches to grasp something, the hand automatically (as well as manually) shapes itself into just the right posture to form the most appropriate grip for that object and for the agent's purpose. If I reach to grab an apple in order to take a bite, the shape of my grasp is different from when I reach to grab a banana, but also different from when I reach to grab the apple to throw it or offer it to someone (Jeannerod 1997; Pacherie 2005; Becchio et al. 2012). This happens without the agent monitoring or being aware of (or seeing) the difference in the shape of the grasp. Still, the hand does not do this blindly; it requires the cooperation of the dorsal visual stream to provide visual information about the shape of the apple and where in the nearby environment it is located.

Similar things can be said about one of Milner and Goodale's patients who has lesions in both temporal lobes, which prevents her from seeing which way an object is oriented. When presented with a disk she is unable to say whether it is presented vertically or horizontally. But when she is handed the disk and asked to put it into a slot (similar to a mailbox), she has no problem orienting the disk to the proper angle (Milner and Goodale 1995). Likewise, Robertson and Treisman (2010, 308) report on a patient of Rafal with visual agnosia who was unable to recognize objects: "When the patient was shown a picture of a clarinet, he hesitated in naming it, suggested it was a 'pencil,' but meanwhile his fingers began to play an imaginary clarinet." There is much to say about gesture in this regard as well. As Andy Clark (this vol.) points out, one may unconsciously express something in gesture that one is unable to articulate in conscious speech (also see Gallagher 2005, ch. 6).

These examples tell us something important about how we should understand brain function. On the enactive view, the brain is not composed of computational machinery locked away inside the head, representing the external world to provide knowledge upon which we can act. Rather, in action—whether reaching and grasping, pointing, or gesturing—the brain partners with the hand and forms a functional unit that properly engages with the agent's environment. For example, one can show experimentally that in action (e.g., reaching and grasping) in contrast to passive perception (e.g., estimating the distance between two sensations on one's skin) the felt differentiation between hand and arm across the wrist is reduced (de Vignemont et al. 2009). That is, in action, the hand is treated not as a body part differentiated from the arm, but as continuous with the arm. Likewise for the arm and the shoulder. In action, the body schema functions in a holistic way (in contrast to the perceptual and articulated aspects of body image; see Gallagher 2005). In the same way, it seems right to say that the brain is part of this holistic functioning.[3] It's not a top-down regulation of movement, brain to hand; nor is it a bottom-up emergence of rationality, hand to brain. Rather, neural processes coordinate with and can be entrained by hand movements, forming a single integrated cognitive system (Iverson and Thelen 1999). This implies a reciprocal unity of feedforward–feedback processes in which the hand and the brain form a dynamic system that reaches into the world—and perhaps beyond. Horst Bredekamp (2007), in his art-historical study of Galileo, comments on the fact that Galileo's drawings of Jupiter were more accurate than the images he could see through his telescope. Galileo's drawing hand, as part of his trained sensorimotor system (he studied art at the Accademia delle Art in Florence), was providing more information than his eyes could consciously see. In some sense—in its own motoric sense—his hand was smarter than his eye. Bredekamp suggests that this was a kind of manual thinking (*manuelles Denken*). Again, however, we should think of this as a holistic thinking that involves eye, brain, and hand. Does the dorsal visual pathway directly inform the drawing hand so that Galileo could go beyond conscious vision? Other possibilities include the idea that Galileo's hand interpolated detail between what vision told him, and certain of his assumptions about smoothness and continuity. But this could not account for his accuracy, which can be verified by more sophisticated telescopes today. Might it be that Galileo received more information through the motion of the telescope, since, in some cases, more details may be discerned in perceptions across moving images? The detail may not appear in a single view from the telescope but may emerge across multiple views. It's not clear, however, how much movement of the telescope was possible. Galileo's friend the artist Cigoli suggested something closer to the enactive story, namely, an interaction between visual perception and the motor ability that comes with practiced drawing:

For Cigoli, Galileo could see better, because he was better prepared by his artistic training and knew how to draw. In an autodidactic process taking place between hand and eye, Galileo was better able to attain knowledge, both because he had learned to perceive the unusual and because he could demonstrate it in the medium of drawing. (Bredekamp 2001, 180)

Practicing one's drawing, of course, will result in plastic changes in the brain. It seems that these plasticity effects accompany whatever habits one forms with one's hands. The famous experiments by Merzenich et al. (1983) demonstrate this. The experimenters tied down certain fingers of monkeys so that they were forced to use only select fingers in habitual movement of the hand. The difference in this habitual movement altered details of the brain's functional maps of the hand. Patterns of hand use physically shape those sensory and motor parts of the brain that register and control hand movement in monkeys, as well as in humans (see Rossini et al. 1994). This brain reorganization also happens as we train our hands to play piano or other musical instruments (Pascual-Leone et al. 1995). Since the general rule is high intermodal connectivity in the brain, it is easy to think that such plastic changes in multiple (motor and sensory) maps related to hand, affecting touch and proprioception, will also inform visual modalities. Even if one is not clinically synesthetic, haptic touch informs the eye and vice versa.

2 The Enactive Manipulatory Area

The hands lead us toward things. As Handy et al. (2003) show, graspable objects grab our attention. When we see different hand postures, they automatically direct our attention toward congruent target objects (Fischer, Prinz, and Lotz 2008). Moreover, the position of one's hands has an effect on visual attention. Objects located near one's hands receive enhanced visual attention. In a study of several classic visual attention tasks (visual search, inhibition of return, and attentional blink)—participants held their hands either near the stimulus display, or far from the display. The position of the hands altered visual processing so that subjects shifted attention more slowly between items when their hands were located near the display (Abrams et al. 2008). The results suggest that the hands facilitate the evaluation of objects for potential manipulation.

The basic principle of "active vision," which is vision in the service of motor control, is summarized by Cagli et al.: "eye movements depend on the task at hand, and if the task is a sensorimotor one, it is reasonable to expect a dependence on body movements as well" (2007, 1016). In this regard, the hands help to define a pragmatic area around the body that has significance for movement, action, attention, and accomplishing tasks. George Herbert Mead called this reachable peripersonal space around the body the "manipulatory area" and suggested that what is present in perception is not a copy of the perceived, but "the readiness to grasp" what is seen (1938, 103). The perception of objects outside of the manipulatory area is always relative to "the readiness of the organism to act toward them as they will be if they come within the manipulatory area. . . . We see the objects as we will handle them. . . . We are only 'conscious of' that in the perceptual world which suggests confirmation, direct or indirect, in fulfilled manipulation" (ibid., 104–105). On this enactive account of perception, the manipulatory area, defined in part by the hands, is the index of how something pragmatically counts as a percept. Perceptual consciousness arises in the spatial and temporal distances between a possibility of action in the manipulatory area and the distant object outside of that area.

Pragmatic engagement with the world is primary and priming for more explicitly cognitive graspings. Heidegger (1962) is famous for making this point. Primarily and for the most part, things are ready-to-hand (*Zuhanden*). That is, they are things that we pick up and use, or integrate into our practical projects. In this regard the world presents us with specifiable affordances (Gibson 1977). Only when the affordance is blocked, or when a tool breaks, or something disrupts our action do we shift gears and start to consider things in more theoretical ways. Things then become, as Heidegger puts it, "present-at-hand" (*Vorhanden*). On this view, it's not so much that we carve out a manipulatory area from the surrounding world as that we discover the surrounding world within an already established manipulatory area.

The very first manipulatory area may be the mouth. One of the earliest coordinated and systematic movements to be found in the fetus is the movement of hand to mouth (de Vries, Visser, and Prechtl 1984; Nillsson and Hamberger 1990). The very same synergetic movement is found in early infancy where the mouth opens to anticipate the hand and where the hand, especially the thumb (Newton's proof for the existence of God), finds a handy place to link up with the gustatory sense (Butterworth and Hopkins 1988; Lew and Butterworth 1997; Rochat 1993; Rochat and Senders 1991). If you allow an infant to grasp your finger, it too will likely end up in the infant's mouth, as do many other things that you put into the infant's hands; it's well known that the infant explores the world orally, but always with the hand involved. As the child learns to reach and grab for itself, and the fine motor skills of the hand are improved, the manipulation becomes more haptic and the exploratory skills become finer (Needham, Barrett, and Peterman 2002; Rochat 1989). Hand-mouth coordination gives way to hand-eye coordination.

Tools and technologies allow us to expand the manipulatory area (e.g., Farnè et al. 2005; Iriki et al. 1996; Witt, Proffitt, and Epstein 2005), and this is reflected in our use of demonstratives ("this," "that") (Coventry et al. 2008). Generally, "this" comes to signify anything reachable; "that" indicates something outside of peripersonal space. We can grasp *this*, or at least touch it; we can only point to *that*, although we can also point (in a different way) to *this*. Goldstein (1971) distinguishes between these two manual capacities: grasping (which is "concrete") and pointing (which is "abstract" or categorical). These distinctions are not strict, however, and they remain somewhat ambiguous so that even normal grasping capacity may require the categorical attitude (Goldstein 1971, 279–280).

> Although the normal person's behaviour is prevailingly concrete, this concreteness can be considered normal only as long as it is embedded in and codetermined by the abstract attitude. For instance, in the normal person both attitudes are always present in a definite figure-ground relation. (Goldstein and Scheerer 1964, 8)

Despite this ambiguity, the distinction between grasping and pointing has been taken by some phenomenologists to mean that concrete behavior (e.g., grasping) is more basic (it survives certain pathologies where pointing does not), and that it characterizes our normal motor intentionality in its nonrepresentational, nonconceptual form (e.g., Kelly 2000, 2004). In pathological cases, for example, these two capacities can come apart, as in the very complex case of Schneider, who was diagnosed with visual agnosia and a form of apraxia after suffering brain damage from a wound (Goldstein and Gelb 1920; Merleau-Ponty 1962). It was reported that Schneider could find his nose in concrete situations, for example, when he wanted to scratch it; but he could not point to his nose on command, or in abstract situations. There are several important qualifications to be made here, however. The extent of Schneider's brain damage is unclear, and as Tony Marcel (2003) notes, we need to be careful to distinguish between normal functions that manifest themselves more clearly in pathological cases, and functions that emerge as compensatory within the pathology. Kelly's assumption that Schneider's intact concrete capacities are normal may be unwarranted. Also, one should clarify what kind of pointing is at stake. There are various forms of noncommunicative pointing (e.g., touching X in an experimental situation) that contrast with communicative (deictic) pointing. In addition, one can distinguish between imperative pointing (to something I want) and declarative pointing (for calling another's attention to something). Moreover, in different pathologies, communicative pointing and gestures may be intact when concrete grasping and noncommunicative pointing are impaired (see e.g., Cole, Gallagher, and McNeill 2002).

Hand actions—grasping, pointing in various ways, gesturing—shape our cognitive processes. There is a reflective reiteration of hand aspects in language—but concurrent hand action can also interfere with judgment. In a study by Glenberg and Kaschak (2002), subjects respond faster in judging whether a sentence makes sense when the direction of their hand movement for responding (away vs. toward their body) matches the movement implied by the sentence compared to when there is a mismatch. This is the case for both abstract and concrete meanings (also see Chen and Bargh 1999). Many priming studies, when the prime is a picture of a hand, for example, show specific effects on perception and cognition (see, e.g., Setti, Borghi, and Tessari 2009).

3 Handling Others

The hand's capacities for pointing and grasping depend not only on the brain, and not only on motoric function, but also on the situation, which may be abstract, instrumentally concrete, or social and communicative. Gallagher and Marcel (1999) distinguished between these three characterizations of situations. Situations may be

- Noncontextualized (e.g., relatively abstract—experimental situations)
- Instrumentally contextualized (e.g., practical or concrete situations)
- Socially contextualized

Consider, for example, an apraxic patient who, in the clinic or testing room, is unable to lift a small block of wood to her cheek (a relatively meaningless and abstract action). In her home, however, she is able to make similar movements (drinking tea) in a close to fluid manner when she is entertaining guests (a socially contextualized situation), and to a lesser extent when she is clearing up dishes (an instrumentally contextualized situation). Her hands behave differently in these different contexts.

Social forces also shape the development of how we use our hands. Imperative pointing is not only a social signal; it depends on other people to develop:

The child may attempt to grasp an object that is out of reach and leave the hand hanging in the air. At this point the mother comes to aid and interprets the gesture. A motor act becomes a gesture for another person, who reacts to the child's attempt and ascribes a specific meaning to the grasping movement. (Sparaci 2008, 210)

Accordingly, as Vygotsky (1986) contends, as the child learns that this gesture motivates that particular response in the other person, the failed reach for grasping is transformed into communicative pointing.

The extent of one's manipulatory area is also modulated by others, and, as mentioned above in regard to the use of tools and technologies, this is reflected in use of demonstratives. Whether any particular object on a table is referred to as "this" or "that" will depend on whether I or another person puts it there, regardless of how close it is to me (Coventry et al. 2008).

Merleau-Ponty points to a certain reversibility associated with hands touching. We can approach this idea by thinking about what happens when we touch some object. We not only feel the surface and shape of the object but we can feel the surface and shape of our own fingers. Tactile perception is ecological in the sense that it tells us something about our own body as well as about the object. Merleau-Ponty borrows an example from Husserl who thinks about the phenomenon of my one hand touching my other hand and the ambiguity that emerges as the touched hand can easily become the touching hand. Merleau-Ponty tries this experiment while one hand is touching an object.

We spoke summarily of a reversibility . . . of the touching and the touched. It is time to emphasize that it is a reversibility always imminent and never realized in fact. My left hand is always on the verge of touching my right hand touching the things, but I never reach coincidence; the coincidence eclipses at the moment of realization, and one of two things always occurs: either my right hand really passes over to the rank of the touched [i.e., becomes an object], but then its hold on the world is interrupted [it is no longer a subject]; or it retains its hold on the world [remains a subject], but then I do not really touch it—my right hand touching; I palpate with my left hand only its outer covering. (1968, 147–148)

This reversibility becomes a principle that is also applicable to our relations with others—relations that never reach complete coincidence. Merleau-Ponty calls this "intercorporeity" and describes it as follows: "between this phenomenal body of mine, and that of another as I see it from the outside, there exists an internal relation which causes the other to appear as the completion of the system" (1962, 352).

The completion remains imperfect, not only in our hands, but also in our relations with others—it remains always imminent and never fully realized, since we either continue to interact or cease to do so, and in the former case the other person experiences something completely different than I do. He experiences *me* in this inner–outer reversibility. We are two embodied perspectives who gaze back at each other, or join our gazes as we look toward something else, ecologically touching and being touched, two hands clapping, but to a tango rhythm. Such interactions have room to develop as long as they do not reach a coincidence of the absurd sort that one finds in the fascinating experiment conducted by Petkova and Ehrsson (2008), where, through the magic of virtual reality, the eye is quicker than the hand and the subject ends up standing in front of himself seemingly shaking hands with himself. The subject wears virtual reality goggles in which he sees a live video feed generated by a camera worn on the head of another person who is standing directly in front of him, with the camera directed at the subject. In effect, he sees himself. When he reaches out and shakes hands with the other person it feels as if he is shaking hands with himself. In this experiment, the hand that I see myself shaking is visually mine, but tactilely (and really) belongs to someone else. So while normally we could agree with Donn Welton, that when "the hand touching our hand is not ours but that of another, when our hand is grasped by a hand not my own, the circuit of reversibility encompasses others and corporeality [and] becomes intercorporeal" (2000, 97), when, in the experiment, the hands touch, the reversibility is short-circuited; it collapses into an ecological involution.

It is also the case that the other person's hands are important for our ability to see their intentions (Becchio et al. 2012). Six-month-old infants react differently to observed grasping actions of human hands than to those of an artificial (mechanical claw). Only the former are perceived as goal directed (Woodward et al. 2001). The infant's early propensity to attend to goals seems to be tied to the specific human aspect of the hand. Infants see meaning in the reaching, grasping, pointing, and gesturing of the other's hands just as they see their own possibilities for action in the actions of and with others.

Such detailed interactions with others are reflected in our hand movements, actions, and gestures, but also in our handed vocabulary: we not only shake hands, we sometimes "lend a hand." Although we sometimes present a fist to others, we also raise our fist in a sign of solidarity. Often, also, we don't know what to do with our hands in the presence of others, unless they are taken up in gesture. That's because we do want to reach out and touch them—either to push them further away, or to share our feelings. Imagine the normative chaos that would be entailed if we let our hands do what they were inclined to do. Anarchic hand syndrome (AHS) presents a good example of this. The result of brain lesions, in AHS the hand seems to have a mind of its own. It's often the case that the anarchic hand ends up doing things that the owner of the hand would never dream of doing. In one case it was reported that the subject's anarchic hand reached over and grabbed food off the plate of a fellow diner (see Della Sala 2000; Della Sala et al. 1994).

One can see power or skill or weakness in the hands of the other, as Michelangelo shows us. We can sometimes see a person's occupation reflected in her hands. My job may even be *to be* a hand, as in "all hands on deck." The ways that human hands are iterated in language and meta-phor reveal some of our central ways of relating to others. But since hands gesture, the very actions that we engage in with our hands can sometimes speak louder than words. We can recognize the authority of others and place ourselves in their hands. Possession is having the thing in hand, and that is some high percentage of the law. We need to have things in hand before we can make a gift or give a handout, or face the demand to stick 'em up and hand it over. The other's hand may be one that you would like to hold, or it may be one that signals that you should keep your dis-tance. Love and war require the use of hands: we touch and caress each other, bind ourselves to each other by slipping rings onto fingers; or we try to get the upper hand, and sometimes destroy one another by wielding swords, pulling triggers, or pressing buttons. "Hands up" signifies surren-der; "hands down" means there is no question. Further thoughts on such matters I leave in your hands.

4 Conclusion

You can count the points I've made in this chapter on one hand. First, hands play an important role in rationality—not by themselves, but as part of a more holistic system that involves the brain and the various sense modalities. The same can be said of the brain—it doesn't work by itself, but in a larger system that includes the hands. This makes rationality in some respects enactive or action-oriented. Second, the enactive possibili-ties that are provided by the fact that we have hands help to define the space around us in terms of a manipulatory area within which we find and help to constitute meaning. This reinforces the idea that rationality is primarily pragmatic. Third, the roles played by hands in enactive rational-ity extend into our relations with others, not only in gestural communica-tion, but in how we recognize their intentions, how we interact, and how we engage in sense-making practices.

Notes

1. According to Aristotle:

Standing erect, man has no need of legs in front, and in their stead has been endowed by nature with arms and hands. Now it is the opinion of Anaxagoras that the possession of these hands is the cause of man being of all animals the most intelligent. But it is more rational to suppose that his endowment with hands is the consequence rather than the cause of his superior intel-ligence. For the hands are instruments or organs, and the invariable plan of nature in distribut-ing the organs is to give each to such animal as can make use of it; . . . For nature adds that which is less to that which is greater and more important, and not that which is more valuable and greater to that which is less. . . . We must conclude that man does not owe his superior intelligence to his hands, but his hands to his superior intelligence. For the most intelligent of animals is the one who would put the most organs to use; and the hand is not to be looked on as one organ but as many; for it is, as it were, an instrument for further instruments. This instrument, therefore,-the hand-of all instruments the most variously serviceable, has been given by nature to man, the animal of all animals the most capable of acquiring the most varied handicrafts. (*On the Parts of Animals*, http://classics.mit.edu/Aristotle/parts_animals.4.iv.html)

2. "It is apparent from the ordinary example of the two hands that the shape of the one body may be perfectly similar to the shape of the other, and the magnitudes of their extensions may be exactly equal, and yet there may remain an inner difference between the two, this difference consisting in the fact, namely, that the surface which encloses the one cannot possibly enclose the other" (Kant 1992, 370).

3. This is a stronger claim than the one made by Aristotle, that the hand, as the "tool of tools" is "analogous" to the soul (*De Anima*, 432a 1–2). There are literally physical connections between brain and hand.

References

Abrams, R. A., C. C. Davoli, F. Du, W. J. Knapp, and D. Paull. 2008. Altered vision near the hands. *Cognition* 107:1035–1047.

Aristotle. 350 BC. *De Anima (On the Soul)*. Trans. J. A. Smith. http://classics.mit.edu/Aristotle/soul.html.

Aristotle. 350 BC. *On the Parts of Animals*. Trans. W. Ogle. http://classics.mit.edu/Aristotle/parts_animals.html.

Becchio, C., V. Manera, L. Sartori, A. Cavallo, and U. Castiello. 2012. Grasping intentions: From thought experiments to empirical evidence. *Frontiers in Human Neuroscience*. doi: 10.3389/fnhum.2012.00117.

Botvinick, M. 2004. Probing the neural basis of body ownership. *Science* 305: 782–783.

Botovnik, M., and J. Cohen. 1998. Rubber hands "feel" touch that eyes see. *Nature* 391:756.

Bredekamp, H. 2007. *Galilei der Künstler: Die Zeichnung, der Mond, die Sonne*. Berlin: Akademie.

Bredekamp, H. 2001. Gazing hands and blind spots: Galileo as draftsman. In *Galileo in Context*, 153–192, ed. J. Renn. Cambridge: Cambridge University Press.

Bronowski, J. 1975. *The Ascent of Man*. New York: Little, Brown.

Butterworth, G., and B. Hopkins. 1988. Hand–mouth coordination in the new-born baby. *British Journal of Developmental Psychology* 6:303–314.

Cagli, R. C., P. Coraggio, P. Napoletano, and G. Boccignone. 2007. What the draughtsman's hand tells the draughtsman's eye: A sensorimotor account of drawing. *International Journal of Pattern Recognition and Artificial Intelligence* 22 (5):1015–1029.

Chen, S., and J. A. Bargh. 1999. Consequences of automatic evaluation: Immediate behavior predispositions to approach or avoid the stimulus. *Personality and Social Psychology Bulletin* 25:215–224.

Cole, J., S. Gallagher, and D. McNeill. 2002. Gesture following deafferentation: A phenomenologically informed experimental study. *Phenomenology and the Cognitive Sciences* 1 (1):49–67.

Coventry, K. R., B. Valdés, A. Castillo, and P. Guijarro-Fuentes. 2008. Language within your reach: Near–far perceptual space and spatial demonstratives. *Cognition* 108:889–895.

Della Sala, S. 2000. Anarchic hand: The syndrome of disowned actions. *Creating-Sparks: The BA Festival of Science*. Available at http://www.creatingsparks.co.uk.

Della Sala, S., C. Marchetti, and H. Spinnler. 1994. The anarchic hand: A fronto-mesial sign. In *Handbook of Neuropsychology*, vol. 9, ed. G. Boller and J. Grafman, 233–255. Amsterdam: Elsevier.

de Vignemont, F., A. Majid, C. Jola, and P. Haggard. 2009. Segmenting the body into parts: Evidence from biases in tactile perception. *Quarterly Journal of Experimental Psychology* 62:500–512.

de Vries, J. I. P., G. H. A. Visser, and H. F. R. Prechtl. 1984. Fetal motility in the first half of pregnancy. In *Continuity of Neural Functions from Prenatal to Post-natal Life*, 46–64, ed. H. F. R. Prechtl. London: Spastics International Medical Publications.

Farnè, A., A. Iriki, and E. Làdavas. 2005. Shaping multisensory action-space with tools: Evidence from patients with cross-modal extinction. *Neuropsychologia* 43:238–248.

Fischer, M. H., J. Prinz, and K. Lotz. 2008. Grasp cueing shows obligatory attention to action goals. *Quarterly Journal of Experimental Psychology* 61 (6):860–868.

Gallagher, S. 2005. *How the Body Shapes the Mind*. Oxford: Oxford University Press.

Gallagher, S., and A. J. Marcel. 1999. The self in contextualized action. *Journal of Consciousness Studies* 6 (4):4–30.

Gibson, J. J. 1977. The theory of affordances. In *Perceiving, Acting, and Knowing*, 67–82, ed. R. Shaw and J. Bransford. Hillsdale, NJ: Erlbaum.

Glenberg, A. M., and M. P. Kaschak. 2002. Grounding language in action. *Psychonomic Bulletin & Review* 9:558–565.

Goldstein, K. 1971. Über Zeigen und Greifen. In *Selected Papers/Ausgewählte Schriften*, ed. A. Gurwitsch, E. M. Goldstein Haudek, and W. E. Haudek. The Hague: Martinus Nijhoff.

Goldstein, K., and A. Gelb. 1920. Über den Einfluss des vollständigen Verlustes des optischen Vorstellungsvermögens auf das taktile Erkennen. In *Psychologische Analysen hirnpathologischer Fälle II*, 157–250, ed. A. Gelb and K. Goldstein. Leipzig: Johann Ambrosius Barth.

Goldstein, K., and M. Scheerer. 1964. *Abstract and Concrete Behavior. An Experimental Study with Special Tests*. Evanston, IL: Northwestern University. Reprint of *Psychological Monographs* 53(2), 1941.

Goodale, M. A., and A. D. Milner. 1992. Separate visual pathways for perception and action. *Trends in Neurosciences* 15 (1):20–25.

Handy, T. C., S. T. Grafton, N. M. Shroff, S. Ketay, and M. S. Gazzaniga. 2003. Graspable objects grab attention when the potential for action is recognized. *Nature Neuroscience* 6:421–427.

Heidegger, M. 1962. *Being and Time*. Trans. J. Macquarrie and E. Robinson. New York: Harper & Row.

Iriki, A., M. Tanaka, and Y. Iwamura. 1996. Coding of modified body schema during tool use by macaque postcentral neurones. *Neuroreport* 7:2325–2330.

Iverson, J., and E. Thelen. 1999. Hand, mouth, and brain: The dynamic emergence of speech and gesture. *Journal of Consciousness Studies* 6:19–40.

Jeannerod, M. 1997. *The Cognitive Neuroscience of Action*. New York: Wiley-Blackwell.

Kammers, M. P., J. A. Kootker, H. Hogendoorn, and H. C. Dijkerman. 2010. How many motoric body representations can we grasp? *Experimental Brain Research* 202:203–212.

Kammers, M. P., F. de Vignemont, L. Verhagen, and H. C. Dijkerman. 2009. The rubber hand illusion in action. *Neuropsychologia* 47:204–211.

Kant, I. 1992. Concerning the ultimate ground of the differentiation of directions in space. In *The Cambridge Edition of the Works of Immanuel Kant: Theoretical Philosophy, 1755–1770*, 365–372, ed. D. Walford and R. Meerbote. Cambridge: Cambridge University Press.

Kelly, S. D. 2000. Grasping at straws: Motor intentionality and the cognitive science of skilled behavior. In *Heidegger, Coping, the Cognitive Sciences: Essays in Honor of*

Hubert L. Dreyfus, vol. 2. ed. M. Wrathall and J. Malpas, 161–177. Cambridge, MA: MIT Press.

Kelly, S. D. 2004. Merleau-Ponty on the body. In *The Philosophy of the Body*, 62–76, ed. M. Proudfoot. London: Blackwell.

Lew, A. R., and G. Butterworth. 1997. The development of hand–mouth coordination in 2- to 5-month-old infants: Similarities with reaching and grasping. *Infant Behavior and Development* 20:59–69.

Marcel, A. 2003. The sense of agency: Awareness and ownership of action. In *Agency and Self-Awareness*, 48–93, ed. J. Roessler and N. Eilan. Oxford: Oxford University Press.

Mead, G. H. 1938. *The Philosophy of the Act*. Chicago: University of Chicago Press.

Merleau-Ponty, M. 1962. *Phenomenology of Perception*. Trans. C. Smith. London: Routledge & Kegan Paul.

Merleau-Ponty, M. 1968. *The Visible and the Invisible*. Trans. A. Lingis. Evanston, IL: Northwestern University Press.

Merzenich, M. M., J. H. Kaas, J. T. Wall, R. J. Nelson, M. Sur, and D. J. Felleman. 1983. Topographic reorganization of somatosensory cortical areas 3b and 1 in adult monkeys following restricted deafferentation. *Neuroscience* 8:33–55.

Milner, D. A., and M. A. Goodale. 1995. *The Visual Brain in Action*. New York: Oxford University Press.

Napier, J. R. 1980. *Hands*. London: Allen & Unwin.

Needham, A., T. Barrett, and K. Peterman. 2002. A pick-me-up for infants' exploratory skills: Early simulated experiences reaching for objects using "sticky mittens" enhances young infants' object exploration skills. *Infant Behavior and Development* 25:279–295.

Nillsson, L., and L. Hamberger. 1990. *A Child Is Born*. New York: Delacorte Press.

Pacherie, E. 2005. Perceiving intentions. In *A Explicação da Interpretação Humana*, ed. J. Sàágua. Lisbon: Edições Colibri.

Pascual-Leone, A., D. Nguyet, L. G. Cohen, et al. 1995. Modulation of muscle responses evoked by transcranial magnetic stimulation during the acquisition of new fine motor skills. *Journal of Neurophysiology* 74:1037–1045.

Pélisson, D., C. Prablanc, M. A. Goodale, and M. Jeannerod. 1986. Visual control of reaching movements without vision of the limb. *Experimental Brain Research* 62 (2):303–311.

Petkova, V. I., and H. H. Ehrsson. 2008. If I were you: Perceptual illusion of body swapping. *PLoS ONE* 3 (12):e3832. doi:10.1371/journal.pone.0003832.

Rohde, M., M. Di Luca, O. Marc, and M. O. Ernst. 2011. The rubber hand illusion: Feeling of ownership and proprioceptive drift do not go hand in hand. *PLoS ONE* 6 (6):e21659. doi:10.1371/journal.pone.0021659.

Robertson, L. C., and A. Treisman. 2010. Consciousness: Disorders. In *Encyclopedia of Perception*, ed. E. B. Goldstein. New York: Sage.

Rochat, P. 1993. Hand–mouth coordination in the newborn: Morphology, determinants, and early development of a basic act. In *The Development of Coordination in Infancy*, 265–288, ed. G. J. P. Savelsbergh. Amsterdam: North-Holland.

Rochat, P. 1989. Object manipulation and exploration in 2- to 5-month-old infants. *Developmental Psychology* 25:871–884.

Rochat, P., and S. J. Senders. 1991. Active touch in infancy: Action systems in development. In *Newborn Attention: Biological Constraints and the Influence of Experience*, 412–442, ed. M. J. S. Weiss and P. R. Zelazo. Norwood, NJ: Ablex.

Rossini, P. M., G. Martino, L. Narici, A. Pasquarelli, M. Peresson, V. Pizzella, F. Tecchio, G. Torrioli, and G. L. Romani. 1994. Short-term brain "plasticity" in humans: Transient finger representation changes in sensory cortex somatotopy following ischemic anesthesia. *Brain Research* 642 (1–2):169–177.

Setti, A., A. M. Borghi, and A. Tessari. 2009. Moving hands, moving entities. *Brain and Cognition* 70 (3):253–258.

Sparaci, L. 2008. Embodying gestures: The social orienting model and the study of early gestures in autism. *Phenomenology and the Cognitive Sciences* 7 (2):203–223.

Vygotsky, L. S. 1986. *Thought and Language*. Cambridge, MA: MIT Press.

Welton, D. 2000. Touching hands. *Veritas* 45 (1):83–102.

Witt, J. K., D. R. Proffitt, and W. Epstein. 2005. Tool use affects perceived distance, but only when you intend to use it. *Journal of Experimental Psychology: Human Perception and Performance* 31 (5):880–888.

Woodward, A. L., J. A. Sommerville, and J. J. Guajardo. 2001. How infants make sense of intentional action. In *Intentions and Intentionality: Foundations of Social Cognition*, 149–169, ed. B. Malle, L. Moses, and D. Baldwin. Cambridge, MA: MIT Press.

10 Radically Enactive Cognition in Our Grasp

Daniel D. Hutto

The question is not to follow out a more or less valid theory but to build with whatever materials are at hand. The inevitable must be accepted and turned to advantage.
—Napoleon Bonaparte

1 Reckoning with Radically Embodied Cognition

For those working in the sciences of the mind, these are interesting times. Revolution is, yet again, in the air. This time it has come in the form of new thinking about the basic nature of mind of the sort associated with radically embodied or enactive approaches to cognition (REC, for short). REC approaches are marked out by their uncompromising and thoroughgoing rejection of intellectualism about the basic nature of mentality. As Varela, Thompson, and Rosch (1991) saw it, the defining characteristic of this movement is its opposition to those theories of mind that "take representation as their central notion" (172). The most central and important negative claim of REC is its denial that all forms of mental activity depend on the construction of internal models of worldly properties and states of affairs by means of representing its various features on the basis of retrieved information.

Not since the ousting of behaviorism with the advent of the most recent cognitive revolution has there been such a root and branch challenge to widely accepted assumptions about the very nature of mentality. In a remarkable reversal of fortune, it is now a live question to what extent, *if any*, representational and computational theories of the mind—those that have dominated for so long—ought to play a fundamental role in our explanatory framework for understanding intelligent activity. Defenders of REC approaches argue that representation and computation neither are definitive of nor provide the basis of *all* mentality.

From the sidelines, interested onlookers might be forgiven for thinking the revolution is already over; embodied and enactive ways of thinking are already comfortably ensconced, having established deep roots in a number of disciplines. Far from merely being at the gates, the barbarians are, it seems, now occupying the local cafes and wine bars in the heart of the city. Even those who most regret this development are prepared to acknowledge that, *de facto*, there has been a major sea change. Lamenting the rise of a pragmatist trend in cognitive science, Fodor (2008) acknowledges that REC-style thinking in cognitive science is now "mainstream." He puts this down to an infectious disease of thought ("a bad cold," as he puts it, 10). Others are nervously aware of the specter of REC approaches "haunting the laboratories of cognitive science" (Goldman and de Vignemont 2009, 154). "Pervasive and unwelcome" is the verdict of these authors: REC may be everywhere but it is something to be cured or exorcised, as soon as possible.

Despite their growing popularity, which some hope is nothing more than a short-lived trend, REC approaches remain hotly contested. Certainly, it is true that there has yet to be a definitive articulation of the core and unifying assumptions of embodied and enactive approaches to cognition—EC approaches—radical or otherwise. Indeed, there is some reason to doubt that it will be possible to group together all of the offerings that currently travel under the banner of EC by identifying their commitment to a set of well-defined core theoretical tenets (see Shapiro 2011, 3). Nevertheless, if REC approaches in particular are to maintain credibility and avoid charges of simply riding the crest of a fashionable wave, at a bare minimum, serious objections from the old guard should be convincingly answered.

Some criticisms are easier to deal with than others. One line of argument draws on observations about the proper order and requirements of cognitive explanations. Fodor (2008) hopes to dispatch REC with one fell blow by observing that positing representationally based thinking is the minimum requirement for explaining any and all activity that deserves the accolade "intelligent responding"—an observation predicated on the assumption that we can draw a principled line between what is properly cognitive and what is not. Accordingly, he insists that we have no choice but to accept that "the ability to think the kind of thoughts that have truth-values is, in the nature of the case, prior to the ability to plan a course of action. The reason is perfectly transparent: Acting on plans (as opposed to, say, merely behaving reflexively or just thrashing about) requires being able to think about the world" (Fodor 2008, 13).

In a nutshell, this is Fodor's master argument for thinking that pragmatist REC-style approaches to the mind *must* be false: for to think the kinds of thoughts that have truth-values is to think thoughts with representational content and, presumably, to make plans requires manipulating these representations (and their components) computationally.[1] So, in short, if all bona fide intelligent action involves planning, and all bona fide planning involves computing and representation, then this is bad news from the front line for REC rebels.

Without a doubt, some problems, indeed, perhaps whole classes of problems, are best addressed through advanced careful planning—planning of the sort that requires the rule-governed manipulation of truth-evaluable representations. Sometimes it is not only advisable, but utterly necessary to stand back and assess a situation in a relatively detached manner, drawing explicitly on general background propositional knowledge of situations of a similar type and using that knowledge to decide— say, by means of deduction and inference—what would be the correct or most effective approach. This can be done before initiating any action or receiving any live feedback from the environment. This is the preferred strategy for dealing with situations, such as defusing bombs, in which a trial-and-error approach is not advisable. Making use of remote representations also works equally well even for more mundane types of tasks—those that include, for example, figuring out the best route from the train station to one's hotel in a foreign city where one seeks to do this in advance from the comfort of one's office, long before boarding a plane.

Such intelligent preplanning can be done at a remove, and reliably, if it is possible to exploit and manipulate symbolic representations of the target domain on the assumption that one has the requisite background knowledge and can bring that knowledge to bear. This will work if the domain itself is stable over time since that will ensure that any stored representations remain up to date and accurate. By using representations of a well-behaved domain's features and properties, and having a means of knowing, determinately, what to expect of it if it changes under specific modifications and permutations, it is possible for a problem solver to plan how to act within it without ever having to (or indeed ever having had to) interact with it in a first-hand manner. This is, of course, the ideal end state of high theoretical science.

As linguistic beings, humans are representation mongers of this sort and thus regularly adopt this basic strategy to solve problems. Our cultural heritage provides us with a store of represented knowledge—in many and various formats—that enables us to do so, successfully, under the sorts of conditions just mentioned. But it hardly follows that this type of cognitive engagement is the basis of, required for, or indeed suitable for, all sorts of tasks—*always and everywhere*. Echoing Ryle (1949), Noë hits the nail on the head when he states that "the real problem with the intellectualist picture is that it takes rational deliberation to be *the most basic kind of cognitive operation*" (2009, 99; emphasis added).

Intellectualism of this unadulterated kind—of the sort that assumes the existence of strongly detached symbolic representations of target domains—has fallen on hard times. It finds only a few hard-core adherents in today's cognitive sciences. Indeed, if anything has promoted the fortunes of REC it has been the dismal failure of this sort of rules-and-representation approach when it comes to dealing with *the most basic forms* of intelligent activity. This is the headline-grabbing lesson of recent efforts in robotics and artificial intelligence, which have provided a series of existence proofs against strong representationalism about basic cognition.

Pioneering work by Rodney Brooks (1991a,b), for example, reveals that intellectualism is a bad starting point when thinking about how to build robots that actually work. We can learn important lessons by paying attention to architectonic requirements of robots that are able to complete quite basic sorts of tasks, such as navigating rooms while avoiding objects or recognizing simple geometrical forms and shapes. Inverting standard intellectualist thinking, Brooks famously rejected the sense-model-plan-act approach, and built robots that dynamically and frequently sample features of their local environments in order to directly guide their responses, rather than going through the extra steps of generating and working with *descriptions* of those environments. These first-generation behavior-based robots, and those that followed them, succeeded precisely because the robots' behaviors are guided by continuous, temporally extended interactions with aspects of their environments rather than working on the basis of represented internal knowledge about those domains, knowledge that would presumably be stored somewhere wholly in the robots' innards. The guiding principle behind Brooks's so-called subsumption architectures is that sensing is *directly connected* with appropriate responding without *representational* mediation. Crucially, the great success of these artificial agents demonstrates that it is possible for a being to act intelligently without creating and relying on internal representation and models. Very much in line with theoretical worries raised by the frame problem, it may even be that, when it comes to basic cognition, this is the only real possibility.

Not just artifice but nature too provides additional support for the same conclusion. Cricket *phonotaxis* (Webb 1994, 1996) is a vivid example of how successful online navigation takes place in the wild, apparently without the need for representations or their manipulation. Female crickets locate mates by attending to the first notes of male songs, frequently adjusting the path of their approach accordingly. They only manage this because the male songs that they attune to have a unique pattern and rhythm—one that suits the particular activation profiles of the female interneurons. The capacity of these animals to adjust their behavior when successfully locating mates requires them to be in a continuous interactive process of engagement with the environment. In doing so they exploit special features of their nonneural bodies—including the unique design of their auditory mechanism—as well as special features of the environment—the characteristic pattern of the male songs. In this case, a beautiful cooperation arises because of the way the cricket's body and wider environment features enable successful navigational activity—activity that involves nothing more than a series of dynamic and regular embodied interactions.

For reasons of space, I will not rehearse in full detail precisely how behavior-based robots or insects make their way in the world. These cases are well known and much discussed (for excellent summaries in greater detail, including other examples, see Wheeler 2005, ch. 8, and Shapiro 2011, ch. 5). For the purposes of this essay, it suffices to note that when bolstered by the articulation of a supporting theoretical framework, one easily provided by dynamical systems theory, these observations offer a serious and well-known challenge to the representationalist assumptions of intellectualist cognitive science (Beer 1998, 2000; Thompson 2007; Garzón 2008; Chemero 2009).

In sum, what the foregoing reflections teach us is that in some cases, bodily and environmental factors play ineliminable and nontrivial parts in making certain types of cognition possible. A familiar intellectualist response to these sorts of examples is to try to cast these wider contributions as playing no more than causal supporting roles that, even if necessary to enable cognition, do not constitute or form part of it. For reasons that should be obvious from the foregoing discussion, it is not clear how one might motivate this interpretation and make it stick with respect to the sorts of cases just described.

In rejecting representationalism, REC takes at face value what attending to the architectonic details of how these agents work suggests—namely, that the specified bodily and environmental factors are fully *equal partners* in constituting the embodied, enactive intelligence and cognition of these artificial and natural agents. Accordingly, although for certain practical purposes and interventions it may be necessary to carve off and focus on specific causally contributing factors in isolation, the cognitive activity itself cannot be seen as other than "a cyclical and dynamic process, with no nonarbitrary start, finish, or discrete steps" (Hurley 2008, 12; see also Garzón 2008, 388). Or, put otherwise, when it comes to understanding cognitive acts, "the agent and the environment are non-linearly coupled, they, together constitute a nondecomposable system" (Chemero 2009, 386).

In promoting this sort of line, REC flags up the "real danger" that "the explanatory utility of representation talk may evaporate altogether" (Wheeler 2005, 200). As Shapiro (2011) notes, the interesting question is whether an antirepresentationalist paradigm has real prospects of *replacing* intellectualist cognitive science altogether. And, as indicated above, he is right to suppose that the two main complementary "sources of support for Replacement come from (i) work that treats cognition as emerging from a dynamical system and (ii) studies of autonomous robots" (Shapiro 2011, 115). While this is a potentially powerful cocktail, it remains to be seen just how far it might take us. To make a convincing case for their far-reaching revolutionary ambitions, proponents of REC must take the next step and "argue that *much or most cognition* can be built on the same principles that underlie the robot's intelligence" (ibid., 116; emphasis added).

Rather than claiming that there can be no such thing as nonrepresentational cognition, intellectualists might take heart from this challenge and agree to split the difference, allowing that very basic forms of cognition—of the sort exemplified by robot and insect intelligence—might be suitable for REC treatment, but not the rest. This is to adopt a kind of containment strategy—a kind of theoretical kettling or corralling. Intellectualists might be tempted to concede that supporters of the radical left have a point, up to a point—allowing that "representations are not necessary for coordinated responses to an environment with which one is dynamically engaged" (ibid., 153). But this concession would be made in the secure knowledge that it "would support only the conclusion that agents do not require representations for certain kinds of activities. However, a stronger conclusion, for instance that cognition never requires representational states, does not follow" (ibid.).

REC approaches would be, accordingly, of limited value on the assumption that they won't scale up. Call this the scope objection. It allows one to accept certain lessons learned from the lab and nature while remaining safe in the knowledge that even if representations are not needed to explain the most basic forms of cognition this in no way poses an interesting threat to intellectualism since the sorts of cases in question "represent too thin a slice of the full cognitive spectrum" (ibid., 156). This is in line with the oft-cited claim that some behavior is too offline and representation hungry to be explained without appeal to the manipulation of symbolic representations. In particular, nonrepresentational cognition, which might do for simple robots and animals, isn't capable of explaining properly world-engaging, human forms of cognition. But should that assessment prove mistaken—if REC approaches were to make substantial inroads in this latter domain—then the boot might just be on the other foot. It might turn out that representationally hungry tasks make up only a very small portion of mental activity; representationally based cognition might be just the tip of the cognitive iceberg.

2   A Helping Hand

This is where reflection on the special prowess of the human hand comes in handy. It cannot be denied that a great deal of human manual activity is connected with sophisticated forms of cognition.

Milner and Goodale's (1995) famous experiments reveal that humans can perform remarkably demanding manual acts with precision—acts requiring the exercise of very fine-grained motor capacities, such as posting items through slots with changing orientations—even when they lack the capacity to explicitly report upon or describe visual scenes they are dealing with.

Nor is it credible, with only rare exceptions, that humans normally learn how to use their hands in these sorts of ways by means of explicit, representationally mediated instruction, the rules for which only later becoming submerged and tacit. It is not as if children are taught by their caregivers through explicit description how to grasp or reach for items. Far more plausible is the hypothesis that we become "handy" through a prolonged history of interactive encounters—through practice and habit. An individual's manual know-how and skills are best explained entirely by appeal to a history of previous engagements and not by the acquisition of some set of internally stored mental rules and representations. To invoke the favorite poetic motto of enactivists, this looks, essentially, to be a process of "laying down a path in walking," or in this case, "handling."

It is possible that the special manual abilities of humans are sophisticated enough to have provided the platform and spurred on other major cognitive developments. Some very strong claims have been made about the critical importance of the ways in which we use our hands in this regard—ways that some believe are responsible for enabling the emergence of other distinctive forms of human cognition, consciousness, and culture. For example, Tallis (2003) regards our special brand of manual activity as the ultimate source of our awakening to self-consciousness. He tells us: "Herein lies the true genius of the hand: out of fractionated finger movements comes an infinite variety of grips and their combinations. And from this variety in turn comes choice—not only in what we do . . . but in how we do it . . . [and with] choice comes consciousness of acting" (Tallis 2003, 175).

If Tallis is to be believed, "between the non-stereotyped prehensions of hominid hand and the stereotyped graspings of the animal paw there is opened a gap which requires, and so creates, the possibility of apprehension to cross it" (ibid., 36). These claims are tempered by the remark that "we may think of the emergence of distinctive capabilities of the human hand as *lighting a fuse on a long process* that entrained many other parts of the human body and many other faculties as it unfolded" (ibid., 6; emphasis added). This allows for the possibility that "the crucially important differences between human and non-human hands do not alone account for the infinitely complex phenomenon, unique in the order of the universe, of human culture. It is not so much the differences—which are very important—but the ability to make much of the differences" (ibid., 33).[2]

Whatever is ultimately concluded about the defensibility of this last set of claims, the point is that if it should turn out that much human manual activity is best explained without appeal to the manipulation of rules or representations then defenders of REC will have made significant progress in addressing the scope challenge. REC approaches will have shown the capacity to advance well beyond dealing with the antics of behavior-based robots and insects, having moved deep in the heart of distinctively human cognition.

Are there further grounds for thinking that manual activity is best explained in a representation-free way? Tallis's (2003) philosophically astute and empirically informed examination of the hand provides an excellent starting point for addressing this question. He claims that "the hand [is] . . . an organ of cognition," and is so "in its own right" (ibid., 28, 31). This is not to say that the hand works in isolation from the brain, indeed, Tallis stresses that the hand—for him, the tool of tools—is the "brain's most versatile and intelligent lieutenant" (ibid., 22). Of course, this way of putting things suggests that the hand is, when all goes well, in some way nothing but a faithful subordinate—one that works under top-down instruction and guidance from above. This underestimates the bidirectional interplay between manual and brain activity—interplay of the sort that explains why the distinctive manual dexterity of *Homo sapiens*, which sets us apart even from other primates who also have remarkable abilities in this regard, was likely one of the "main drivers" of the growth of the human brain (ibid.).

These ideas can be taken much further if one fully rejects what Tallis calls the standard ploy:

While it is perfectly obvious that voluntary activity must be built up out of involuntary mechanisms, there are *profound problems* in understanding this. There are particular problems with *the standard ploy* invoked by movement physiologists: proposing that the automation incorporates "calculations" that the brain (or part of it) "does," which permit customisation of the programs to the singularities of the individual action. (Ibid., 65; emphases added)

Invoking the standard ploy amounts to a sort of hand waving and making an anthropocentric appeal to representational contents so as to specify and fill out hypothesized motor plans and motor programs that supply instructive orders from on high, lending intelligence to and directing manual activity. For example, on this view, motor plans, intentions, and programs are understood as "propositional attitudes with contents of the form 'let [my] effector E perform motor act M with respect to goal-object G'" (Goldman 2009, 238).[3] The trouble is that even if we imagine that such representational contents exist, it is difficult to see how they could do the required work. The only chance such contents could have of specifying what is to be done, and how it is to be done, would be if they go beyond issuing very general and abstract instructions of the sort that Goldman gestures at above.

Only very fine-grained instructions would be capable of directing or controlling specific acts of manual activity successfully. This raises a number of questions. How do brains decide which general kind of motor act, M, is the appropriate sort of motor act to use in the situation at hand? This, alone, is no simple business—given the incredible variety of possible manual acts.[4]

And, even if we put that concern aside, proponents of the view that brains can initiate and control manual acts by traditional intellectualist means are left with the problem of explaining how and on what basis the brain decides how to execute any given act. A major problem for traditional forms of intellectualism is that the requirements for successfully performing any particular motor act are tied to a unique and changing context. For example, even if everything else remains static, the speed, angle of approach, and grip aperture need to be altered appropriately at successive stages when one does something as simple as picking up a coffee cup:

A particular challenge . . . has been to explain how cognition and perception processes regulate complex multi-articular actions in dynamic environments. The problem seeks to ascertain how the many degrees of freedom of the human motor system (roughly speaking the many component parts such as muscles, limb segments, and joints) can be regulated by an internally represented algorithm . . . and how the motor plan copes with the ongoing interaction between the motor system and energy fluxes surrounding the system, e.g., frictional forces and gravitational forces. . . . Not even the attempt to distinguish between the motor plan and the motor program has alleviated the problem in the literature (Summers and Anson 2009). (Araújo and Davids 2011, 12)

Successful manual activity requires bespoke and on-the-fly alterations. Hence, it is deeply implausible that the brain can simply identify what is required for successfully completing a certain type of activity and simply issuing general instructions to be carried out in the form of ordering preprogrammed routines. The implausibility of this suggestion is underscored by the fact that "most of the things we do are unique even though they may have stereotyped components" (Tallis 2003, 67). Not surprisingly, human manual activity—despite its unique complexities—seems to depend on interactions between the brain, body, and environmental interactions that involve essentially the same kinds of dynamic interactive feedback and temporally extended engagements needed to explain the intelligent antics of behavior-based robots and insects.

It must be noted that concomitant with abandoning the standard ploy in favor of a REC-based approach comes the admission that what we are dealing with in most cases of manual activity are not strictly speaking actions. This will surely be so if we operate with a strict concept of action—one that insists on a constitutive connection between actions and intentional states, where the latter are conceived of as requiring the existence of propositional attitudes of some sort. But all that follows from this, as Rowlands observes, is that "most of what we do does not count as action" (2006, 97).

Respecting the stipulated criterion on what is required for action, many philosophers acknowledge the existence of nonintentional doings, motivated activities, and/or deeds. For example, Velleman recognizes the need to "define a category of ungoverned activities, distinct from mere happenings, on the one hand, and from autonomous actions, on the other. This category contains the things one does rather than merely undergoes, but that one somehow fails to regulate in the manner that separates autonomous human action from merely motivated activity" (2000, 4).

On the face of it, the great bulk of animal doings takes the form of sophisticated forms of highly coordinated, motivated activity that falls well short of action if acting requires forming explicit, if nonconscious, intentions and deliberate planning, at any level. Picking up on Fodor's earlier remark, far from being mere "thrashings about" or "reflexive behaviors," such unplanned engagements appear to be quite skillful, and sometimes even expert, dealings with the world. If REC has the right resources for explaining the wide class of such doings, then it has the potential to explain quite a lot of what matters to us when it comes to understanding mind and cognition.

3   The Nonstandard Ploy: Representationalism Rescued?

Despite all that has been said in favor of REC, many will balk at going so far. There are weaker and much more conservative and conciliatory ways of taking on board what is best in embodied and enactive ideas without abandoning intellectualism in a wholesale way. For example, intellectualists can happily accept that various facts about embodiment are causally necessary in making certain types of intelligent responding possible and in shaping its character without this concession in any way threatening the idea that cognition is wholly constituted by representational facts or properties.

Trivially, it is clearly true that what a creature perceives depends on contingent facts about the nature of its sensory apparatus—thus, bats, dolphins, and rattlesnakes perceive the world differently and perceive different things because they are differently embodied. Moreover, no one denies that what and how we perceive causally depends on what we do—thus, it is only by moving my head and eyes in particular ways that certain things become visible and audible. Obviously, these truisms in no way threaten intellectualism.

Things can be taken further still without rocking the boat too much. A more daring thesis, one that several authors have lighted upon, is that extended bodily states and processes might—at least on occasion—serve as representational or information-carrying vehicles. As such, they can play unique computational roles in enabling some forms of cognition (Clark 2008a). Or, in the lingo of Goldman and de Vignemont (2009), perhaps those attracted to embodied and enactive accounts of cognition should be taken as claiming that some mental representations are encoded in essentially bodily formats. These renderings of what enactive and embodied accounts have to offer are conservative with respect to a commitment to representationalism. They are perfectly compatible with asserting that "the mind is essentially a thinking or representing thing" (Clark 2008b, 149), or that "the manipulation and use of representations is the primary job of the mind" (Dretske 1995, xiv).

Without breaking faith with intellectualism, conservative embodied/enactive cognition, or CEC, still allows one to recognize "the profound contributions that embodiment and embedding make" (Clark 2008a, 45). For those who endorse only CEC and not REC, the new developments in cognitive science, far from posing a threat to the existing paradigm, can be seen as supplying new tools or "welcome accessories" of considerable potential value that could augment intellectualist accounts of the mind.

CEC-style thinking is best exemplified by a recent bid to save the representationalist baby from the embodied bathwater, by arguing for the existence of action-oriented representations, or AORs. According to Wheeler (2008), who has done more than most to promote this view, an action-oriented representation is one that is:

(i) action-specific (tailored to a particular behavior and designed to represent the world in terms of specifications for possible actions);
(ii) egocentric (features bearer-relative content as epitomized by spatial maps in an egocentric co-ordinate system); and
(iii) intrinsically context-dependent (the explicit representation of context is eschewed in favor of situated special-purpose adaptive couplings that implicitly define the context of activity in their basic operating principles) (see also Wheeler 2005, 199).

Believing in AORs is consistent with accepting the neural assumption—an assumption that pays homage to the intuition that neural states and processes have a special cognitive status. Those attracted to this assumption believe it should be respected because, even though nonneural factors can qualify as representational vehicles, as it turns out, in most cases, they do not. As such, the great majority of cognitive explanations only involve representations that are wholly brain-bound. This is so even in cases in which it is necessary to making appeal to extraneural but nonrepresentational causal factors in order to explain the particular way that some particular intelligent activity unfolds. By accepting this last caveat, defenders of CEC allow that the full explanation of a given bout of intelligent behavior need not be strongly instructional in character in the way demanded by the standard ploy.

Wheeler (2005) highlights the core features of CEC-style thinking, illustrating the role of AORs by appeal to the architecture of a simple behavior-based robot created by Francheschini, Pichon, and Blanes (1992).

The robot has a primary visual system made up of a layer of elementary motion detectors (EMDs). Since these components are sensitive only to movement, the primary visual system is blind at rest. What happens, however, is that the EMD layer *uses* relative motion *information*, generated by the robot's own bodily motion during the previous movement in the sequence to build a temporary *snap map* of the detected obstacles, constructed using an egocentric coordinate system. Then, in an equally temporary *motor map, information* concerning the angular bearing of those detected obstacles is *fused with information* concerning the angular bearing of the light source (supplied by a supplementary visual system) and a directional heading for the next movement is generated. (Wheeler 2005, 196; first, second, fifth and sixth emphases added)

Wheeler (2005) considers and dismisses a number of possible minimal criteria for being an AOR—including appeal to selectionist strategies and decoupleability. After careful review, he settles on the idea that what is necessary and sufficient to distinguish behavior-based systems that operate with AORs from those that do not is that the former systems exhibit arbitrariness and homuncularity. A system exhibits arbitrariness just when the equivalence class of different inner elements is fixed "by their capacity, when organized and exploited in the right ways, to carry specific items of information or bodies of information about the world" (ibid., 218). A system is homuncular just when (a) it can be compartmentalized into a set of hierarchically organized communicating modules, and (b) each of those modules performs a well-defined subtask that contributes toward the collective achievement of the overall adaptive solution.

For Wheeler, the linchpin holding this account of AORs together is that some cognitive systems are information-processing systems. Thus: "The connection between our two architectural features becomes clear once one learns that, in a homuncular analysis, the *communicating subsystems* are conceptualized as *trafficking in the information that the inner vehicles carry.* So certain subsystems are interpreted as producing information that is then consumed downstream by other subsystems" (ibid.; emphases added).

We can legitimately describe a cognitive system as employing AORs just in case it is a genuine source of adaptive richness and flexibility and it turns out that its subsystems use "information-bearing elements to stand in for worldly states of affairs in their communicative dealings" (ibid. 219). Satisfaction of the above conditions is all that is required for the existence of weak or minimal representations. In the end, all of the weight in this account is placed on the idea that it suffices for minimal representations to be present in a system, S, if it manipulates and makes use of informational content in well-defined ways.

This minimal notion of representation is, no doubt, attractive to cognitive scientists. For anyone in the field it is utterly textbook to be told that information is a kind of basic commodity—the raw material of cognition. After all, "minds are basically processors of information; cognitive devices [for] receiving, storing, retrieving, modifying and transmitting information of various kinds" (Branquinho 2001, xii–xiii).

There is great latitude for thinking about the processes that enable this. Thus: "Mental representations might come in a wide variety of forms; there being no commitment in the claim itself to a specific kind of representation or to a particular sort of representational vehicle . . . mental representations might be thought of as images, schemas, symbols, models, icons, sentences, maps and so on" (ibid., xiv).

Accordingly, representations or representational vehicles "are items in the mind or brain of a given system that in some sense 'mirror,' or are mapped onto, other items or sets of items . . . in the world" (ibid., xiv). But what makes something into a vehicle, the essence of representing, is that they bear or possess content. Content is key. Thus: "The whole thrust of cognitive science is that there are sub-personal contents and sub-personal operations that are truly cognitive in the sense that these operations can be properly *explained only in terms of these contents*" (Seager 1999, 27; emphasis added).

Dietrich and Markman define representations as "any internal state that mediates or plays a mediating role between a system's inputs and outputs *in virtue of that state's semantic content. We define semantic content in terms of information* causally responsible for the state, and in terms of the use to which that information is put" (2003, 97; emphasis added). In sum, believing in AORs only requires acceptance of "the general idea of inner states that bear contents" (Clark 2002, 386).

Given this, it might be thought that accepting that at least some cognitive systems employ AORs is a no-brainer. Certainly, it seems that this must be true of human manual activity of the sort described in section 2—activity of the sort that, if my argument goes through, would enable fans of REC to answer the scope objection. After all, even in cases in which that sort of activity is not supported by focused, conscious perception, "the motor activity of the hand—reaching, gripping and manipulation—cannot function in the absence of what is usually called 'sensory information'" (Tallis 2003, 27). Indeed, as Tallis puts it: "the information the hand needs to support its manipulative function is most clearly evident in the first stage . . . in reaching out prior to grasping, shaping, etc. Here the hand is under primarily visual control: the target is located, the relationship to the body determined, the motion initiated to home in on the target—these are all regulated [or, better, assisted] by sight, which measures what needs to be done and the progress of the doing" (ibid.).

With this in mind, it looks like manual activity is surely dependent on information-processing activity of the sort that would qualify as involving AORs—hence it is the sort of activity better suited to CEC rather than a REC treatment. If so, any ground gained by supporters of REC in section 2 would be lost.

4   The Hard Problem of Content

Despite some obvious attractions, the AOR story, and the support it lends to CEC over REC, isn't beyond question. Indeed, as I argue in this concluding section, anyone who favors CEC must face up to the hard problem of content—and I suggest that REC is a small price to pay to allow one to avoid that problem.

Before turning to that issue, it is worth highlighting an immediate concern about CEC and its reliance on AORs. Appeal to AORs seems to secure the fate of minimal representations—winning a key metaphysical battle—only at the cost of losing a wider explanatory war. On the assumption that the AORs need not be decoupled in order to qualify as representations—an assumption Wheeler (2005, 2008) explicitly defends, and with good reason (see Chemero 2009, ch. 3), defenders of CEC face the charge that "talk of representations in coupled systems may be too cheap, or too arbitrary, and thus adds little or nothing to an explanation of how these systems work" (Shapiro 2011, 147).

Chemero also voices this exact worry, noting that "the representational description of the system does not add much to our understanding of the system . . . [thus] despite the fact that one can cook up a representational story, once one has the dynamical explanation, the representational gloss does not predict anything about the system's behavior that could not be predicted by the dynamical explanation alone" (2009, 77). Although initially cast as a purely explanatory concern, it is clear that this issue cannot be kept wholly free of metaphysical considerations. For instance, Chemero goes on to note that "in terms of the physics of the situation, the ball, the outfielder, and the intervening medium are just one connected thing. In effective tracking, that is, the outfielder, the ball, and the light reflected from the ball form a single coupled system. No explanatory purchase is gained by invoking representation here: in effective tracking, any internal parts of the agent that one might call mental representations are causally coupled with their targets" (ibid., 114).

Part of the trouble here is that there does not appear to be any clean-cut way to decide, with precision, which systems actually satisfy the relevant conditions for being minimally representational systems. For example, opinions diverge about whether Watt's much discussed centrifugal governor—a device originally designed to ensure a constant operating speed in rotative steam engines—qualifies as a representational device. This is despite the fact that the relevant parties in the debate are fully agreed about the characteristics of the governor's internal design, which are quite elegant and simple. The positions of the device's spindle arms interact with and modify the state of a valve that controls the engine's speed—when the arm is high the valve slows the engine, and when the arm is low the valve increases engine speed.

In line with the criteria laid out in the previous section, Chemero (2009) concludes that "It is possible to view the governor's arms as [noncomputational] representations. It is the function of particular arm angles to change the state of the valve (the representation consumer) and so adapt it to the need to speed up or slow down. The governor was so designed . . . to play that role . . . it is both a *map* and a *controller*. It is an *action-oriented representation*, standing for the current need to increase or decrease speed" (71; emphases added).[5]

Shapiro (2011) dissents. After careful review of this issue, he concludes that "Watt's design of the centrifugal governor *requires that* the angle of the flyball arms carry information about the flywheel speed in order to regulate the speed. Still, the function of the flyball arms is not to carry information, but rather to play a role in regulating the speed of the flywheel" (147).

The point is that in order to carry out the governor's control function, the spindle arms must covary with the relevant changes in engine speed. That they carry information in this sense is an unavoidable, collateral feature of their design that enables them to perform their regulatory work. Apparently, this is not sufficient to qualify as being a true information processor in the relevant sense—hence, the governor is misdescribed as making use of AORs.

Shapiro (2011) goes on to contrast the governor with other kinds of information using devices: "Some devices surely do include components that have the function to carry information. The thermostat . . . is such a device. Thermostats contain a bimetal strip because the behavior of this strip carries information about [i.e. covaries with] temperature, and the room" (148).

The important difference is that although the governor's arms carry information about the flywheel arms, the governor does not use that information in order to perform its control tasks. This is meant to mark the subtle but utterly critical difference between merely complex systems and properly cognitive complex systems. Apparently, the thermostatic systems are designed to use information about the temperature *as such* in carrying out their work.

Put as starkly as this, one might be forgiven for failing to see what changes so dramatically when it comes to the operation of thermostats as opposed to Watt governors. A thermostat regulates the system's temperature, maintaining it at a desired point. Its mechanisms exploit the properties of the bimetallic strip that—when all is well—responds in reliable ways to temperature changes, bending one way if heated and the opposite way if cooled.

The important difference between the two types of systems is not that more mechanisms or steps are involved in regulating temperature by this means. Rather, the crucial distinction is meant to be that the bimetallic strips in thermostats have the systemic function of indicating specific desired temperatures to other subsystems that use those indications to regulate their behavior. It is because they function in this special way that devices of this general type are representational—they exploit preexisting indication relations because they have the function to indicate how things stand externally and use those indications in particular ways.[6] Following Dretske (1995), if such devices were naturally occurring then they would "have a content, something *they say or mean*, that does not depend on the existence of our purposes and intentions. . . . [They would] have original intentionality, something they represent, say, or mean, that they do not get from us" (8; emphasis added).

To qualify as representational, an inner state must play a special kind of role in a larger cognitive economy. Crudely, it must, so to speak, have the function of *saying or indicating* that things stand *thus and so* and of being consumed by other systems because it says or indicates in that way. Only then will an internal state or structure meet Ramsey's (2007) job description challenge.

It is plausible that many of the states (or ensembles of states) of systems that enable basic cognition are merely (1) reliably caused by (or nomically depend on) the occurrence of certain external features, (2) disposed to produce certain effects (under specific conditions), and (3) have been selected because of their propensities for (1) and (2). Yet states or structures that only possess properties 1–3 fail to meet the job description challenge. They fail to qualify as truly representational mental states having the proper function of saying "things stand thus and so"; rather, they—like the Watt governor—only have the proper function of guiding a system's responses with respect to specific kinds of worldly offerings.

Exactly what else is required to be a representation-using system? Wheeler speaks of the need for communicative transactions between homuncular subsystems. This informational dealing is the basis of true cognition—nonetheless, he stresses that this does not imply that the subsystems "in any literal sense understand that information" (2005, 218). Fair enough, but even if they literally lack understanding, it might be thought that at least such subsystems must be literally trading in informational content—using and fusing it—even if they don't understand what it says. But talk of using and fusing contents, although quite common, cannot be taken literally either. It is not as if informational content is a kind of commodity that gets moved about and modified in various ways; information is not "like a parcel in the mail" (Shapiro 2011, 35).

This being so, it seems that *bona fide* cognitive systems are not special because they *literally* use and manipulate informational content (not even content that they don't understand). They are special because it is their function to convey informational content without actually manipulating it as such.

We are now getting down to brass tacks. For this story to work, there must at least be content that these subsystems have the special function to convey; there must be something that it is their function to say even if they don't understand what they are saying or what is said. But exactly what is informational content?

Dretske speaks about informational content as "the what-it-is-we-can-learn from a signal or message in contrast to how-much-we-can-learn" (1981, 47). He makes clear that he understands a signal's informational content to be a kind of propositional content of the *de re* variety. Propositions or propositional contents have special properties—minimally, they are bearers of truth. Assuming that informational contents are propositional is presumably what allows Dretske to hold that when signals carry information to the senses they tell "us truly about another state of affairs" (ibid., 44).

This view is, of course, quite compatible with holding that informational content lacks full-fledged representational properties. Thus one can hold that informational content is supplied by the senses, which is not representational content, and that more is required for informational content to be properly representational.

It is at this juncture that defenders of AORs and CEC—at least those who subscribe to an explanatory naturalism—face a dilemma. Since so much hangs on this, it is worth going very slowly over some familiar ground. In the opening passage of Dretske's *Knowledge and the Flow of Information*, we find the *locus classicus* and foundational statement on how to understand information-processing systems in a way that is both required by the CEC story and that expresses a commitment to explanatory naturalism:

In the beginning there was information. The word came later . . . information (though not meaning) [is] an *objective commodity*, something whose generation, transmission and reception does not require or in any way presuppose interpretative processes. One is therefore given a framework for understanding how meaning can evolve, how genuine cognitive systems—those with the resources for interpreting signals, holding beliefs, acquiring knowledge—can develop out of *lower-order, purely physical, information-processing mechanisms*. . . . Meaning, and the constellation of mental attitudes that exhibit it, are manufactured products. The raw material is information. (1981, vii; emphases added)

Any explanatory version of naturalism seeks to satisfy what Wheeler charmingly calls the Muggle constraint: "One's explanation of some phenomenon meets the Muggle constraint just when it appeals only to entities, states and processes that are wholly nonmagical in character. In other words, no spooky stuff" (2005, 5).

It is widely supposed that the informational theory of content comfortably meets this constraint. At least, its defenders have attempted to convince us that, when they promote it, there is nothing up their sleeves. This is because, as Jacob emphasizes,

the relevant notion of information at stake in informational semantics is the notion involved in many areas of scientific investigation as when it is said that a footprint or a fingerprint carries information about the individual whose footprint or fingerprint it is. In this sense, it may also be said that a fossil carries information about a past organism. The number of tree rings in a tree trunk carries information about the age of the tree. (1997, 45; emphasis added)

This picks out the relevant notion by means of examples. We can call it the notion of information-as-covariance. Although theorists quibble about the strength and scope of the degree of covariance required for informational relations to exist, there is consensus that s's being F "carries information about" t's being H if and only if the occurrence of these states of affairs lawfully, or reliably enough, covary.

But here's the rub. Anything that deserves to be called content has special properties—truth, reference, implication—that make it logically distinct from and irreducible to mere covariance relations holding between states of affairs. Although the latter notion is surely scientifically respectable, it isn't able to do the required work of explaining content. Put otherwise, if information is nothing but covariance then it is not any kind of content—at least not if content of the sort defined in terms of its truth bearing properties. The number of a tree's rings can covary with its age; this does not entail that the first state of affairs says or conveys anything true about the second, nor vice versa. The same goes for states that happen to be inside agents and that reliably correspond with external states of affairs—these too, in and of themselves, do not "say" or "mean" anything in virtue of instantiating covariance relations. Quite generally, covariation in and of itself neither suffices for nor otherwise constitutes content, where content minimally requires the existence of truth bearing properties. Call this the "covariance doesn't constitute/confer content" (or CCC) principle.

The CCC principle undermines the assumption that covariation is the worldly source of informational content. There is no doubt the idea of information-as-covariance is widely used in sciences; hence, it is not a hostage to fortune for explanatory naturalists. But if the CCC is true, there is a gaping explanatory hole in the official story propounded by those who follow Dretske's lead. Anyone peddling such an account is surely violating the Muggle constraint and ought to be brought to the attention of the Ministry of Magic.[7]

One might opt for the first horn of this dilemma and retain the scientifically respectable notion of information-as-covariance, and thus retain one's naturalistic credentials while relinquishing the idea there is such a thing as informational content. That is the path I recommend, but it requires giving up on CEC since—as argued in the previous section—the minimal requirement for distinguishing information-*processing* systems is that they make use of AORs, which are defined as content-bearing vehicles. But the distinction between vehicles and contents falls apart, at least at the relevant level, if there are no informational contents to bear.

To avoid this, one might opt to be impaled on the second horn. This would be to accept that contentful properties exist even if they don't reduce to, or cannot be explained in terms of, covariance relations. If contentful properties and covariance properties are logically distinct, they might still be systematically related. Hence, it might be hoped that contentful properties can be naturalistically explained by some other means (e.g., by some future physics). Alternatively, they could be posited as explanatory primitives—as metaphysical extras that might be externally related to covariance properties. Thus they might have the status that Chalmers (2010) still assigns to qualia—they might require us to expand our understanding of the scope of the natural. Contentful properties might pick out properties that—like phenomenal properties—are irreducible to and exist alongside basic physical properties. If so, the explanatory project of naturalism with respect to them would look quite different—it would be to discover the set of fundamental bridging laws that explain how contentful properties relate to basic physical properties. That would be the only way to solve what we might call the hard problem of content.

Of course, one might try to avoid both horns by demonstrating the falsity of the CCC by showing how contentful properties—for example, truth-bearing properties—reduce to covariance properties (good luck with that!). A more plausible dilemma-avoiding move would be to show that the notion of information that is in play in these accounts is, in fact, meatier than covariance but is nonetheless equally naturalistically respectable.

After all, Dretske talks of indication relations, not covariance relations, though the two are often confused. Tellingly, in continuing the passage cited above Jacob remarks that "in all of these cases, it is not unreasonable to assume that the informational relation holds between an indicator and what it indicates (or a source) independently of the presence of an agent with propositional attitudes" (1997, 45). In making this last point, he stresses that "the information or indication relation is going to be a relation between states or facts" (ibid., 49–50).

However, following Grice, Dretske is wont to think of indication as natural meaning—as in "smoke" means "fire." But smoke *means* fire only if it indicates fire to someone. It makes no sense to talk of indication in the absence of a user. Indication is, at least, a three-place relation, whereas covariance, by contrast, is a two-place relation. To think of indication as the basis for informational semantics therefore is already to tacitly assume that there is more going on than mere covariation between states of affairs.[8] This raises questions about how exactly the notion of information-as-indication relates to its scientifically respectable cousin, the notion of information-as-covariance. Moreover, we might wonder if this notion has independent naturalistic credentials of its own. Until these questions are answered, promoters of AORs and CEC—those who rely on the existence of informational content to distinguish genuine cognitive systems from all others—haven't really left the starting blocks with their theorizing.

5 Epilogue: Decoding Information

The "code" metaphor is rife in the cognitive sciences, but the cost of taking it seriously is that one must face up to the hard problem of content. In light of the problems with CEC-style stories highlighted above, we have reason to think that online sensory signals "carry information" (in one sense) but not that they "pass on" meaningful or contentful messages—contentful information that is used and fused to form inner representations. Unless we assume that preexisting contents exist to be received through sensory contact, the last thread of the analogy between basic cognitive systems and genuinely communicating systems breaks down at a crucial point.

In line with REC, there are alternative ways to understand cognitive activity as involving a complex series of systematic—but not contentfully mediated—interactions between well-tuned mechanisms (see, e.g., Hutto 2011a,b; Hutto and Myin forthcoming). Supporters of REC press for an understanding of basic mentality as literally constituted by, and to be understood in terms of, concrete patterns of environmentally situated, organismic activity, nothing more nor less. If they succeed, the above arguments should encourage more cautious CEC types—those trying to occupy the mid-left—to take a walk on the wild side.[9]

Notes

1. This intellectualist way of understanding the basic nature of minds taps into a long tradition stretching back at least as far as Plato; it was revived by Descartes in the modern era, and regained ascendency, most recently, through the work of Chomsky during the most recent cognitive revolution. As Noë observes: "What these views have in common—and what they have bequeathed to cognitive science—is the idea that we are, in our truest nature, thinkers. It is this intellectualist background that shapes the way cognitive scientists think about human beings" (2009, 98).

2. There are clear connections that might be forged between this view and Donald's claim that when it comes to understanding human cognition, "the most critical element is a capacity for deliberately reviewing self-actions so as to experiment with them. . . . It would be no exaggeration to say that this capacity is uniquely human,

and forms the background for the whole of human culture, including language" (1999, 142).

3. It is perhaps understandable that in seeking to make sense of this cognitive activity we are naturally inclined to assume the existence of representations that "include not only 'commands' and 'calculations,' but also 'if-then' and other logical operations. This shows how it seems impossible to make sense of cerebral control—requisition and modification—of motor programs, to describe them in such a way that they deliver what is needed while avoiding anthropomorphisms" (Tallis 2003, 65). The problem is that "attributing to the brain, or parts of it, or neural circuits, the ability to do things that we, whole human beings, most certainly cannot do, seems unlikely to solve the puzzle" (ibid.).

4. At a pinch, one could give a short list of these, which could include: "grasping, seizing, pulling, plucking, picking, pinching, pressing, patting, poking, prodding, fumbling, squeezing, crushing, throttling, punching, rubbing, scratching, groping, stroking, caressing, fingering, drumming, shaping, lifting, flicking, catching, throwing, and much besides" (Tallis 2003, 22).

5. Notably, Chemero holds that the centrifugal governor is not a computer even though it can be regarded as a representational device; in this respect, he does not break faith with the conclusion of van Gelder's original analysis when he first introduced the example into the literature (van Gelder 1995).

6. Thus, "if we suppose that, through selection, an internal indicator acquired a biological function, the function to indicate something about the animal's surroundings, then we can say that this internal structure represents" (Dretske 1988, 94).

7. To make vivid what is at stake, it is worth noting that early analytic philosophers were at home with the view that the world is ultimately and literally composed, at least in part, by "propositions." These were conceived of as bedrock Platonic entities—mentionable "terms" that, when standing in the right complex relations, constitute judgeable objects of thought. In commenting on Russell's version of this idea, Makin underscores the features that parallel many of the properties that Dretske demands of informational content. He stresses that "with propositions, it is crucial to bear in mind that they are not, nor are they abstracted from, symbolic or linguistic or psychological entities. . . . On the contrary, *they are conceived as fundamentally independent of both language and mind. Propositions are first and foremost the entities that enter into logical relations of implication, and hence also the primary bearers of truth* . . . 'truth' and 'implication' apply, in their primary senses, to propositions and only derivatively to the sentences expressing them" (Makin 2000, 11; emphasis added).

8. Others have noticed this as well. For example, Ramsey comments on the peculiar features of the quasi-semantic indication relation as follows: "Dretske and many

authors are somewhat unclear on the nature of this relation. While it is fairly clear what it means to say that state A nomically depends upon state B, it is much less clear how such a claim is supposed to translate into the claim that A is an indicator of B, or how we are to understand expressions like 'information flow' and 'information carrying'" (Ramsey 2007, 133).

9. I am not alone in trying to persuade those in the CEC brigade to make this shift; see also Gangopadhyay 2011.

References

Araújo, D., and K. Davids. 2011. What exactly is acquired during skill acquisition? *Journal of Consciousness Studies* 18 (3–4):7–23.

Beer, R. D. 1998. Framing the debate between computational and dynamical approaches to cognitive science. *Behavioral and Brain Sciences* 21:630.

Beer, R. D. 2000. Dynamical approaches to cognitive science. *Trends in Cognitive Sciences* 4 (3):91–99.

Branquinho, J. 2001. *The Foundations of Cognitive Science*. Oxford: Oxford University Press.

Brooks, R. 1991a. New approaches to robotics. *Science* 253:1227–1232.

Brooks, R. 1991b. Intelligence without representation. *Artificial Intelligence* 47: 139–159.

Chalmers, D. 2010. *The Character of Consciousness*. Oxford: Oxford University Press.

Chemero, A. 2009. *Radical Embodied Cognitive Science*. Cambridge, MA: MIT Press.

Clark, A. 2002. Skills, spills, and the nature of mindful action. *Phenomenology and the Cognitive Sciences* 1:385–387.

Clark, A. 2008a. Pressing the flesh: A tension in the study of the embodied, embedded mind? *Philosophy and Phenomenological Research* 76 (1):37–59.

Clark, A. 2008b. *Supersizing the Mind: Embodiment, Action, and Cognitive Extension*. Oxford: Oxford University Press.

Dietrich, E., and A. Markman. 2003. Discrete thoughts: Why cognition must use discrete representations. *Mind & Language* 18:95–119.

Donald, M. 1999. Preconditions for the evolution of protolanguages. In *The Descent of Mind: Psychological Perspectives on Hominid Evolution*, ed. M. C. Corballis and S. E. G. Lea. Oxford: Oxford University Press.

Dretske, F. 1981. *Knowledge and the Flow of Information*. Cambridge, MA: MIT Press.

Dretske, F. 1988. *Explaining Behavior: Reasons in a World of Causes*. Cambridge, MA: MIT Press.

Dretske, F. 1995. *Naturalizing the Mind*. Cambridge, MA: MIT Press.

Fodor, J. A. 2008. *LOT 2: The Language of Thought Revisited*. Oxford: Oxford University Press.

Francheschini, N., J.-M. Pichon, and C. Blanes. 1992. From insect vision to robot vision. *Philosophical Transactions of the Royal Society of London, Series B: Biological Sciences* 337:283–294.

Gallagher, S. 2008. Are minimal representations still representations? *International Journal of Philosophical Studies* 16 (3):351–369.

Gangopadhyay, N. 2011. The extended mind: Born to be wild? A lesson from action understanding. *Phenomenology and the Cognitive Sciences* 10 (3):377–397.

Garzón, F. C. 2008. Towards a general theory of antirepresentationalism. *British Journal for the Philosophy of Science* 59 (3):259–292.

Goldman, A. I. 2009. Mirroring, simulating, and mindreading. *Mind & Language* 24 (2):235–252.

Goldman, A. I., and F. de Vignemont. 2009. Is social cognition embodied? *Trends in Cognitive Sciences* 13 (4):154–159.

Hurley, S. 2008. The shared circuits model: How control, mirroring, and simulation can enable imitation and mindreading. *Behavioral and Brain Sciences* 27:1–58.

Hutto, D. D. 2011a. Enactivism: Why be radical? In *Sehen und Handeln*, 21–44, ed. H. Bredekamp and J. M. Krois. Berlin: Akademie.

Hutto, D. D. 2011b. Philosophy of mind's new lease on life: Autopoietic enactivism meets teleosemiotics. *Journal of Consciousness Studies* 18 (5–6):44–64.

Hutto, D. D., and E. Myin. Forthcoming. *Radicalizing Enactivism*. Cambridge, MA: MIT Press.

Jacob, P. 1997. *What Minds Can Do*. Cambridge: Cambridge University Press.

Makin, G. 2000. *The Metaphysics of Meaning*. London: Routledge.

Milner, D., and M. Goodale. 1995. *The Visual Brain in Action*. New York: Oxford University Press.

Noë, A. 2009. *Out of Our Heads*. New York: Hill & Wang.

Ramsey, W. M. 2007. *Representation Reconsidered*. Cambridge: Cambridge University Press.

Rowlands, M. 2006. *Body Language*. Cambridge, MA: MIT Press.

Ryle, G. 1949. *The Concept of Mind*. London: Hutchinson.

Shapiro, L. 2011. *Embodied Cognition*. London: Routledge.

Seager, W. 1999. *Theories of Consciousness: An Introduction*. London: Routledge.

Summers, J. J., and G. A. Anson. 2009. Current status of the motor programme revisited. *Human Movement Science* 28:566–577.

Tallis, R. 2003. *The Hand: A Philosophical Inquiry into Human Being*. Edinburgh: Edinburgh University Press.

Thompson, E. 2007. *Mind in Life: Biology, Phenomenology, and the Sciences of the Mind*. Cambridge, MA: Harvard University Press.

van Gelder, T. 1995. What might cognition be, if not computation. *Journal of Philosophy* 92:345–381.

Varela, F. J., E. Thompson, and E. Rosch. 1991. *The Embodied Mind: Cognitive Science and Human Experience*. Cambridge, MA: MIT Press.

Velleman, J. D. 2000. *The Possibility of Practical Reason*. Oxford: Oxford University Press.

Webb, B. 1994. Robotic experiments in cricket phototaxis. In *From Animals to Animats 3: Proceedings of the Third Annual Conference on Simulation of Adaptive Behavior*, ed. D. Cliff, P. Husbands, J. A. Meyer, and S. W. Wilson. Cambridge, MA: MIT Press.

Webb, B. 1996. A cricket robot. *Scientific American* 275 (6):62–67.

Wheeler, M. 2005. *Reconstructing the Cognitive World*. Cambridge, MA: MIT Press.

Wheeler, M. 2008. Minimal representing: A response to Gallagher. *International Journal of Philosophical Studies* 16 (3):371–376.

IV THE GIST OF GESTURES

11 Gesture as Thought?

Andy Clark

1 Why Gesture?

Following a sustained empirical enquiry into the nature and organization of human gesture, the psychologist Susan Goldin-Meadow (2003) asks an intriguing question: Is gesture simply about the expression of fully formed thoughts, and thus mainly a prop for interagent communication (listeners appreciating meanings through others' gestures), or *might gesture function as part of the actual process of thinking*? Some clues (ibid., 136–149) that it might be more than merely expressive include:

that we do it when talking on the phone;
that we do it when talking to ourselves;
that we do it in the dark when no one can see;
that gesturing increases with task difficulty;
that gesturing increases when speakers must choose between options;
that gesturing increases when reasoning about a problem rather than merely describing the problem or a known solution.

A deflationist might suggest that most of these effects are easily explained by mere association: that gesturing without a viewer is just a habit, installed by our experience of gesturing in the normal communicative context. It turns out, however (ibid., 141–144), that speakers blind from birth, who have never spoken to a visible listener, and never seen others moving their hands as they speak, gesture when they speak. Moreover (see also Iverson and Goldin-Meadow 1998, 2001), they do so even when speaking to others they know to be blind.

Supposing (for the sake of argument) that gesture does play some kind of active causal role in thinking, just what role might that be? One way to find out is to see what happens when actual physical gesture is removed from the mix of available resources. To explore the impact of restricting gesture on thought, Goldin-Meadow and colleagues asked two matched groups of children to memorize a list of words and then to carry out some mathematical problem solving before trying to recall the list (see Goldin-Meadow 2003; Goldin-Meadow et al. 2001). One group (call it the "free-gesture group") could freely gesture during the intervening maths task; the other (call it the "no-gesture group") was told not to gesture. The results were that restricting the use of gesture during the intervening mathe-matical task had a robust and significant detrimental effect on the sepa-rate memory task (remembering the list of words). The best explanation, according to Goldin-Meadow, is that the act of gesturing somehow shifts or lightens aspects of the overall neural cognitive load, thus freeing up resources for the memory task.

Before pursuing this idea, it is necessary to rule out a rather obvious alternative account. According to this alternative account, the effort of remembering *not* to gesture (in the no-gesture group) is *adding* to the load, rather than gesture (in the free-gesture group) lessening the load. If this were so, the no-gesture group would indeed perform less well, but not because gesturing lightens the load. Rather, remembering *not* to gesture increases it. As luck would have it, some children and adults spontaneously chose not to gesture during some of the episodes of mathematical problem solving. This allowed the experimenters to compare the effects of removing gesture by instruction and by spontaneous (hence presumably effortless) inclination. Memory for the initial task turned out to be equally impaired (Goldin-Meadow 2003, 155) even when the lack of gesture was a spontane-ous choice, supporting the claim that the gestures themselves play some active cognitive role.[1]

An important hint as to the nature of this active role emerges, Goldin-Meadow argues, when we look (Goldin-Meadow 2003, ch. 12) at cases of gesture-speech mismatches. These are cases where what you say and what you gesture[2] are in conflict, for example, you gesture a one–one mapping while failing to appreciate the importance of such a mapping in your simultaneous vocal attempts at solving the problem. Many such cases were found, and (importantly) the gestures tended to prefigure the child's con-sciously finding the right solution in speech at a very slightly later point. Even if the right solution was not shortly found, the presence of the apt gesture turned out to be predictive of that child's being able to learn the right solution more easily than others whose gestures showed no such tacit or nascent appreciation.

After much conjecture and experiment, Goldin-Meadow is led to the following story (drawing also, as she clearly notes, on the groundbreaking work of David McNeill 1992; see also McNeill 2005 and discussion in section 2 following). The physical act of gesturing, Goldin-Meadow suggests, plays an active (not merely expressive) role in learning, reasoning, and cognitive change by providing an alternative (analog, motoric, visuospatial) representational format. In this way: "Gesture . . . expands the set of representational tools available to speakers and listeners. It can redundantly reflect information represented through verbal formats or it can augment that information, adding nuances possible only though visual or motor formats" (Goldin-Meadow 2003, 186).

Encodings in that visuomotor format enter, it is argued, into a kind of ongoing coupled dialectic with encodings in the verbal format. Gesture thus continuously informs and alters verbal thinking, which is continuously informed and altered by gesture, that is, the two form a genuinely coupled system. This coupled dialectic creates points of instability (conflict) whose attempted resolutions move forward our thinking, often (though of course not always) in productive ways. The upshot, according to Iverson and Thelen, is "a dynamic mutuality such that activity in any one component of the system can potentially entrain activity in any other component" (1999, 37).

But is it (at least partly) the actual physical gestures that matter here, or do they merely reflect the transfer of load between two different neural stores? Does gesturing simply shift the burden from a neural verbal store to a neural visuospatial store? If so, then it should be harder to perform a separate spatial memory task when freely gesturing than when not. This was tested (Goldin-Meadow et al. 2001; Wagner et al. 2004) by replacing the original word-recall task with a spatial one: that of recalling the location of dots on a grid. The results were unambiguous. The availability of gesture still helps (still yields improved performance on the memory task) even when the second task is itself a spatial one.

The act of gesturing, all this suggests, is not simply a motor act expressive of some fully neurally realized process of thought. Instead, the physical act of gesturing is part and parcel of a coupled neural-bodily unfolding that is itself usefully seen as an organismically extended process of thought. In gesture, we plausibly confront a cognitive process whose implementation involves machinery that loops out beyond the purely neural realm. This kind of cognitively pregnant unfolding need not stop at the boundary of the biological organism.

Something very similar may occur, as frequently remarked, when we are busy writing and thinking at the same time. It is not always the case that fully formed thoughts get committed to paper. Rather, the paper provides a medium in which, this time via some kind of coupled neural-scribbling-reading unfolding, we are enabled to explore ways of thinking that might otherwise be unavailable to us. Just such a coupled unfolding was eloquently evoked in a famous exchange between Richard Feynman and the historian Charles Weiner:[3]

Weiner once remarked casually that [a batch of notes and sketches] represented "a record of [Feynman's] day-to-day work," and Feynman reacted sharply.

"I actually did the work on the paper," he said.

"Well," Weiner said, "the work was done in your head, but the record of it is still here."

"No, it's not a *record*, not really. It's *working*. You have to work on paper and this is the paper. Okay?"

Feynman is right. If, following McNeill and Goldin-Meadow, we allow that actual gestures (not simply their neural pre- or postcursors) can form part of an individual's cognitive processing, there seems no principled reason to suddenly stop the spread the moment skin meets air.

2 Material Carriers

In recent work, David McNeill (2005) offers a clear expression of the view that the physical gestures are actually elements in the cognitive process itself. McNeill's work is grounded in extensive empirical case studies of the use of gesture in free speech. The key idea that McNeill uses to understand and organize these studies is the notion (briefly mentioned above) of an ongoing imagery-language dialectic in which physical gesture acts as a "material carrier." The phrase "material carrier," due to Vygotsky (1962/1986), is meant to convey the idea of a physical materialization that has systematic cognitive effects. But once more, we should not be misled by the image of cognitive effects. For according to McNeill, "the concept [of a material carrier] implies that the gesture, *the actual motion of the gesture itself*, is a dimension of thinking" (2005, 98; emphasis in original). Our free (i.e., spontaneous, nonconventional) gestures are not, McNeill argues, merely expressions of or representations of our fully achieved "inner" thoughts but are themselves "thinking in one of its many forms" (ibid., 99). Notice that this is not to say that the gestures do not follow from, and lead to, specific forms of neural activity. They do, and McNeill (see his 2005, chs. 7 and 8) has much to say about the neural systems preferentially involved in the generation and reception of spontaneous gesture. Rather, it is to see the physical act of gesturing as *part of a unified thought-language-*

hand system whose coordinated activity has been selected or maintained for its specifically cognitive virtues.

There are important differences between McNeill's account and that of Goldin-Meadow, but they are united in seeing the physical gestures as genuine elements in the cognitive process. McNeill stresses the idea of "growth points," described as "the minimal unit of an imagery-language dialectic" (ibid., 105). A growth point is a package of imagistic and linear propositional (linguistic) elements that together form a single idea (e.g., both conveying the concept of an antagonistic force as a speaker describes some series of events). The points of productive conflict stressed by Goldin-Meadow are not growth points in this technical sense (see, e.g., McNeill 2005, 137). But they are growth points in another, quite routine sense: They are collisions in meaning space, crucially mediated by gestural loops into the physical world, that are able to move our thinking along in productive ways. These differences in emphasis do not amount, as far as I can tell, to any deep inconsistency in the underlying models of the cognitive virtues of gesture. In each case, the loop into gesture creates a material structure that is available to both speaker and listener. And just as that material structure may have a systematic cognitive effect on the listener, so too it may have a systematic cognitive effect on the speaker. Here, McNeill invokes an evolutionary hypothesis that he dubs "Mead's Loop."[4] The background to McNeill's suggestion is the discovery of so-called mirror neurons. These are neurons, first discovered in the frontal lobes of macaques, that fire both when an animal performs some intentional action and when it sees another animal performing the same action (see Rizzolatti et al. 2001). McNeill's suggestion is that our own gestures activate mirror-neuron-dominated neural resources so that: "One's own gestures [activate] the part of the brain that responds to intentional actions, including gestures, by someone else, and thus treats one's own gesture as a social stimulus" (McNeill 2005, 250).

Whether this is the correct evolutionary and mechanistic account is unimportant for present purposes. What matters is rather the guiding idea that by materializing imagistic thought in physical gesture we create a stable physical presence that may productively affect and constrain the neural elements of thought and reason. The role of gesture, if this is correct, is closely akin to that of certain forms of self-directed, overt or covert speech or (looping outside the organismic shell) to certain forms of writing-for-thinking (see McNeill 2005, 99).

Shaun Gallagher, in a very rewarding recent discussion of gesture and thought, writes that "Even if we are not explicitly aware of our gestures, and even in circumstances where they contribute nothing to the com-municative process, they may contribute implicitly to the shaping of our cognition" (2005, 121). Gallagher approaches the topic of gesture in the larger framework of his account of the "prenoetic" role of embodiment. This is a term of art that Gallagher uses to signify the role of the body in structuring mind and consciousness. The idea is that facts about the body, and about bodily orientation and the like, set the scene for conscious acts of perception, memory, and judgment (the "noetic" factors) in various important ways. A prenoetic performance, we are told, is "one that helps to structure consciousness but that does not explicitly show itself in the contents of consciousness" (ibid., 32). Thus, to take a very simple example, embodied agents perceive the world from a certain spatial perspective. That perspective shapes what is given to us explicitly in phenomenal experience, but it is not itself part of what we experience. Instead, it "shapes" or "structures" experience (for this example, see Gallagher 2005, 2–3). In this manner, Gallagher speaks of the role of gesture in "shaping" cognition and (following Merleau-Ponty's usage in describing the cognitive role of speech) in the "accomplishment of thought." Such locutions neatly (though only temporarily) sidestep the thorny issue of whether to see gesture as part of the *actual machinery* of thought and reason. In a footnote to the quoted passage Gallagher is less cautious, suggesting that "It may be . . . that certain aspects of what we call the mind just are in fact nothing other than what we tend to call expression, that is, occurrent linguistic practices ('internal speech'), gesture, and expressive movement" (ibid., 121). Gesture, Gallagher here suspects, is both a means by which thought is "accomplished" *and* an aspect of mind—an aspect of the think-ing[5] itself.

What finally emerges from our consideration of the role of gesture in thought is that gestures form part of an integrated "language-thought-hand"[6] system, and that it is activity in this whole interlinked system that has been selected for its specifically cognitive virtues. Neural systems coor-dinate with, help produce, exploit, and can themselves be entrained by those special-purpose bodily motions that constitute free gestures. Such a system forms an organic whole, with problem-solving virtues not reducible to the virtues of any of its individual parts.

3   Loops as Mechanisms

At this point, a critic may attempt a more deflationary reading, suggesting that gesturing (or scribbling, etc.) is not *itself* part of the thinking or the cognitive process but rather merely *affects* it (perhaps by, e.g., "lightening the load" on the real (inner-neural) cognitive processes). To think otherwise, according to critics like Adams and Aizawa (2001) and Rupert (2004), is to make a causal–constitutive error, where that means to mistake inputs that causally act on a cognitive system for parts of the cognitive system itself. In this section, I show why such a deflationary move seems inappropriate in the case at hand.

We can start with a simple observation. A single integrated system can have a variety of distinct parts whose contributions to some overall process are hugely different. Some of those parts, moreover, may be cognitive processes in their own right (that is to say, they remain cognitive processes even when considered in isolation from the others) and others not. Just so, a sequence of physical gestures *alone* could never implement a cognitive state. It is only in coordination with crucial forms of neural activity that the cognitive role of the gestures can emerge and be maintained. By contrast, some set of neural goings-on is often sufficient for the presence of some cognitive state or other. But this (genuine) asymmetry provides no reason to reject the notion that gestures form part of the machinery of cognition. To see this, we need only remind ourselves that the activity of a single neuron is (likewise) never sufficient for the existence of a cognitive state, yet that activity can, in the proper context, still form part of the machinery that implements a cognitive state or process.

In addition, it may (or may not) also be true that for any gesture-involving cognitive unfolding, there is a pure sequence of neural events such that *if* they were somehow held in place, or ushered into being without the loop through physical gesture, the cognitive states of the embodied agent would be the same. It does not follow from this that the gestures play only a causal role and do not help constitute the machinery of cognition, for the same may also be true of a sequence of neural states held together by some internal operation. Achieve that very sequence some other way and the chain of thoughts, let's assume, will come out the same. It doesn't follow (and see Hurley 1998, ch. 8, for a full demonstration of this) that the inner or the outer operations involved are thereby not (as things actually unfold) genuine aspects of the cognitive process. Thus Susan Hurley (2010) usefully cautions against what she dubs "the 'causal-constitutive error' error," where this is

the error of objecting that externalist explanations give a constitutive role to external factors that are "merely causal" while assuming without independent argument or criteria that the causal/constitutive distinction coincides with some external/internal boundary. To avoid thus begging the question, we should not operate with prior assumptions about where to place the causal/constitutive boundary, but wait on the results of explanation. (Hurley 2010, 103)

In trying to get a grip on these matters, we are easily misled by various inessential features of many common cases where biological, external factors and forces impact thought and reason. Thus suppose the rhythmic pulse of rain on my Edinburgh window somehow helps the pace and sequencing of a flow of my thoughts. Is the rain now part of my cognitive engine? No. It is merely the backdrop against which my cognizing takes shape. But this, I submit, is not because the rain is outside the bounds of skin and skull. Rather, it is because the rain is not part of (it is not even a side effect or a "spandrel" within) any system either selected or maintained for the support of better cognizing. It is indeed *mere* (but as it happens helpful) backdrop. Compare this with a robot *designed* to use raindrop sounds to time and pace certain internal operations essential to some kinds of problem solving. Such a robot would be vulnerable to (non-British) weather. But it is not clear (at least to me) that the whole drop-based timing mechanism is not usefully considered as one of the robot's cognitive routines. Consider finally the Self-Stimulating Spitting Robot. This is a robot that evolved to spit stored water at a plate on its own body for the same purpose, so as to use the auditory signal as a kind of "virtual wire" (Dennett 1991, 196) to time other key operations. Those self-maintained cognition-supporting signals are surely part of the cognitive mechanism itself. A neural clock or oscillator would count, after all.

What these simple examples show is that (as we might expect) coupling alone is not enough. *Sometimes*, all coupling does is provide a channel allowing externally originating inputs to drive cognitive processing along. But in a wide range of the most interesting cases there is a crucially important complication. These are the cases where we confront a recognizably cognitive process, running in some agent, that creates outputs (speech, gesture, expressive movements, written words) that, recycled as inputs, drive the cognitive process along. In such cases, any intuitive ban on counting *inputs* as parts of *mechanisms* seems wrong. Instead, we confront something rather like the cognitive equivalent of a forced induction system. A familiar example is the turbo-driven automobile engine. The turbocharger uses exhaust flow from the engine to spin a turbine that spins an air pump that compresses the air flowing into the engine. The compression squeezes more air into each cylinder, allowing more fuel to be combined, leading to more powerful explosions (that drive the engine that creates the exhaust flow that powers the turbo). This self-stimulating automotive arrangement provides up to 40 percent more power on demand.

The exhaust flow is an engine output in good standing that also serves as a reliable, self-generated input. There can surely be little doubt that the whole turbo-charging cycle should count as part of the automobile's own overall power-generating mechanism.

The same is true, I submit, in the case of gesture. In gesture, as when we write or talk, we materialize our own thoughts. We bring something concrete into being, and that thing (in this case, the arm motion) can systematically affect our own ongoing thinking and reasoning. This is what Dennett (1991) calls the power of "cognitive self-stimulation": we can now see it as a kind of cognitive turbo-drive. The speech–neural processing–gesture triad allows gesture (like speech itself, so this is really a twin [differing] turbo system) to play the role of turbo for gesture is both a systemic output and a self-generated input[7] that plays an important and at time transformative role, as we saw, in an extended neural-bodily cognitive economy. The wrong image here is that of a central reasoning engine that merely uses gesture to clothe or materialize preformed ideas. Instead, gesture and overt or covert speech emerge as interacting parts of a distributed cognitive engine, participating in cognitively potent self-stimulating loops whose activity is as much an *aspect* of our thinking as its *result*.

This resolves an apparent tension highlighted earlier in the text. The tension concerned the delicate question of whether gesture "merely" "materialized thought" or actually helped to "accomplish" thought, to bring thought about. But we can now see that this is a false dichotomy. What the turbo-drive model helps show is that the very thing that materializes a thought can be the thing that helps move the cognitive process along. Gesture here emerges as both a means of bringing a thought to completion and an operator that helps transform one state of understanding into another.

4 An Open Question: True Neural "Screening Off"

The turbo-drive model of gesture is appealing, and it certainly helps explain the range of empirical data we scouted earlier in the chapter. It shows that the causal–constitutive objection is not necessarily valid in this case. But for all that, it remains possible that neural activity is actually doing all the cognitive work that I have been ascribing to the actual gestures themselves. This would be so if it turned out that all the cognitive benefits of gesture are actually secured by a covert, fully neural route. There are at least two reasons why folk might suspect this.

First, gesture is experienced by some subjects born with no arms (Ramachandran and Blakeslee 1998). Might they be gaining the "cognitive bonus" too? If so, then the bonus is secured not by the action of a gross-bodily material carrier that materializes thought, but by that of a more covert, fully inner circuit.

Second, consider patient IW (Cole, Gallagher, and McNeill 2002; see also Cole, this vol.). IW has no feeling below the neck and receives no proprioceptive feedback from his arms. But he now produces (even in an artificially blind condition) delicately time-linked expressive, noniconic gestures in what seems to be a spontaneous, often unconscious fashion. If IW gains cognitive benefits in the artificially blind condition, it must be via some covert, purely neural route.

Such a route might involve the direct action of an efference copy of the nonfunctional motor command (Wolpert et al. 1998), or it might reflect "pre-motor preparatory processes involved in the generation of the gestural movement rather than from the gestural movement itself" (Cole, Gallagher, and McNeill 2002, 62). The same kind of deflationary possibility attends McNeill's appeal to "Mead's Loop" discussed earlier. Is the mirror-neuron stimulation thus posited actually achieved by the real loop into physical gesture, or might it be supported by a wholly neuro-internal route, for example, via some internal emulation circuit (Grush 2004) that predicts what the gesture would be like (including any proprioceptive feedback that might arise) were we to make it?

This looks to be an open empirical question. If gesture arises as a result of our urge to express ourselves, that fact alone explains the presence of gesture in the phantom case and the reemergence of gesture in IW. But if actual physical gesturing then confers additional cognitive benefits via the "turbo-drive" route I suggested, then we should not see these same benefits in those cases. At the very least, we should not see them to the same degree (unless, of course, some *other* gross-bodily strategy rushes in to compensate, such as mouth gestures, or micro-body motions, etc.).

The situation is complex, however, as the following simple comparison shows. Consider the case of writing while thinking. It may help our thinking to try to write things down, even in the dark, and even if we couldn't feel the motion of our hands. But the benefits may increase if (unlike IW) we can feel our hands moving, thus adding a self-stimulating loop. And they may increase still further if we can also see what we are writing as we write it, completing another self-stimulating loop (but imagine doing this using rapidly disappearing ink). Finally, we may get still further benefits if the traces persist longer, so that self-stimulating loops can be spread out in various ways across time and space.

The upshot is that there remains a need to check experimentally whether IW and/or phantom gesturers really do get some cognitive bonus despite lack of limbs or propriceptive feedback. If they do get it, do they get it to the same degree? And if this is the case, does that establish the inner-route model or does it merely reflect the emergence of some alternative, compensatory, but still gross-body-involving loop (e.g., signers have been shown to use some mouth motions as a form of spontaneous noniconic gesture; see Fontana 2008)?

5 Conclusions: Minds, Watches, and Gestures

I'd like to end with a small confession: I don't really much care how we end up using the term "cognitive" or "cognitive system." What matters is to explore the many ways in which human thought and reason are potentially altered and empowered by the complex and temporally nuanced interplay between what we do with our bodies and what we do with our brains. I don't want to claim that minds (unlike brains) have parts or (a fortiori) that my arms (not to mention my iPhone; see Fodor 2009) are parts of my mind.

Rather, I would follow the lead of Gilbert Ryle (1949/1990) and Daniel Dennett (1987) in considering the concept of mind as being more like the concept of timekeeping than like the concept of a thing that keeps time (e.g., a watch). Timekeeping does not have parts—neither internal nor external. Watches have parts. Brains have parts. Bodies have parts. And parts of brains, and possibly parts of nonbrainy bodies, contribute, often in subtle and unexpected ways, to the skills and capacities we identify as mental and as cognitive.

The real moral, then, is just that we should be alert to the vast potential for cognitively potent self-stimulation created by the fact that we are:

mobile active beings;
replete with limbs/effectors;
richly endowed with sensory systems;
with vast swathes of neural tissue devoted to the detection (interoception) of our own bodily states.

As such, and given that we evolved by natural means hugely insensitive to demands of neatness and explanatory hygiene, we should consider gesture on its own terms, uninfected by the meme that "minds are simply what brains do."

Acknowledgments

This chapter includes some material that first appeared in Clark 2008. Thanks to Oxford University Press for permission to use that material here. Thanks to Zdravko Radman and all the participants at "The Hand: An Organ of the Mind: Pointing, Touching, Grasping, Gesturing, and Beyond," a meeting held in 2009 at the Interuniversity Centre, Dubrovnik, Croatia. Special thanks to Michael Wheeler and Shaun Gallagher for many invaluable discussions of these issues.

Notes

1. There is also rumored to be some older work in which children were simply told to sit on their hands, thus effectively removing the gestural option without adding to the memory load.

2. Much of Goldin-Meadow 2003 is devoted to the task of systematically attributing meaning to spontaneous free gestures. See also McNeill 1992, 2005.

3. The exchange is quoted in Gleick 1993, 409. Thanks to Galen Strawson for drawing this material to my attention.

4. After G. H. Mead. See Mead 1934.

5. There is, unfortunately, substantial ambiguity in the notion of "thinking" invoked in many of these discussions of gesture, since it can sometimes mean (1) "verbal thought," which is conceived, by Goldin-Meadow, as distinct from (though intertwined with) (2) the kinds of holistic, imagistic thinking specifically accomplished by gesture. Finally, there is (3) the overall cognitive state achieved by an agent who has engaged in some ongoing process involving both gestural and verbal elements. To say that gesturing is part of the process that constitutes thinking is thus to say both that it helps mediate and inform the verbal thinkings, and that in so doing it forms part of a larger integrated cognitive system.

6. The use of "thought" here is misleading (see note 5). It reflects common usage rather than the actual model than McNeill and others develop. There is a similar ambiguity in the use of "language," since gesturing, on McNeill's account, is actually part of language. McNeill is aware of these infelicities but thinks the usage will do no harm: see "Terminological Tango" on page 21 of McNeill 2005.

7. In the case of gesture, the relation between the self-created inputs and other processing elements also looks to involve the full complexities of "continuous reciprocal causation" as discussed in Clark 1997, 163–166.

References

Adams, F., and K. Aizawa. 2001. The bounds of cognition. *Philosophical Psychology* 14 (1):43–64.

Clark, A. 1997. *Being There: Putting Brain, Body, and World Together Again.* Cambridge, MA: MIT Press.

Clark, A. 2008. *Supersizing the Mind: Embodiment, Action, and Cognitive Extension.* New York: Oxford University Press.

Clark, A., and D. Chalmers. 1998. The extended mind. *Analysis* 58 (1): 7–19. Reprinted in *The Philosopher's Annual*, vol. 21, ed. P. Grim, 1998. Also reprinted in *Philosophy of Mind: Classical and Contemporary Readings*, ed. D. Chalmers. Oxford: Oxford University Press, 2002. Also reprinted in *The Extended Mind*, ed. R. Menary. Cambridge, MA: MIT Press, 2010.

Cole, J., S. Gallagher, and D. McNeill. 2002. Gesture following deafferentation: A phenomenologically informed experimental study. *Phenomenology and the Cognitive Sciences* 1 (1):49–67.

Dennett, D. 1987. *The Intentional Stance.* Cambridge, MA: MIT Press.

Dennett, D. 1991. *Consciousness Explained.* Boston: Little, Brown.

Fodor, J. 2009. Where is my mind? Review of *Supersizing the Mind* by Andy Clark. *London Review of Books* 31 (3):13–15.

Fontana, S. 2008. Mouth actions as gesture in sign language. *Gesture* 8 (1): 104–123.

Gallagher, S. 2005. *How the Body Shapes the Mind.* Oxford: Oxford University Press.

Gleick, J. 1993. *Genius: The Life and Times of Richard Feynman.* New York: Vintage.

Goldin-Meadow, S. 2003. *Hearing Gesture: How Our Hands Help Us Think.* Cambridge, MA: Harvard University Press.

Goldin-Meadow, S., H. Nusbaum, S. Kelly, and S. Wagner. 2001. Explaining math: Gesturing lightens the load. *Psychological Science* 12:516–522.

Goldin-Meadow, S., and S. Wagner. 2004. How our hands help us learn. *Trends in Cognitive Sciences* 9 (5):234–241.

Grush, R. 2004. The emulation theory of representation: Motor control, imagery, and perception. *Behavioral and Brain Sciences* 27:377–442.

Hurley, S. 1998. *Consciousness in Action.* Cambridge, MA: Harvard University Press.

Hurley, S. 2010. The varieties of externalism. In *The Extended Mind*, 101–154, ed. R. Menary. Cambridge, MA: MIT Press.

Iverson, J., and E. Thelen. 1999. Hand, mouth, and brain. In *Reclaiming Cognition: The Primacy of Action, Intention, and Emotion*, 19–40, ed. R. Núñez and W. J. Freeman. Bowling Green, OH: Imprint Academic.

Iverson, J., and S. Goldin-Meadow. 1997. What's communication got to do with it? Gesture in blind children. *Cognitive Development* 9:23–43.

Iverson, J., and S. Goldin-Meadow. 1998. Why people gesture when they speak. *Nature* 396:228.

Iverson, J., and S. Goldin-Meadow. 2001. The resilience of gesture in talk. *Developmental Science* 4:416–422.

Mead, G. H. 1934. *Mind, Self, and Society*. Ed. C. W. Morris. Chicago: University of Chicago Press.

McNeill, D. 1992. *Hand and Mind*. Chicago: University of Chicago Press.

McNeill, D. 2005. *Gesture and Thought*. Chicago: University of Chicago Press.

Ramachandran, V. S., and S. Blakeslee. 1998. *Phantoms in the Brain: Probing the Mysteries of the Human Mind*. New York: Morrow.

Rizzolatti, G., L. Fogassi, and V. Gallese. 2001. Neurophysiological mechanisms underlying the understanding and imitation of action. *Nature Reviews. Neuroscience* 2:661–670.

Rupert, R. 2004. Challenges to the hypothesis of extended cognition. *Journal of Philosophy* 101 (8):389–428.

Ryle, G. 1949/1990. *The Concept of Mind*. London: Penguin.

Vygotsky, L. S. 1962/1986. *Thought and Language*. Trans. A. Kozulin. Cambridge, MA: MIT Press.

Wolpert, D. M., R. C. Miall, and M. Kawato. 1998. Internal models in the cerebellum. *Trends in Cognitive Sciences* 2:338–347.

Wagner, S. M., H. C. Nusbaum, and S. Goldin-Meadow. 2004. Probing the mental representation of gesture: Is handwaving spatial? *Journal of Memory and Language* 50:395–407.

12   Is Cognition Embedded or Extended? The Case of Gestures

Michael Wheeler

1   Locating Thought

When we perform bodily gestures, are we ever literally thinking with our hands (arms, shoulders, etc.)? In the more precise, but correspondingly drier, technical language of contemporary philosophy of mind and cognition, essentially the same question might be asked as follows: Are bodily gestures ever among the material vehicles that realize cognitive processes? More precisely still, is it ever true that a coupled system made up of neural activity and bodily gestures counts as realizing a process of thought, in such a way that the gross bodily movements concerned should be granted cognitive status along with, and in essentially the same sense as, the neural activity?

This question, however it is phrased, is, I think, acutely interesting in itself, but it enjoys the added value of bearing on the truth or otherwise of an increasingly prominent (although still very much minority) view in contemporary thinking about thinking, a view known as the *extended cognition (or extended mind) hypothesis* (henceforth ExC). ExC states that there are actual (in this world) cases of intelligent action in which thinking and thoughts (more precisely, the material vehicles that realize thinking and thoughts) are spatially distributed over brain, body, and world, in such a way that the external (beyond-the-skull-and-skin) factors concerned are rightly accorded cognitive status. In this formulation, the term "cognitive status" signals whatever status it is that we usually grant the brain within mainstream (internalistic, nonextended) cognitive theory, in relation to the production and explanation of psychological phenomena. According to ExC, then, under the right circumstances, the material vehicles that realize thinking and thoughts include not only neural states and processes, but also nonneural bodily structures and movements and, crucially (see next paragraph), certain environmentally located elements. Paradigmatic

candidates for such environmental elements include a range of technological artifacts from notebooks and slide-rules to smartphones and other
examples of contemporary mobile and pervasive computing. (The canonical statement of ExC is due to Clark and Chalmers 1998. Clark's own recent
presentation and defense can be found in Clark 2008b. For a timely collection that places the original Clark and Chalmers paper alongside a
number of developments, criticisms and defenses of the view, see Menary
2010.)

For what it's worth, I am inclined to think that ExC is ultimately defen-
sible (see e.g., Wheeler 2010a) and so far more likely to be true than any
knee-jerk first impressions of implausibility might suggest. However, even
in the more revolutionary enclaves of contemporary cognitive theory, it is
the fan of ExC who bears the burden of proof. To see why, we need to
contrast ExC with a still radical but ultimately less revisionist position on
mind, which construes thinking not as extended, but as *embedded* (or *situated*). According to the embedded view, the distinctive adaptive richness
and flexibility of intelligent behavior is regularly, and perhaps sometimes
necessarily, causally dependent on (a) various nonneural bodily structures
and/or movements, and/or on (b) the bodily exploitation of certain environmental props or scaffolds. Thus, to adapt an example from Clark (1997),
consider an intelligent agent whose strategy for solving a jigsaw puzzle
involves the systematic deployment of a range of bodily actions. For
example, she might physically manipulate the pieces, either by rotating
them to help visually pattern-match for possible fits or by testing them
in the target position. Such problem-solving manipulations succeed by
exploiting certain as-it-happens useful environmental factors, such as the
geometric properties of the pieces. In the process, the external environment itself is effectively transformed into a readily available problem-
solving resource, one whose elements restructure the piece-finding problem
and thereby reduce the information-processing load placed on the inner
mechanisms in play.

As the jigsaw example shows, the embedded theorist seeks to register
the important, and sometimes perhaps even necessary, contribution made
by nonneural bodily and environmental factors to many cognitive outcomes. That said, the embedded position, as I shall understand it, is that
the actual thinking in evidence in such cases remains *either* a purely neural
phenomenon (one that is given a performance boost by its embodied
context and its technological ecology) *or* a phenomenon that is realized in
the brain and the nonneural body (but which is given a performance boost
by its technological ecology). These two variations on the embedded view

will be disentangled further in a moment. Right now, it is the distinction with ExC that concerns us. Here is the key point: If what we confronted in our jigsaw-solving scenario were a *full-blown* case of cognitive extension, as opposed to one of cognitive embeddedness, then not only the piece-manipulating bodily movements, but also the manipulated pieces themselves, would count, alongside and in the same fundamental sense as the active neural structures, as being among the material realizers of the thinking.

The case for embedded cognition has been made over and over again. (For what I now take to be my own book-length contribution to the cause, see Wheeler 2005, although in hindsight I realize that, in line with the philosophical *Zeitgeist* of that era, my treatment of the issues in that text fails to distinguish between the embedded view and ExC in a sufficiently clear or consistent manner.) Given that ExC remains controversial in a way that the embedded view mostly does not, I think it is fair to say that it is the latter view that currently deserves to be treated as the default position in the debate. That's why the burden of proof rests with the advocate of ExC. Against this background, and taking it largely (although not entirely) for granted that the contribution of bodily gestures to thinking will surely indicate at least the embeddedness of cognition, my goal in what follows will be to determine whether or not a careful consideration of gestures and their role in thought mandates the more radical conclusion that cognition is extended. Putting gestures center stage in this way immediately highlights a subtle but important complication (or ambiguity) that was touched on in the previous paragraph, and which is sometimes lost sight of in the literature. To bring this complication into clear view, consider the following, hypothetical turn of events: it turns out, when all the data and arguments are in, that although the properly cognitive part of our behavior-generating system includes nonneural as well as neural bodily factors, this process of cognitive spread stops at the skin. Under these circumstances, it seems to me that we would readily conceptualize cognition as a phenomenon that is embodied in previously unexpected ways, but not as a phenomenon that is extended in any truly interesting sense. That's why the formulation of ExC that I gave earlier is couched partly in terms of the cognitive status of beyond-the-skull-*and-skin* external factors.

If I am right about our inclinations here, then although the result that bodily gestures are among the realizing vehicles of cognition would certainly depose (what I shall call) the *conservative* embedded view, according to which extraneural bodily factors and environmental structures are no more than powerful props for wholly neural cognitive activity, it would not *necessarily* be enough to establish ExC. For, without further argument, what that result would establish would be "only" (what I shall call) the *radical* embedded view, according to which, although nonneural bodily factors sometimes count as being among the realizing material vehicles of cognition, the same is not true of any of the behavior-shaping factors located beyond the skin. What the fan of ExC needs to do, then, if she wants to build her case on gestures, is show that a robust implication of the very considerations that, when extracted from a careful analysis of gestures, bestow cognitive status on those gestures also bestow cognitive status on a range of extrabodily environmental elements. Only then, it seems, would the cognitive status of bodily gestures provide proper support for a genuinely interesting and distinctive thesis of extended cognition. (Of course, what I am calling the radical embedded view might equally be dubbed minimal ExC, or alternatively be classified as a thesis of profoundly embodied or body-centric cognition; for more on body-centrism in relation to ExC, see Clark 2008a; Wheeler 2011a. One's choice of terminology here depends on what aspect of the target position one wants to emphasize. In the end, what is important is that proper conceptual space is made for the position, not what we call it.)

With the substantive and somewhat lengthy scene-setting now over, it is time to get down to business. Here is where the rest of this chapter will take us. I shall begin (in section 2) by pointing to certain empirical studies of bodily gestures that, on the face of things, provide a hefty experimental nudge in the direction of the claim that such gestures are indeed material realizers (more accurately, partial material realizers) of thought processes. I shall argue, however, that once we focus with due critical caution on the conceptual borders between the embedded view (in whatever form it manifests itself) and ExC, it becomes far less obvious that these empirical considerations alone mandate the eviction of the default (i.e., the embedded) view. To open the door to considerations that might conceivably complete the transition to a full-strength extended cognition framework here, I shall first say rather more about how my favored version of ExC works (section 3). Then, on the basis of that improved understanding of ExC, I shall proceed (in sections 4 and 5) to examine an argument due to Clark (2008b; this vol.), which concludes explicitly that gestures are not merely props for subtly embedded but wholly neural cognitive activity, but instead are themselves literally part of cognitive processing, and then another argument, due to Gallagher (2005), which at least toys with that conclusion. In each case, I shall say why it is that even though the considerations on offer exhibit the right theoretical profile to make cognitive extension a genuine possibility, they ultimately fail even to secure the cognitive status of bodily gestures, which means that they don't tell in favor of ExC in contrast to the embedded view. Indeed, they leave us firmly in the grip of the conservative embedded view, rather than its radical cousin.

2 Experiments at the Edge of Embeddedness

What sort of empirical data on bodily gestures might lead us in the direction of ExC? Here is a brief selection of some intriguing observations and experimental results. (Andy Clark's contribution to the present volume already contains a detailed summary of the first of my three examples, so I shall not repeat all its finer points here. Readers who hanker after more nuances are referred to Clark's chapter. Clark also discusses work that is closely related to my second example.)

Goldin-Meadow et al. (2001; see also Goldin-Meadow 2003) took two matched groups of children and asked them first to memorize a list of words, then to carry out a mathematical problem-solving task, and then finally to recall the list. The children in one group were allowed to gesture freely during the mathematical problem-solving phase. The children in the other group were instructed not to gesture. The outcome was that the children in the gesturing group did significantly better at recalling the list.

In another experiment, Broaders et al. (2007; see also Goldin-Meadow 2009) asked children to gesture while explaining their answers to novel mathematical problems. Children who could not solve the problems, and who did not give any spoken indication of having appropriate problem-solving strategies available to them, nevertheless indicated the possession of such strategies via their gestures. In this case, speech and gestures are vehicles for different meanings, with gestures carrying problem-solving content that, in an intuitive sense, is in advance of that carried by speech. This divergence seems to promote, or at least can be used to predict, later learning, since when the children in this gesturing group were later given an appropriate mathematics lesson, they were more likely to learn how to solve the problems in question than were children who, when explaining their preliminary, preteaching answers, had been asked not to gesture.

As a final example, consider Hutchins's (2010) analysis of a video record of an embodied communicative interaction between a company instructor pilot and his student. The context for this interaction is a training exercise in a flight simulator, in which the task is to land an airplane using the Instrument Landing System approach. After the student has made a poor landing, the instructor turns to him and says "Want to do it again?" while performing a hand gesture on the words "do it." This gesture is interesting because it is not positioned over any of the controls or instruments on the simulated flight deck. It is a movement that starts at the current simulated position of the airplane on the runway and moves toward the back of the simulator, with the hand cupped and the palm facing backward. Hutchins's analysis concludes that the gesture is in fact coupled to the simulated flight deck and thereby to the imaginary environment of an imaginary airplane in a partly simulated and partly imagined airspace. To explain: The referent of the anaphora "it" in the spoken verb phrase "do it again?" is of course the flying of a simulated approach; but the referent of the accompanying gesture is a fictional event that would need to happen *before* another landing could be attempted, namely, taxiing back up to the approach end of the runway in order to take off again. Indeed, the simulated airplane, unlike a real airplane, can be instantaneously repositioned at the start of an approach path, so there is no need for even a simulated taxiing and take-off phase. The direction of the gestural movement thus indicates the direction that a real airplane would need to travel in order to get back in the air, while the orientation of the hand corresponds to the orientation of the simulated approach path, in relation to the simulated flight deck of the imaginary airplane, as parked on the runway after the previous landing.

How are we to interpret these data? Goldin-Meadow (2003) suggests that the best explanation for the first result reported above is that the gestures performed during the mathematical problem-solving task contribute actively to its execution. In this way, some of the cognitive load imposed by that task is transferred from the brain to the gestures, which leaves more of the available neural resources free for deployment on the accompanying memory assignment. In discussing the second result, Goldin-Meadow writes that "telling children to gesture . . . encourages them to convey previously unexpressed (and correct) ideas, which, in turn, makes them receptive to instruction that leads to learning" (Goldin-Meadow 2009, 108). Perhaps most striking of all, Hutchins argues that his instructor pilot's gesture "indexes a conceptual construct that is a precondition for the concept indexed by talk" (2010, 96), and does so by "bring[ing] into being a conceptual space [a kind of scale model of the space of the approach path] in which [the gesture] acquires its own meaning" (ibid., 98).

These various attempts to characterize the cognitive impact of gestures bring us tantalizingly close to the conclusion that at least some gestures are among the material realizers of thinking. In other words, we are currently located on the threshold of the radical embedded view and therefore firmly in the vicinity of ExC. Without further argument, however, it is not clear that the theoretical glosses in question actually force us to go beyond the conservative embedded position, for even if it is true that gestures contribute to thought in precisely the ways that Goldin-Meadow and Hutchins suggest they do, that still wouldn't compel us (even though it might warmly invite us) to go beyond the view that gestures are "no more than" powerful, noncognitive causal influences on good old-fashioned neurally instantiated thought. Here are some reflections that indicate why.

First, it seems clear that a processing or storage burden on my neural cognitive resources may be offloaded onto a nonneural resource without that latter resource *necessarily* acquiring cognitive status as a result. Consider, for example, the burden that used to be placed on my neural resources through the need to store important phone numbers in my brain. That load has more recently been relocated to my smartphone. And maybe that transferal has freed up some of my neural resources to do other things. But these facts *alone* surely don't endow my smartphone with cognitive status, although they might be part of such a story. This deflationary (with respect to ExC) moral would seem to apply in just the same measure to the case of a processing or storage load being shifted onto bodily gestures.

Second, perhaps the solution-anticipating gestures performed by the children in Broaders et al.'s experiment are, in truth, bodily expressions of problem-solving strategies that have been unconsciously arrived at and stored in those children's brains, but which the children concerned cannot yet express verbally. If this is indeed what is going on, then one might think that the gestures themselves are best understood as noncognitive windows on the state of those children's neural cognitive resources. As Clark (this vol.) puts this kind of point, perhaps "all the cognitive benefits of gesture are actually secured by a covert, fully neural route."

Finally, and here I confess that one has to squint with a certain amount of enthusiasm in order to ignore certain potentially disruptive details, compare (a) the way in which the gesture performed by Hutchins's instructor pilot creates a conceptual space for a shared imaginary scenario with (b) the way in which a software games designer or a contemporary film animator may use certain computational tools to create an imaginary world for public access. One intuitive gloss on (b), roughly analogous to the previously sketched offloading scenario, takes it that the computational tools available to the software world-builder constitute useful noncognitive props and scaffolds that enable him to reduce the cognitive burden on his neural resources as he designs his imaginary space. A second gloss on (b), roughly analogous to the previously highlighted view that gestures are merely public expressions of fully formed neurally realized cognitive strategies, takes it that the world-builder, whether consciously or not, has the design of his imaginary space fully represented in his brain, in advance of that design merely being transferred to, by virtue of being implemented in, the appropriate software environment. Arguably, either gloss provides us with a deflationary yet satisfying interpretation of (b). And it is surely plausible that whatever goes for our software world-builder goes for Hutchins's gesturing instructor pilot.

From what we have seen so far, the interim conclusion is this: The (interesting and significant) contributions that gestures undoubtedly make to thought remain amenable to the conservative embedded picture according to which cognition is entirely neurally realized, even though it is intimately situated in the wider body and the world. In short, gestures help us to think, but (in the relevant sense) they are not themselves part of thinking. So much the worse, it seems, for the attempt to build a case for cognitive extension on the basis of gestures. Perhaps, however, in our rush to find empirical evidence for gesture-based cognitive extension, we are simply looking in the wrong place for the crucial considerations. To investigate that possibility, we need to bring ExC itself—or at least my favored version of it—into better focus. This demands that we temporarily turn our attention away from our main topic of gesture and toward more general questions of cognitive extension. The eventual payoff from this detour, however, will be the uncovering of a theoretical structure, the so-called *mark of the cognitive*, that will enable us to see precisely what is needed by any gestures-based argument for ExC.

3   Riding the Waves

Broadly speaking, there are two prominent lines of thought in the ExC literature. Sutton (2010) dubs these the *first-wave* and the *second-wave* versions of the view. First-wave ExC is standardly characterized (by Sutton, among others) as emphasizing and defending the kinds of arguments for cognitive extension that were to the fore in the original Clark and Chalmers (1998) paper. Almost all of the attention here is concentrated on the much discussed (and much misunderstood) *parity principle*. Clark's recent formulation of this principle is as follows: "if, as we confront some task, a part of the world functions as a process which, were it to go on in the head, we would have no hesitation in accepting as part of the cognitive process, then that part of the world is (for that time) part of the cognitive process" (Clark 2008b, 77; drawing on Clark and Chalmers 1998, 8). So, the parity principle begins by asking us to consider an actual system that generates some psychologically interesting outcome (e.g., some example of intelligent action) and whose operation involves an important functional contribution from certain externally located physical elements. It then encourages us to imagine a hypothetical scenario in which exactly the same functional contribution, to an equivalent outcome, is made by certain internally located physical elements. Having taken this imaginative step, if we then judge that the internal realizing elements in the hypothetical case count as bona fide parts of a genuinely cognitive system, we are driven to conclude that the very same (i.e., cognitive) status should be granted to the external realizing elements in the actual, environment-involving case. After all, by hypothesis, nothing about the functional contribution of the target elements has changed. All that has been varied is the spatial location of those elements. And if someone were to claim that being shifted inside the head is alone sufficient to result in a transformation in status, from noncognitive to cognitive, he would, it seems, be guilty of begging the question against ExC.

By contrast, so-called *second-wave* ExC rejects, or at least downplays, the parity principle, in favor of considerations of either *complementarity* (Sutton 2010; Kiverstein and Farina 2011) or, in a closely related vein, *cognitive integration* (Rowlands 1999; Menary 2007). As Sutton (2010, 194) argues, "in extended cognitive systems, external states and processes need not mimic or replicate the formats, dynamics, or functions of inner states and processes," so "different components of the overall (enduring or temporary) system can play quite different roles and have different properties while coupling in collective and complementary contributions to flexible thinking and acting." Adding a further dimension to, or perhaps making explicit an existing dimension of, complementarity, the integrationists emphasize the processes by which internal and external elements with different properties may be combined into a single, although essentially hybrid, cognitive whole. Thus, they foreground factors such as the completion of cognitive tasks through the skilled manipulation of external elements, the transformation of our cognitive abilities through the learning of the manipulative skills just mentioned, and the application of norms of manipulation with a distinctively cognitive character (Menary 2007). The key, second-wave-defining commitment, however, is shared by complementarity theorists and their integrationist cousins. Put crudely, for the second-wavers, it's difference, not sameness, that matters.

Now, for my own part, I always did prefer the original punk bands to the so-called new wave bands that, in the context of the UK music scene anyway, came into being later (give me the Clash over the Cure any day), and things are not so different when it comes to ExC. So, although this is not the place to present a detailed critical discussion of second-wave ExC, here is the kernel of an objection. According to the second-waver, it is precisely the differences between certain internal and certain external elements that explain how many cognitive tasks are performed. For example, following Bechtel (1994), one might explain how some examples of mathematical problem solving are accomplished, by citing a complementary combination of internally located pattern-sensitive connectionist networks and externally located combinatorial symbol systems. One might even highlight the skilled embodied manipulation of the symbols in question, according to learned normative rules that, when mastered, transform what we can do. But none of this, as far as I can see (and here I am echoing the deflationary attitude to gestures that appeared earlier in our discussion), compels us to adopt ExC. At root, the problem for the second-wave theorist is this: Once we remind ourselves that it is the embedded view that is the default position in the debate, and once we allow the inner elements in some proposed complementary combination of interest to have cognitive status (which, given my understanding of the notion of cognitive status, is not to say that the inner is necessarily always a self-standing cognitive system in its own right or that the cognitive is defined by whatever the inner does—see below), it seems that the second-wave emphasis on the existence of theoretically significant differences between the internal and the external elements in question creates a conceptual ravine between, on the one hand, the undoubtedly important phenomena of complementarity and integration and, on the other, cognitive extension. As Rowlands (2010, 90) puts it at the culmination of a similar line of reasoning, "given that there are significant differences between internal cognitive processes and external processes involved in cognition, why not simply suppose that the latter are part of the extraneous scaffolding in which the real, internal cognitive processes are *embedded*?" (In the interests of completeness, I should note that while Rowlands's [1999] earlier work is standardly, and rightly, identified as one important source for integrationism, and thus as one wellspring of second-wave ExC, his more recent position [e.g., Rowlands 2010] has seen him argue that parity considerations [properly understood] and complementarity-integrationist thinking have equal weight in the justification for cognitive extension. Rowlands's recent approach to parity is an issue to which I shall return briefly below.)

With at least the shape of an objection to second-wave ExC duly recorded, let's refocus our attention on the first-wave version and the parity principle. Notice that, as stated above, the parity principle depends on the notion of *multiple realizability*, the idea that a single type of mental state or process may enjoy a range of different material instantiations. This dependence becomes visible (Wheeler 2011a) once one recognizes that the all-important judgment of parity is based on the claim that it is possible for the very same cognitive state or process to be available in two different generic formats—one nonextended and one extended. Thus, in principle at least, that state or process must be realizable in either a purely organic medium or in one that involves an integrated combination of organic and nonorganic structures. In other words, it must be multiply realizable. So, if we are to argue for cognitive extension by way of parity considerations, the idea that cognitive states and processes are multiply realizable must make sense. Now, one of the first things that undergraduate students taking philosophy of mind classes get taught is that the philosophical position known as *functionalism* provides a conceptual platform for securing multiple realizability. According to functionalism, as I shall understand it here, what matters when one is endeavoring to identify the specific contribution of a state or process *qua* cognitive contribution is not the material constitution of that state or process, but rather the functional role it plays in generating cognitive phenomena, by intervening causally between systemic inputs, systemic outputs, and other functionally identified, intrasystemic states and processes. This is the kind of functionalism that, I think, still deserves to be called the house philosophy of mind in cognitive science. For example, computational explanations of mental phenomena, as pursued in, say, most branches of cognitive psychology and artificial intelligence are functionalist explanations, in this sense.

As an aside, notice that I am not proposing functionalism as a way of specifying the constitutive criteria that delineate the mental states that figure in our prescientific, folk (i.e., commonsense) psychology. Thus, for example, I am not advocating functionalism as a way of specifying what it is for a person to be in pain, as we might ordinarily think of that phenomenon. That ambitious brand of functionalism faces a range of well-documented objections that are widely thought to be fatal to the view. (For an introduction to the main lines of argument, see, e.g., Levin 2010.) It strikes me, however, that the functionalism that I have advocated in relation to ExC—call it *cognitive-scientific functionalism*—is pretty much immune to the objections traditionally leveled at the more ambitious project. Proper analysis and argument would undoubtedly be needed to make this point stick, but good empirical evidence for such an immunity is provided by the fact that the functionalist (in my sense) brands of artificial intelligence and cognitive psychology have not ground to a halt in response to the (alleged) failure of the folk-psychology-oriented version of the position.

Back to the main plot: Because a function is something that enjoys a particular kind of independence from its implementing material substrate, a function must, in principle, be multiply realizable, even if, in this world, only one kind of material realization happens to exist for that function. What this tells us is that functionalism is sufficient for some kind of multiple realizability. But is it *the right sort* of multiple realizability to support ExC? Historically, of course, the guiding cognitive-scientific assumption has been that the economy of functionally identified material states and processes that causally explain psychological phenomena will be realized by the nervous system or, in hypothetical cases of minded robots or aliens, whatever the differently realized counterpart of the nervous system inside the bodily boundaries of those cognitive agents turns out to be. In other words, functionalism has standardly been used to secure (what we might call) *narrow* (within-the-skin) multiple realizability. In truth, however, there isn't anything in the letter of functionalism as a generic philosophical framework that mandates this exclusive focus on the inner (Wheeler 2010a,b). After all, what the functionalist schema demands of us is only that we specify the causal relations that exist between some target element and a certain set of systemic inputs, systemic outputs, and other functionally identified, intrasystemic elements. There is no essential requirement that the outer boundary of the system of interest must fall at the organic sensorimotor interface. In other words, in principle at least, functionalism straightforwardly allows for the existence of material cognitive systems whose borders are located at least partly outside the skin. To put a different slant on this point, functionalism straightforwardly allows for (what we might now call) *wide* (both within-the-skin and beyond-the-skin) multiple realizability. And that is precisely what the first-wave ExC theorist needs in order to run the argument from parity. So, although I am happy to adopt Clark's term *extended functionalism* for the version of ExC that I favor (Clark 2008a,b; see also Wheeler 2010a,b, 2011a), it is instructive to note that the term "extended" in extended functionalism refers to the spatial limits of the material cognitive system of interest and not, as might perhaps be thought, to the theory of functionalism. In other words, the possibility of cognitive extension is not something that requires any theoretical enhancement of (any theoretical extension to) conventional cognitive-scientific functionalism. Rather, the possibility that ExC is true is a straightforward consequence of such functionalism. Put another way, if conventional cognitive-scientific functionalism is true, then ExC is not in any way conceptually confused, even though it may be empirically false. (For an argument that concludes that extended functionalism leads to a deadlock between ExC and its critics, see Rowlands 2010; for an argument that concludes that extended functionalism is committed to an excessively liberal notion of cognitive extension, see Sprevak 2009; for replies to both these arguments, see Wheeler 2010a.)

So far we have uncovered two conceptual components of (what I take to be) the most plausible form of ExC. But neither the parity principle nor functionalism, nor even the two of them combined, can carry the case for cognitive extension. What is needed, additionally, is an account of precisely which functional contributions count as cognitive contributions and which do not. After all, as the critics of ExC have often observed, and as the officially pro-ExC complementarity arguments canvassed above point out, there will undoubtedly be *some* functional differences between extended cognitive systems (if such things exist) and purely inner cognitive systems. For instance, our purely inner organic memory systems ordinarily exhibit primacy and recency effects that extended memory systems plausibly would not, or at least need not (Adams and Aizawa 2008; see Rupert 2009 for similar observations, and Wheeler 2010a, 2010b for discussion). So, faced with the task of deciding some putative case of parity, we will need to decide which, if any, of those functional differences matter. I see no secure way of doing this, except by providing what Adams and Aizawa (2008) have dubbed a *mark of the cognitive*, a scientifically informed account of what it is to be a proper part of a cognitive system that, so as not to beg any crucial questions, is fundamentally independent of where any candidate element happens to be spatially located (Wheeler 2010a,b, 2011a,b). This way of explicating the basic idea of a mark of the cognitive specifies certain general conditions of adequacy that any particular suggestion for such a mark would need to meet. Of course, once a candidate mark of the cognitive has been placed on the table (i.e., once we have an account of what it is to be cognitive that meets the proposed adequacy conditions), further philosophical and empirical legwork will be required to find out (i) whether that account is independently plausible, and (ii) just where cognition (so conceived) falls—in the brain, in the brain and the nonneural body, or, as ExC predicts will sometimes be the case, in a system that extends across brain, body, and world.

In contrast to the foregoing picture, Clark has argued that the fan of ExC should shun the idea of a mark of the cognitive (as I have characterized it), in favor of "our rough sense of what we might intuitively judge to belong to the domain of cognition" (Clark 2008b, 114). Since this disagreement will figure, albeit in a minor way, later in our discussion, it is worth just pausing to introduce its main currents. According to Clark, judgments about whether or not some distributed system counts as an extended cognitive system should not be constrained by any scientific account of cognition, since such accounts are standardly "in the grip of a form of theoretically loaded neurocentrism" (Clark 2008b, 105). Rather, those judgments should be constrained by our everyday, essentially prescientific sense what counts as cognitive, since the "folk [i.e., commonsense] grip on mind and mental states . . . is surprisingly liberal when it comes to just about everything concerning machinery, location, and architecture" (ibid., 106). As a strategy for identifying cognitive structures and systems that doesn't immediately beg the question against ExC, Clark's appeal to folk intuitions strikes me as misguided. Indeed, as far as I can see, our ordinary folk practices of mental state attribution strongly presume the within-the-skin internality of our cognitive machinery (see Wheeler 2011b for the detailed argument; and see Clark 2011 for a response). Moreover, there is every reason to believe that it is possible to extract, from the functionalist varieties of artificial intelligence and cognitive psychology, certain candidate marks of the cognitive that, in their fundamental theoretical commitments, if not in the way they have actually been applied in practice, have not been seduced by any theoretically loaded neurocentrism (see, e.g., Wheeler's 2011a interpretation of Bechtel's aforementioned hybrid models of mathematical competence, as realizing extended physical symbol systems). This is, of course, precisely as one would expect, if my foregoing reflections on cognitive-scientific functionalism are on the right track.

At this point in the proceedings, something interesting happens. We are currently courting one of those Wittgensteinian ladder-discarding moments, in which we are invited to throw away one of the very theoretical supports that got us to where we are. To explain: The role that is now being played, in our explication of ExC, by the idea of a mark of the cognitive threatens to remove the need for any appeal to parity. After all, if a candidate extended solution satisfies some agreed mark of the cognitive, it looks as if the argument for ExC is already complete and parity considerations are made redundant, meaning that the parity principle itself may be jettisoned. (Thanks to Peter Sullivan for pressing me on this point.) One response here would be to agree that, strictly speaking, the parity principle is not required as the engine room of first-wave ExC, but to point out that nevertheless it may continue to be a useful heuristic mechanism that helps to ensure equal treatment for different spatially located systems judged against an unbiased and theoretically motivated standard of what counts as cognitive. In short, it is a helpful bulwark against what Clark (2008b, 77) calls "biochauvinistic prejudice." Another response would be to point out that a perfectly reasonable notion of parity may continue to figure in the method by which we appeal to our mark of the cognitive. Thus, some purely inner solution to a cognitive problem and some alternative, extended solution to the same problem may be judged to enjoy *parity with respect to a particular mark of the cognitive* (Wheeler 2010a, 2011a,b; cf. Rowlands 2010, 90).

With our updated understanding of parity in hand, the first-wave ExC theorist is well positioned to accommodate, within her theory, the kinds of differences in the formats, dynamics, and functions of purely inner and extended systems that so impress both the second-wavers and some of ExC's critics. The only differences between internal and external elements that, in the light of our revised parity considerations, would count against ExC would be those that, according to some particular mark of the cognitive in force, are stationed at the conceptual boundary between the cognitive and the noncognitive (Wheeler 2010a,b; cf. again Rowlands 2010, 90, although it should be noted that Rowlands doesn't frame his project as a defense of first-wave ExC, but as the development of a view that he calls the "amalgamated mind," which is designed to subsume embodied and extended approaches).

On the strength of the foregoing reflections, the notion of a mark of the cognitive can now be added alongside functionalism and the (properly understood) parity principle, in order to give us the conceptual profile of (the most plausible form of) ExC. Within this profile, a certain theoretical priority needs to be granted to the notion of a mark of the cognitive (or perhaps to the notion of a structured set of marks of the cognitive—one should admit that things are likely to get complicated). ExC might well be able to survive without the parity principle (hence our ladder-discarding moment) and without functionalism (which is certainly sufficient but arguably not necessary for multiple realizability). But it is less clear that ExC could survive without the notion of a mark of the cognitive, given the way in which, without that notion in play, deflationary conservatism always favors an embedded view. With that in mind, we can now be quite specific in stating what, in the overall context of this chapter, is the key issue. Any case for ExC based on gestures needs to provide a mark of the cognitive that has the consequence that bodily gestures are rightly counted among the material vehicles that realize cognitive states and processes. Moreover, to move us beyond the radical embedded view to full-blown cognitive extension (see above), that mark of the cognitive must plausibly allow various environmental elements to be accredited with cognitive status. In the next two sections, I shall consider two accounts from the recent embedded-extended literature that, although they do not frame things in quite the way I just have, may ultimately be interpreted as offering us just such a mark of the cognitive. In each case, however, I shall argue that even though the account in question is in the right theoretical ballpark for ExC—in the sense that if it did establish the cognitive status of gestures, it would generalize so as to secure, in addition, the cognitive status of various environmentally located elements—it fails to reach first base, because it does not establish the cognitive status of gestures.

4 Cognitive Self-Stimulation

Clark (2008b, this vol.) argues that the best way to understand the contribution of gestures to thought is to depict such activity as a form of cognitively potent *self-stimulation*. Here is the idea. In gesturing activity, neural systems are (partly) causally responsible for producing certain special-purpose bodily movements that are then recycled as inputs to those and/ or other neural systems. This feedback process sustains sophisticated brain-body loops of exploitation, coordination, and mutual entrainment, with various problem-solving benefits. In short, gestures are self-generated inputs to neural processing that (to borrow one of Clark's own analogies, more on which below) turbo-charge thought. In the interests of completeness, it is worth noting that Clark usefully distinguishes between three different species of cognitive self-stimulation: *fully anarchic, semi-anarchic,* and *centrally controlled* (see Clark 2008b, 131–135). Cognitive self-stimulation is fully anarchic when it is realized by "a vast parallel coalition of more or less influential forces" whose unfolding is "largely self-organizing" (ibid., 131). It is semi-anarchic when it is realized by a system of loosely coupled autonomous mechanisms that, through purely local control protocols, are capable of exploring divergent trajectories without destructively interfering with each other. And it is centrally controlled when the self-stimulating activity of the system is orchestrated by some privileged executive system, canonically located in the brain. Clark argues that gestural cognitive self-stimulation could not be fully anarchic, because the gestural and verbal reasoning systems often require the "protection" afforded by loose coupling, in order to explore different spaces of thought. (We saw experimental evidence of this effect earlier, in the experiment conducted by Broaders et al. [2007], in which children gestured while explaining their answers to novel mathematical problems.) Having ruled out the anarchic option, Clark allows that cognitive self-stimulation may be either semi-anarchic or centrally controlled, with neither of these possibilities having any detrimental implications for the claim that gestures are among the material vehicles of cognition. In particular, he writes that it is "open to even the staunchest fan of central control to endorse [ExC]" (Clark 2008b, 244, n. 27). Since Clark is, at root, neutral between the semi-anarchic and centrally controlled options, I shall not distinguish between them here and so will refer henceforth simply to the process of cognitive self-stimulation.

Clark's compelling image of self-generated bodily movements acting as components in cognitively self-stimulating loops nicely captures what is distinctive about the kinds of gestures that, as we saw earlier, may plausibly enable the reallocation of neural resources, by soaking up some of the overall processing load, or may encode verbally inexpressible problem-solving strategies in ways that promote future learning. Less obviously, perhaps, the notion of cognitive self-stimulation also helps us to understand the case of Hutchins's gesturing instructor pilot. Recall that, in this example, while the instructor's speech serves to index a future simulated approach flight (that's the reference of the word "it" in the sentence "Want to do it again?"), the gesture that accompanies those words indexes the wholly imaginary event of the fictional airplane taxiing back up the runway in order to take off again. This imaginary event constitutes a precondition for the future simulated approach. Of course, unlike our other two examples of gesturing, the instructor's bodily motion has a communicative function. But that shouldn't blind us to the fact that the shared conceptual space it establishes also creates a platform for some otherwise unavailable individual reasoning. Hutchins touches on this very point when he notes that the "gesture depicts a fictional event that facilitates reasoning" (2010, 97), and that it does so by bringing online complex visualization skills that instrument-rated pilots possess, and which support thinking and planning not only in flight simulators but also when, for example, the view out of a real airplane is obscured by clouds. The instructor's gesture may thus be seen as reflexively engaging his own visualization skills, as well as communicatively engaging those of his student.

What seems clear, then, is that gestures may act as self-generated aids that enhance thought in subtle and powerful ways. The question for us, though, is this: Does the fact that gestures realize the phenomenon of cognitive self-stimulation provide ExC-related support for the claim that gestures are among the material vehicles that instantiate psychological processes? There seems little doubt that Clark thinks the answer to this question is "yes." Indeed, he introduces his treatment of gesture as a "worked out example of extended cognizing in action" (Clark 2008b, 123). For Clark, then, cognitive self-stimulation is supposed to be the route by which gestures attain cognitive status. Unfortunately, once we look closely at the details of Clark's argument, it is far from obvious that he delivers on this promise.

First, let's deal with what is ultimately a misguided objection to Clark's position. It wouldn't do for a critic here to insist that, because a self-generated input that figures in a self-stimulating loop remains an input, it cannot count as cognitive. Without further argument, this would only beg the question against Clark, who maintains that, *within the kind of self-stimulating loops at issue*, the fact that a self-generated input has the status of an input to a self-standing cognitive system (the neural system) is not a barrier to that input enjoying cognitive status. As Clark himself puts it, in a passage quoted in full below, "in such cases [i.e., of cognitive self-stimulation], any intuitive ban on counting *inputs* as parts of [cognitive] *mechanisms* seems wrong" (Clark 2008b, 131). This point is reinforced by the analogy that Clark draws with a turbo-driven car engine, in which the exhaust flow is both an output and a self-generated input. There "can be little doubt that the whole turbocharging cycle [including the exhaust flow] should count as part of the automobile's own overall power-generating mechanism" (ibid.). Similarly, we are invited to think, there should be little doubt that the whole cognitive self-stimulation loop, including the self-generated input (e.g., a pattern of bodily gestures), counts as part of the thinker's own cognitive mechanism. Interpreted carefully, as the specific claim that the self-generated input's standing as an input is, in itself, no impediment to it being awarded cognitive status, this point seems to me to be well taken. The problem with Clark's treatment of cognitive self-stimulation must reside elsewhere.

To see where the problem lies, we need to pay detailed attention to precisely what Clark says about the character and the theoretical consequences of cognitive self-stimulation. Clark's cornerstone claim is that the "key distinction between 'merely impacting' some inner cognitive process and forming a proper part of an *extended* cognitive process looks much less clear . . . in cases involving the systematic effects of *self-generated* external structure on thought and reason" (Clark 2008b, 126). To keep things in line here, we need to situate this claim in relation to the three-way distinction that I have been adopting, between the conservative embedded view, the radical embedded view, and ExC. The first thing to note is that, in the target passage from Clark, the term "external structure" refers to extra*neural* factors, so cognitive extension will be automatically secured if extraneural factors count among the material realizing vehicles of cognition. One might think that this simply elides the distinction, on which I have been insisting, between radical embeddedness and cognitive extension. However, because Clark is a good functionalist, he takes cognitive self-stimulation to be a process with respect to which it is fundamentally *irrelevant* whether the self-generated structure is a nonneural bodily factor or an environmental element. What this means is that, *if* the considerations he introduces support the radical embedded view, they will thereby support ExC, which is, of course, just another way of saying that Clark's appeal to cognitive self-stimulation is in the right theoretical ballpark to underpin ExC. Because Clark's position has this structure, it will be safe for us to explicate and interrogate that position largely in terms of a two-way distinction between the conservative embedded view and ExC. For the rest of this section, then, I shall refer to the radical embedded view only when it is helpful to do so.

Using the conceptual machinery that we have just set out, we can see how Clark's cornerstone claim applies to bodily gestures in relation to thought. Clark's suggestion is that once the phenomenon of cognitive self-stimulation is brought into correct theoretical view, the idea that there is a clear boundary between the conservative embedded account of gestures (as noncognitive factors that have an impact on neurally realized thought) and the extended account of gestures (as material vehicles of cognition) becomes far less compelling. It is at this point that the problem with Clark's position comes into focus. Strictly speaking, what Clark argues is not that, where cognitive self-stimulation is in evidence, we have a case of cognitive extension, but instead that, where cognitive self-stimulation is in evidence, *there is no clear distinction between the conservative version of cognitive embeddedness and cognitive extension*. Now, one might reasonably wonder why this constitutes an argument for the conclusion that ExC is to be preferred over the conservative embedded view, rather than an argument for the conclusion that there are circumstances in which whether one describes an action as a case of conservative cognitive embeddedness or as a case of cognitive extension is ultimately a matter of intellectual temperament rather than metaphysical correctness. Of course, Clark's own description of his treatment of gesture (see above) strongly indicates that this sort of even-handed outcome is not what he intends. But, at first sight anyway, that seems to be what he offers us.

Maybe we are missing something. Perhaps the right reconstruction of the target reasoning is this: in cases of cognitive self-stimulation, (i) the distinction between conservative embeddedness and cognitive extension is eroded in such a way that whatever evidence there is that tells in favor of the conservative embedded view tells equally in favor of ExC, and (ii) under such circumstances, we are theoretically permitted to adopt ExC. Let's consider subclaims (i) and (ii) in turn. One worry about subclaim (i) is that it threatens to flout a point that has been made forcibly and repeatedly by Adams and Aizawa (e.g., 2008) in the ExC literature, namely that the bare causal dependence of thought and reason on external factors—even when that dependence is of a necessary kind or indicates that the systems in question are closely causally coupled—is simply not sufficient for genuine cognitive extension. What is needed, in addition, is a relation of *constitutive* dependence. That is, it must be that external factors don't merely exert a causal influence on, but rather partly constitute, the realizing base of cognition. Given this, the worry for Clark's subclaim (i) is that the explicitly argued-for erosion of the distinction between conservative embeddedness and cognitive extension is being purchased using the dubious currency of a tacit and un-argued-for erosion of the more fundamental distinction between causal and constitutive dependence.

How might Clark respond? There is textual evidence that suggests that he believes the very same empirical evidence of cognitive self-stimulation that supposedly undermines the embeddedness-extension distinction also undermines the causal-constitutive distinction. Thus, in a passage from which we have already quoted, he writes:

Sometimes, all coupling does is provide a channel allowing externally originating inputs to drive cognitive processing along. But in a wide range of the most interesting cases, there is a crucially important complication. These are the cases where we confront a recognizably cognitive process, running in some agent, that creates outputs (speech, gesture, expressive movements, written words) that, recycled as inputs, drive the cognitive process along. In such cases, any intuitive ban on counting *inputs* as parts of [cognitive] *mechanisms* seems wrong. (Clark 2008b, 131)

As I interpret this passage, Clark's contention is that the empirical fact of cognitive self-stimulation establishes a context in which the causal-constitutive (or coupling-constitution) distinction has no force. If cognitive self-stimulation did indeed have this effect, my objection to subclaim (i) would lose its force. But surely, as it stands, Clark's contention threatens to gets things precisely back to front, for, given that we are in the business of pursuing a distinction-collapsing strategy, it is only by collapsing the

causal-constitutive distinction that it becomes possible to take the empirically identified causal contribution of the self-generated inputs in question to be evidence of their constitutive status as material vehicles that
partly realize thought. If my reasoning here is on track, then what Clark
owes us, but doesn't give us, is an argument for collapsing the causal-
constitutive distinction *that is independent of the empirical fact of cognitive
self-stimulation.*

Even if we ignore the problems with subclaim (i) of our reconstructed
version of Clark's reasoning, subclaim (ii) is questionable. In effect, this is
the claim that, if the available evidence tells equally in favor of both the
conservative embedded view and ExC, then we are theoretically permitted
to adopt ExC. That would be true, of course, *if* the default position in the
debate were ExC, but it is hard to see just what, in the present philosophical and scientific climate, might justify that assessment of the relative
standings of the two theses. I pointed out earlier that, relative to ExC, the
case for the embedded view is already well established in the recent literature. And even if Clark responded to this observation by insisting that any
empirical support for the conservative embedded view ought to be inherited by ExC in cases of cognitive self-stimulation (which is a putative payoff
from subclaim (i)), the fact remains that ExC undoubtedly demands the
more significant revision to our scientific and philosophical approaches to
mind. In virtue of this genuine asymmetry, a perfectly reasonable theoretical inertia currently favors counting the less revisionist, embedded option
as the default view, with the conservative embedded view emerging as even
harder to shift than its radical embedded cousin. This is all bad news for
subclaim (ii).

If I am right, our reconstructed version of Clark's reasoning falls doubly
short. So, how might the appeal to cognitive self-stimulation be recon-
ceived, so as to avoid the difficulties just highlighted? The reasoning that
we have recently rejected fails partly because it either ignores or engages
unsuccessfully with the causal-constitutive distinction. What would it be,
therefore, to pay proper heed to, and thus to engage successfully with,
that distinction? It would be to provide a set of constitutive criteria for a
behavior-shaping element to count as a genuine material realizer of cognition. Readers who have been paying attention will right now be crying out
that we have met this idea already, in the guise of the mark of the cognitive. So, here is a suggestion that would revitalize the appeal to cognitive
self-stimulation: We should adopt the view that being the kind of self-
generated input that supports a process of cognitive self-stimulation is a
mark of the cognitive. If this understanding of cognitive self-stimulation

is warranted, then, given the experimental and observational evidence that we have already reviewed, to the effect that gestures are indeed best understood as self-generated inputs that support a process of cognitive self-stimulation, gestures will count among the material realizers of cognition.

As we have noted, the idea of a mark of the cognitive is something from which Clark has sometimes displayed a noticeable tendency to distance himself. It is worth just pausing, then, to register the fact that Clark's notion of what it is to be a self-generated input within a cognitively self-stimulating loop already meets the structural adequacy criteria identified earlier for being a mark of the cognitive. Thus, consider the following features of cognitive self-stimulation, as Clark thinks of it. First, it provides us with a *scientifically informed* proposal for identifying the cognitive: Clark motivates the notion by citing the kinds of experimental data on gesture that have been discussed in this chapter. Second, it's impressively *neutral regarding where the material vehicles that realize thought might be located*: Clark argues that not only bodily gestures, written words, and overt speech (see earlier quotation from Clark 2008b, 131) but also neurally realized inner speech (see Clark 2008b, 135) may be vehicles for such self-stimulation. Finally, it is deployed by Clark precisely to *distinguish the cognitive from the noncognitive*. In my mind at least, that settles it: If the payoff here were to be a persuasive gesture-based argument for ExC, Clark himself is well positioned to, and would have much to gain from, endorsing the view that being a self-generated input that supports a process of cognitive self-stimulation is a mark of the cognitive.

Unfortunately, a serious obstacle stands in the way of the proposed strategy. It seems that, just because some target element is a self-generated input that figures in a cognitive self-stimulating loop, that fact alone isn't *sufficient* for the element in question itself to count as cognitive, because it may very well make its turbo-charging contribution to thought while remaining noncognitive in character. The problem, then, is not that the self-generated inputs that figure in self-stimulating loops *cannot* be cognitive, but rather that an element may be a self-generated input that figures in a self-stimulating loop and *still not be cognitive*. This is just another way of saying that, even though a careful consideration of cognitive self-stimulation may give us a feature that meets the structural adequacy conditions for being a mark of the cognitive, nevertheless once our attention shifts to the specific content of the proposed candidate feature, it turns out to be an insufficiently robust indicator of the cognitive to play the kind of constitutive role being asked of it.

To bring this point (which is in truth no more than a modulation of the deflationary attitude that we encountered earlier) into proper view, we need to return, yet again, to Clark's text. Given the way in which we are currently understanding the target argument for ExC (i.e., as turning on the claim that being the kind of self-generated input that supports a process of cognitive self-stimulation is a mark of the cognitive), Clark's discussion contains a stark moment of what might be interpreted as either ambiguity or under-determination. Recall once more his claim that, in cases of cognitive self-stimulation, "any intuitive ban on counting *inputs* as parts of [cognitive] *mechanisms* seems wrong." Now notice that the term "cognitive" is not present in Clark's original text. I inserted it for expository reasons, because the result is a passage that more conspicuously expresses what Clark intends. But if we now look at things with a duly critical eye, with Clark's own distinction-collapsing strategy found wanting, and with the causal-constitutive distinction in force, it is desperately unclear that the right to add the term "cognitive" here—a right that Clark ultimately needs—has been properly earned. After all, both sides of the current debate (i.e., the conservative embedded side and the extended side) cheerfully accept that self-generated inputs that support cognitive self-stimulating loops operate within well-defined *mechanisms* that turbo-charge thinking. However, this observation does not do enough to establish that every element within the mechanisms concerned—and, in particular, the self-generated inputs in question— count as *vehicles that realize* thought, as opposed to *causal influences* on thought. For the turbo-charging mechanism at issue may very well be a hybrid system of cognitive and non-cognitive elements that interact causally so as to enhance overall psychological performance. Put another way, the problem is that, although we have an argument for the conclusion that the self-generated inputs at issue are proper parts of certain loop-shaped *mechanisms* that turbo-charge thought, we do not yet have an argument for the conclusion that those loop-shaped mechanisms are, *in their entirety and in their own right, cognitive mechanisms*. Perhaps the properly cognitive mechanisms in play are subsystems of larger, performance-enhancing loops, where the latter are not cognitive mechanisms in their own right, even though they contain cognitive mechanisms. Putting the point yet another way, it is not that the kind of thought-enhancing self-stimulating loops on which we have been concentrating can never be cognitive mechanisms in their entirety and in own right, but rather that, if sometimes they are, it's not their character *as* thought-enhancing self-stimulating loops that makes them so.

This claim may be bolstered if we reflect on an analogous, noncognitive case. Consider: Computer technology is now routinely used in many sports training regimes to improve the performance of elite athletes. Sometimes, this process of skill-enhancement happens by way of self-stimulating loops. For example, Baca and Kornfeind (2006) have designed a system in which individual rowers who are training on indoor rowing machines are monitored for factors that affect their technique. Real-time data displayed on a screen in front of the rowers (regarding, e.g., ground reaction and pulling forces) enable them to improve their rowing movements. In effect, the rowers lay down structures (patterns of bodily movement) that are then recycled as inputs to support them in tuning their bodily capacities in ways that will eventually result in improved performance on the water. This is, I submit, a self-stimulating loop, but one that enhances not reasoning or cognitive learning but the acquisition and honing of bodily skills. Now notice that although the feedback systems in place here, and in particular the self-generated inputs that the rower exploits, are core aspects of the *mechanism* by which the rower's body is tuned for improved performance, there is presumably little temptation to categorize those inputs as *realizers* of the observed bodily adaptation, as opposed to elements that have a *critical causal impact* on that adaptation. And it is not at all obvious why things should carve up any differently when the focus of attention is a self-stimulating loop that enhances thought. This is a shortfall that leaves ample room for the conservative embedded theorist to claim (with the additional weight provided by the default status of her view) that where the self-generated inputs in which we are interested are extraneural, as in the case of bodily gestures, those structures are located outside the boundaries of the properly cognitive mechanisms in play, mechanisms that themselves remain neurally located.

Of course, this is not necessarily the end of the matter. It is open to the fan of cognitive self-stimulation to reply that it is only certain instances of the phenomenon that should be expected to secure cognitive extension. The hunt would then be on for the additional factor or factors that are required. If we then allow ourselves to plug in the proposal that bodily gestures exhibit the extra factor or factors in question, the gestures-based argument for ExC would be squarely back in the game. With this sort of strategy in mind, it might be tempting to appeal to the fact that cognitive self-stimulation mechanisms in human behavior will standardly be the products of design—either by natural selection, development, or learning. Thus Clark (2008b, 130) offers us three intuition-pumping examples: (i) a situation in which the rhythm of the rain on my window happens, by chance, to improve my thinking, by positively affecting the pacing, sequencing, and timing of my thoughts; (ii) a robot that has been designed to exploit the rhythm of the rain to improve the pacing, sequencing, and timing of its reasoning operations; and (iii), following Dennett (1991), a self-stimulating spitting robot that has evolved to spit stored water in a rhythmic fashion at a metal plate on its body, in order to achieve the same outcome as in (ii). In essence, Clark's analysis of these examples is that (i) is (at best) a case of conservative embeddedness, (ii) is usefully categorized as a case of extended cognition (although some hesitation is due), and (iii) is a clear case of extended cognition. The feature that makes example (ii) a (perhaps precarious) case of extended cognition is supposed to be the fact that the robot has been *designed* to exploit the external structures in question. That feature—the designed exploitation of external structures—is then presumably solidified and enhanced through the addition of the (also designed) self-stimulation mechanism described in (iii).

On the strength of Clark's brief discussion of the role that design may play in an argument for ExC, it seems reasonable to give serious consideration to the following thought: If the bare fact of cognitive self-stimulation isn't enough to guarantee ExC, perhaps the recognition that cognitive self-stimulation will standardly be a product of design (by evolution, development, or learning) is. A moment's reflection, however, reveals a serious difficulty with this proposal, namely that *having been designed is not a robust mark of the cognitive*. For example, the heart has been designed by natural selection to pump blood around the body, but its designed character does not in any way make it a realizer of cognitive states or processes. So, merely pointing out that a material system or element that has not already attained cognitive status has been designed to perform its function will not result in the desired conceptual "upgrade." What is still missing is an account of why the designed function of interest counts as a cognitive function. The lesson for the proposed extended account of gestures is seemingly straightforward. Given that merely being causally coupled to an existing cognitive system such as the brain is not sufficient for cognitive status (that's one implication of the causal-constitution distinction), and given that (as I have argued) being a self-generated structure that supports a cognitive self-stimulation loop is not sufficient for cognitive status, simply adding the thought that the loop and its components are designed features is of no relevant consequence. So, even if, as seems likely, bodily gestures are indeed designed structures of the appropriate kind, we still find ourselves some way short of ExC.

5 Prenoetic Constraints

Time for a change of tack. In this section, I shall examine certain central threads in a detailed treatment of bodily gestures due to Gallagher (2005). Gallagher, it must be admitted, does not quite come out and say that gestures are among the material realizing vehicles of cognition, but he does openly flirt with that thought, and it would not be unreasonable to interpret his arguments as warmly recommending an ExC-friendly outcome. Thus, in a passage also quoted in part and commented on by Clark (2008b, this vol.), Gallagher writes as follows:

The question here is . . . about the cognitive effects gestures might have even if we have no conscious access to them. This is an extremely difficult question to answer if we think of cognition (thought) as a completely internal process that happens in a disembodied mind. It may be, however, that certain aspects of what we call the mind just are in fact nothing other than what we tend to call expression, that is, occurrent linguistic practices ("internal speech"), gesture, and expressive movement. (Gallagher 2005, 121, n. 7)

Henceforth, I shall ignore Gallagher's hesitancy and assess the considerations he offers us as a gesture-driven case for ExC.

Gallagher provides (what he dubs) an *integrative* theory of gesture, one that understands gesture to be (i) *embodied*, in the sense of being facilitated and constrained by motor capacities, (ii) *communicative*, in the sense of being used in intersubjective co-ordination and communication, and (iii) *cognitive*, in the sense (or senses—see below) of "contributing to the accomplishment of thought" and "shaping the mind" (ibid., 123). It is of course (iii) and its accompanying locutions that concern us. In accordance with our strategic orientation, I shall treat (iii) as an attempt to offer us a candidate mark of the cognitive. Let's begin, then, by considering the idea that gestures are genuine realizers of cognitive processes because they are elements that *shape the mind*.

Gallagher unpacks the relevant notion of "shaping" by way of his concept of a *prenoetic* contribution to thought. A contribution to thought is prenoetic when it shapes (or structures) thought, but "does not normally enter into the phenomenal content of experience in an explicit way" (ibid., 2). Here are two illustrative examples given by Gallagher (ibid., 2–3). First, because of my nature as an embodied being, my perceptual access to the world necessarily takes place from a particular and limited spatial perspective. That perspective shapes what I experience, but it is not normally something of which I am explicitly conscious in my experience. Second, it is arguable that I always experience the world through an affective lens formed by my current mood. Roughly, if I'm depressed, the world strikes me as a somber place; if I'm euphoric, the same elements that might have had a gloomy hue strike me as no more than mildly troublesome irritants. My affective lens shapes what I experience, but it is not normally something of which I am explicitly conscious in my experience. For Gallagher, bodily gestures are further examples of such prenoetic factors. As he puts it, gestures "shape cognition in a prenoetic manner" (ibid., 123).

To do justice to Gallagher's talk of "shaping" here, and in light of his examples of our limited spatial perspective and our affective lens, one might reasonably conceptualize prenoetic contributions to thought as *constraints on* thought, or at least on thought as we know it. But once this interpretation of the prenoetic is in place, any claim that such factors, wherever they happen to be located, are material realizers of cognition, as opposed to important causal determinants of cognition, is less than compelling. Once again it is expositorily useful to consider an analogous, noncognitive example. It is widely recognized that the biological process of adaptation by natural selection operates against a backdrop of various constraints. For example, developmental constraints are imposed by the heavily conserved Hox genes, which are active determinants in body segmentation and organ development in the anterior-posterior body pattern of many animals, including humans. In other words, basic bodily form—roughly, where the head goes, where the legs go, and so on—is a developmental constraint on adaptation. Practically speaking (i.e., outside of thought experiments), this sort of constraint cannot (or at least can only very rarely) be overcome by selection, even if an obvious adaptive benefit would accrue. So one might picture the set of such constraints as taking a space of conceivable phenotypes (variation that we can imagine) and reducing it to a space of possible phenotypes ("possible" in the sense of variation that is actually available to selection). By channeling selection in this way, the constraints in question may be said to shape the adaptations that actually come about. Roughly, just as our embodiment constrains our spatial perspective and thereby the structure of our experience, our Hox genes may constrain, for example, the number of digits we can evolve on each hand and thereby the structure of our bodily adaptations. Crucially, once we have classified a biologically relevant factor as a constraint on some adapted trait of interest, and so have accepted that there is a fundamental sense in which that factor is simply not available for modification by selection, it would surely be theoretically uncomfortable to think of it as a proper part of, or as realizing, the adapted trait itself. Similarly, once we have classified a psychologically relevant factor as a prenoetic element with regard to some cognitive trait of interest, and if we have accepted that prenoetic elements shape cognition in the sense of being constraints on the structure of cognition, it would surely be equally theoretically uncomfortable to think of the prenoetic factor in question as a proper part of, or as realizing, the cognitive trait itself. The upshot, then, is that even if Gallagher is right that gestures prenoetically shape thought, that does not make gestures realizing material vehicles of cognition.

What about the variant understanding of gestures (as cognitive) that Gallagher offers us? This is the claim that gestures contribute to the *accomplishment* of thought. As Gallagher (2005, 121) makes clear, our source of illumination here is Merleau-Ponty's account of language as accomplishing thought, an account that turns on the somewhat nebulous notion of the *expression of thought as the completion of thought*. As Merleau-Ponty puts it:

> If speech presupposed thought, if talking were primarily a matter of meeting the object through a cognitive intention or through a representation, we could not understand why thought tends towards expression as towards its completion, why the most familiar thing appears indeterminate as long as we have not recalled its name, why the thinking subject himself is in a kind of ignorance of his thoughts so long as he has not formulated them for himself, or even spoken and written them, as is shown by the example of so many writers who begin a book without knowing exactly what they are going to put into it. (1945/1962, 206)

Gallagher (2005, 121) builds on Merleau-Ponty's thinking about language to suggest that gesture, itself conceptualized as a mode of language, assists in the accomplishment of thought. In this Merleau-Pontian register, then, Gallagher's claim may be glossed in the following way: *Gestures are expressions of thinking which complete that thinking.*

One can see how, in general terms, this picture might plausibly have the implication that gestures are material realizers of cognition, for, where this picture applies, an unexpressed thought is an incomplete thought, so the expression (e.g., speech or gesture) is plausibly a proper part of the thinking. This may well be what drives Gallagher's suggestion that "certain aspects of what we call the mind just are in fact nothing other than what we tend to call expression." What is still missing, though, is a proper account of what it is for an element to express thought in such a way that the process of expression accomplishes the completion of the thinking.

Further light may be shed on this matter if we adapt, for our own purposes, Krueger's (forthcoming) useful distinction between three ways in which the notion of "expressing thought" might be understood. As we shall see, Krueger's analysis makes good use of our old friend, the causal-constitutive distinction. The first option that Krueger identifies is that expressive behavior may be understood as expressing thought, because such behavior is the *causal output* of certain internal psychological phenomena. It should be clear that, as it stands, this understanding of what it is to express thought will not be adequate to ground a notion of cognitive accomplishment that succeeds in conferring cognitive status on the expressing elements. For the proposal gives us no reason to classify the relevant causal outputs of some inner psychological phenomenon as anything other than noncognitive structures that are associated with that phenomenon. On the second option that Krueger identifies, the expressed psychological phenomenon is held to be *experientially copresent* with the expressive behavior associated with that phenomenon. The inspiration for this suggestion is the phenomenological point that when I visually perceive, say, a tomato, although my visual access to that entity is aspectual (there is an obvious sense in which, given my embodied spatial perspective, I have perceptual access only to certain portions of it), my experience is of the tomato as an intact, solid, three-dimensional object. The tomato's hidden-from-view aspects are not known inferentially, but are experientially copresent with those aspects that are perceptually present. If we transpose this model to the present context, we get the following model: although what we perceive is the expressive behavior, such as the gesturing, we simultaneously experience the associated psychological phenomena. As Krueger himself notes, there is undoubtedly a *prima facie* disanalogy here, since whereas I could turn the tomato around to bring its currently unperceived aspects into view, no similar action is available to me, within the copresence model, in the case of the psychological phenomena that are associated with expressive behavior. What is interesting about this disanalogy in the present context is that it points us in the direction of a crucial observation that Krueger proceeds to make, namely that, on the copresence model, when I observe another's gesturing I observe not realizing material vehicles of her thinking, but rather movements that are associated with such vehicles. Once again, then, the proposed understanding of what it is to express thought will not be adequate to ground a notion of cognitive accomplishment that succeeds in conferring cognitive status on the expressing elements.

The third and final option that Krueger identifies takes expressive behavior to express thought in the sense that the behavior in question is *partly constitutive* of the psychological phenomenon of which it is an expression. This is a notion of expressing thought that fits with Gallagher's suggestion that expressive movements help to accomplish thinking by expressing thought in the sense of completing the thinking in question.

Of course, given that the constitutive dependence of thought on expressive bodily movements is precisely what is being advocated here, the proposal on the table will certainly support the conclusion that bodily gestures are among the material realizers of cognition. Moreover, there seems to be little doubt that environmentally located elements may partly realize expressive activity, so the account is plausibly ExC-compliant. Perhaps, then, the route to a gestures-based argument for ExC is to be found in Gallagher, via Merleau-Ponty and Krueger.

Regrettably, we are not quite home. By appropriating Krueger's analysis in the way that we have, we have certainly learned that if we want to maintain both the claim that gestures are expressions of thought and the claim that gestures partly realize thought, then we will need to understand the notion of expression in Krueger's constitutive sense. But while that is an important outcome, it is in truth no more than a reminder that a mark of the cognitive—a constitutive account of what makes a state, structure, or process a cognitive one—is required. In the present context, such an account would specify the function or functions that an expressive behavior would need to perform, in order to count as completing, rather than just as influencing, thought. In fact, Krueger's own analysis implicitly reflects, although it does not explicitly make, this very point: he proceeds to complete his account of expressive behavior as partly constitutive of thought by appealing to Clark's notion of cognitive self-stimulation (Krueger forthcoming, 27). Unfortunately, as I argued earlier in this chapter, the fact that the relevant kinds of behavior exhibit the phenomenon of cognitive self-stimulation ultimately fails to plug the causal-constitutive gap. What all this indicates, I think, is that Gallagher's Merleau-Pontian suggestion that gestures help to accomplish thought—that gestures express thought so as to complete thought—is perhaps best seen as a placeholder for a mark of the cognitive, rather than as a mark of the cognitive itself. And that means that we still remain some way short of a gestures-based argument for ExC.

6   An Unfinished Business

The conclusion of our investigation may be put as follows. If we ask ourselves the question "Is cognition embedded or extended?" and we appeal only to the gesture-related arguments from the recent literature canvassed in this chapter, our answer ought to be "embedded." Moreover, it ought to be "embedded" in the conservative sense of that term. But while this is in some ways a disappointing result for those of us who are fans of extended cognition, it's not all doom and gloom. Along the way we have learned some useful lessons. Most notably, we have learned that, once the extended cognition hypothesis is understood, as I think it should be, in terms of a tripartite profile involving the parity principle, extended functionalism, and the mark of the cognitive, certain gesture-related considerations, such as cognitive self-stimulation, prenoetic shaping, and expression as completion, are seemingly unable to carry the weight of argument that has sometimes been placed on them. Of course, other considerations (other marks of the cognitive) may ultimately confer cognitive status on gestures, and yet other considerations (not applicable to gestures) may persuade us (or most of us anyway) that cognitive extension is indeed the way of things. Right now, however, it's an unfinished business—and that's the most interesting kind.

Acknowledgments

Many thanks to Andy Clark, Shaun Gallagher, Zdravko Radman, and audiences in Dubrovnik, Bielefeld, and Milan for important critical feedback on the ideas presented here. Some of the text in section 3 was adapted from passages that appear in Wheeler 2011a,c.

References

Adams, F., and K. Aizawa. 2008. *The Bounds of Cognition*. Malden, MA: Blackwell.

Baca, A., and P. Kornfeind. 2006. Rapid feedback systems for elite sports training. *IEEE Pervasive Computing/IEEE Computer Society [and] IEEE Communications Society* 5 (4):70–76.

Bechtel, W. 1994. Natural deduction in connectionist systems. *Synthese* 101: 433–463.

Broaders, S., S. W. Cook, Z. Mitchell, and S. Goldin-Meadow. 2007. Making children gesture brings out implicit knowledge and leads to learning. *Journal of Experimental Psychology: General* 136 (4):539–550.

Clark, A. 1997. *Being There: Putting Brain, Body, and World Together Again*. Cambridge, MA: MIT Press.

Clark, A. 2008a. Pressing the flesh: A tension in the study of the embodied, embedded mind? *Philosophy and Phenomenological Research* 76 (1):37–59.

Clark, A. 2008b. *Supersizing the Mind: Embodiment, Action, and Cognitive Extension*. New York: Oxford University Press.

Clark, A. 2011. Finding the mind. *Philosophical Studies* (symposium on Clark's *Supersizing the Mind*) 152 (3): 447–461.

Clark, A., and D. Chalmers. 1998. The extended mind. *Analysis* 58 (1):7–19.

Dennett, D. C. 1991. *Consciousness Explained*. Boston: Little, Brown.

Gallagher, S. 2005. *How the Body Shapes the Mind*. Oxford: Oxford University Press.

Goldin-Meadow, S., H. Nusbaum, S. Kelly, and S. Wagner. 2001. Explaining math: Gesturing lightens the load. *Psychological Science* 12:516–522.

Goldin-Meadow, S. 2003. *Hearing Gesture: How Our Hands Help Us Think*. Cambridge, MA: Harvard University Press.

Goldin-Meadow, S. 2009. How gesture promotes learning throughout childhood. *Child Development Perspectives* 3:106–111.

Hutchins, E. 2010. Imagining the cognitive life of things. In *The Cognitive Life of Things: Recasting the Boundaries of the Mind*, 91–101, ed. L. Malafouris and C. Renfrew. Cambridge: McDonald Institute Monographs.

Kiverstein, J., and M. Farina. 2011. Embraining culture: Leaky minds and spongy brains. *Teorema* 32 (2):35–53.

Krueger, J. Forthcoming. Seeing mind in action. *Phenomenology and the Cognitive Sciences* (special issue on empathy and intersubjectivity).

Levin, J. 2010. Functionalism. *The Stanford Encyclopedia of Philosophy* (summer 2010 edition), ed. E. N. Zalta, http://plato.stanford.edu/archives/sum2010/entries/functionalism/.

Menary, R. 2007. *Cognitive Integration: Mind and Cognition Unbounded*. Basingstoke: Palgrave Macmillan.

Menary, R., ed. 2010. *The Extended Mind*. Cambridge, MA: MIT Press.

Merleau-Ponty, M. 1945/1962. *Phenomenology of Perception*. London, New York: Routledge.

Rowlands, M. 1999. *The Body in Mind*. Cambridge: Cambridge University Press.

Rowlands, M. 2010. *The New Science of the Mind: From Extended Mind to Embodied Phenomenology*. Cambridge, MA: MIT Press.

Rupert, R. 2009. *Cognitive Systems and the Extended Mind*. New York: Oxford University Press.

Sprevak, M. 2009. Extended cognition and functionalism. *Journal of Philosophy* 106:503–527.

Sutton, J. 2010. Exograms and interdisciplinarity: History, the extended mind, and the civilizing process. In *The Extended Mind*, 189–225, ed. R. Menary. Cambridge, MA: MIT Press.

Wheeler, M. 2005. *Reconstructing the Cognitive World: The Next Step*. Cambridge, MA: MIT Press.

Wheeler, M. 2010a. In defense of extended functionalism. In *The Extended Mind*, 245–270, ed. R. Menary. Cambridge, MA: MIT Press.

Wheeler, M. 2010b. Minds, things, and materiality. In *The Cognitive Life of Things: Recasting the Boundaries of the Mind*, 29–37, ed. L. Malafouris and C. Renfrew. Cambridge: McDonald Institute Monographs. (To be reprinted in *Action, Perception and the Brain: Adaptation and Cephalic Expression*. J. Schulkin, ed. Basingstoke: Palgrave Macmillan.)

Wheeler, M. 2011a. Embodied cognition and the extended mind. In *The Continuum Companion to Philosophy of Mind*, 220–238, ed. J. Garvey. London: Continuum.

Wheeler, M. 2011b. In search of clarity about parity. *Philosophical Studies* (symposium on Andy Clark's *Supersizing the Mind*) 152 (3): 417–425.

Wheeler, M. 2011c. Thinking beyond the brain: Educating and building, from the standpoint of extended cognition. *Computational Culture* 1, http://computationalculture.net/.

13 Pointing Hand: Joint Attention and Embodied Symbols

Massimiliano L. Cappuccio and Stephen V. Shepherd

One of the most characteristic and notable powers of the hand is its ability to indicate via pointing—a gesture produced to draw the attention of specific observers toward a distal target. When pointing successfully aligns signaler's and recipients' gazes and they become mutually aware of this alignment, one may say it manipulates "joint attention" (JA), the experience of openly sharing with others a common focus of interest (Seemann 2011). JA is crucial in human development; its robustness in nine- to twelve-month-old children predicts the acquisition of cognitive skills, including language (Mundy et al. 2007; Brooks and Meltzoff 2008); its absence is diagnostic of autism (American Psychiatric Association 1994). Though JA is pervasive among humans and crucial to social intelligence, scientists and philosophers have yet to agree on its precise features, mechanisms, and evolutionary and developmental origins. In this chapter, we argue that an understanding of pointing—a gesture that explicitly incarnates and manipulates JA—is key to tackling these questions. In particular, we will discuss how coattenders become aware that they are focusing on the same object and what abilities are required.

Philosophers of mind typically dichotomize representational and dispositional skills in social cognition (Seemann 2007), presenting us with a theory of mind (ToM) on the one hand, whether derived from folk psychology or mental simulation, versus embodied cognition on the other, comprising interactionist, enactivist, and/or narrative-practice accounts. Unsurprisingly, these views favor alternate approaches to JA: Either coattenders match their representations about their attentional focus by inferring shared beliefs or they coordinate their dispositions toward an attentional focus through overtly observable behaviors. Such alternate approaches have distinct cognitive prerequisites: If the representational model is correct, coattenders must be reciprocal mindreaders and infer one another's (possibly propositional) mental contents in order to confirm parity. But if the dispositional approach is correct, coattenders may only engage in forms of reciprocal coordination mediated by embodied intentions (e.g., posture, orientation, motion, expression, and gaze). Both stories have some developmental and primatological support. Supporting the representational account, JA produced by fourteen-month-old infants reportedly involves appreciation of beliefs shared with the recipients (Moll et al. 2007); can convey abstract, absent contents (Liszkowski et al. 2007); and can aim to achieve altruistic rather than purely self-centered goals (Liszkowski et al. 2006). Supporting the dispositional account, infants younger than four years fluently produce and understand pointing despite poorly distinguishing between their own and others' beliefs, confounding the matching of mental representations (Gallagher and Hutto 2008); similarly, many nonhuman species coordinate group behavior toward a target without (it is assumed) possessing propositional attitudes and inferential capabilities (Knoblich and Sebanz 2008).

Individuals coordinate attention in many ways, but pointing is one of the most prototypical. Although the morphological details of this gesture can change across cultures and contexts, it appears to be universally used (Kita 2003) and possibly innate (as evident from blind-born infants). However, while the gesture of pointing is not conventional in the sense of signs, neither is it reducible to a mere epiphenomenon of attention (Kendon 2004), because unlike bodily postures and gaze, it has an explicitly communicative intention.

Whether JA is representational or dispositional depends primarily on the *minimal* cognitive prerequisites needed for nonhuman primates or human infants to engage in pointing. Because adult humans unequivocally have the ability to use JA symbolically (SJA), and because this ability arises (in evolution and development) from precursors that are capable only of basic JA (BJA), we are interested in observable behaviors that seem to scaffold the transition from BJA to SJA (cf. Shepherd and Cappuccio 2011). While some form of this distinction is broadly accepted, past debate has contrasted a declarative mode of pointing (assertive and informative), said to imply JA, with an imperative mode (requesting and ordering), said to not imply it. It remains contested whether nonhumans ever acquire declarative pointing: Some authors believe that even apes lack the representational abilities required (e.g., Povinelli et al. 2002; Tomasello et al. 2005), while others stress that apes at least sometimes point declaratively, at least when enculturated by humans (e.g., Leavens 2011; Leavens and Racine 2009).

We do not take a stand as to whether apes represent minds or engage in declarative pointing. Instead, we argue that declarative pointing does not *require* but, on the contrary, intrinsically *produces* a primitive form of representational intelligence. Any well-formed instance of pointing—that is, any act of pointing that effectively aligns recipients' gaze with the signaler's—incarnates the possibility of communal attention in physical form and is simultaneously recognized by all parties as explicitly produced to coordinate awareness. The hand can thus symbolically represent the coattenders' "jointness" in a minimal, prototypical, and embodied form, and does so through direct perception. Since pointing, so understood, *invites* rather than *requires* inference, and does so without specialized cognitive resources, we argue that although representation can clearly augment JA, it comes late to the party. Specifically, we propose that SJA arises in development and evolution through observable somatic states—most notably through declarative gestures such as pointing—long before the intervention of ToM.

In the following sections, we describe (1) basic and symbolic joint attention and (2) the phenomenology of declarative pointing, arguing (3) that pointing can arise from and mediate BJA and so evidences, to signaler and receiver alike, the relevance of coordinated behavior. Pointing thus (4) scaffolds the acquisition of SJA by physically cuing information not just in the world, but in the common ground of shared knowledge and intentions. Finally, we consider how such embodied manifestations of joint intention build a state of (5) "open knowledge."

1 Basic and Symbolic Joint Attention

Following or cuing another's gaze, even when the target is understood to be the subject of the other's attention, is insufficient evidence for BJA. BJA additionally requires that the coattenders are aware of mutually targeting the same object (Tomasello 1995), which then becomes available for overtly shared consideration. If BJA is initiated whenever a signaler (S) indicates to a recipient (R) her attention toward a certain targeted object (T), it is necessary that there exist:

[I]   *Shared attention* to T:
      S→T & R→T
[II] *Reciprocal attention* between participants:
      S→R & R→S
[III] Occurrent *iterative awareness* of the others' attention:
      R→S→T, S→R→T, S→R→S→T, etc.

The potential for iteration is crucial because it completes the collective attentional state Peacocke (2005) called "Mutual Open-Ended Availability" (MOEA): If attention is entirely symmetrical and overt, then as soon as one subject perceives a certain relation of iterative awareness (e.g., R→S→R→T), the opportunity arises for the other to notice, producing a higher-order iteration (S→R→S→R→T). Were individuals' perceptual and cognitive abilities infinite, the length of iteration could be arbitrary, but in practice only a small number of iterations is likely and indispensable for most social interactions among human adults (four, according to Sperber 2000 and Moore forthcoming). This constraint undermines not the open-endedness of JA, but merely the ability of each partner to analyze its details: By analogy, we can recognize the infinitely nested reflections in a Hall of Mirrors with neither need nor ability to count them.

MOEA is the essential feature of JA, and indeed most authors (e.g., Schiffer 1972; Sperber and Wilson 1986; Wilby 2010) account for the mutually, manifestly overt quality of JA experience through "mutual knowledge"— a reciprocal iterative understanding of the form "S knows that R knows that S knows . . . that X, and R knows that S knows that R knows . . . that X"—but whether this knowledge is dispositional (Campbell 2005; Gallagher and Hutto 2008) or representational (Tomasello 2008) is contested, perhaps signifying a distinction between mutual "know-how" and "know-that" (cf. Ryle [1945] 1971). In our view, while *dispositional* mutual knowledge is a necessary and sufficient condition for fluency in BJA, it is only necessary and *not sufficient* for fluency in SJA.

This is because the crucial issue at stake in JA is the presence of MOEA, irrespective of whether SJA is engaged. Consider that you may notice someone observing you admiring a motorcycle and see that he saw you spot him—and *you can do this purely perceptually*. You needn't ponder his covert thoughts or intentions; you simply directly perceive a social context. In fact, you may automatically smile appreciatively at the motorcycle— or guiltily withdraw from same—based on the other's directly perceived affect. It is surely true that these interactions are informed by a culturally and linguistically rich context—but dispositional MOEA neither precludes a role for these factors (because learned scripts and norms may shape our direct perception of a scene) nor requires explicit reflection on them (Gallagher and Hutto 2008). Thus, dispositional MOEA appears sufficient to account for BJA, and BJA to account for most instances of JA among animals and humans alike. Nonetheless, some instances of JA remain to be explained.

By introducing representations, SJA—but not BJA—can assign meaning to arbitrary locations, explicitly reference absent or abstract contents, engage informative contents and propositional attitudes, and interact with mentalistic attributions of belief. We argue that pointing plays a crucial role in making the evolutionary and developmental transition from BJA to SJA in that it physically embodies joint mental activity. To understand how this may occur we must consider the process of declaration in the context of the pointing gesture.

2 Dynamics of Declarative Pointing

The three most influential views on declaration conceptualize its preverbal function as to either (a) assert informative contents endowed with truth-values (cf. Leonard 1959), (b) direct the recipient's attention instrumentally to obtaining a certain reaction (pointing "proto-declaratively," as in Bates et al. 1975), or (c) share attention simply for its own intrinsic, contemplative, value (Tomasello 2008). All three have caveats when applied to pointing: (a) declarative gestures don't necessarily have a true/false content; (b) gestures can reference symbolic contents in a manner incompatible with purely dispositional or instrumental accounts (McNeill et al. 2008); and (c) infants show dissatisfaction if their "altruistic" pointing fails to evoke an affective response (Liszkowski et al. 2004). We therefore prefer a different definition: a deictic signal is declarative when the immediate interactional features of T (affordances recognizable in the occurrent perceptual context) are irrelevant to understand the gesture's meaning, except instrumentally to make the recipients aware of some background information related to shared attitudes toward T.

In order to unpack this idea, we shall partially follow Tomasello (2008), who builds on Grice's (1957) theory of "nonnatural meaning" to characterize declarative communication as representational. When pointing has an imperative valence its meaning is natural, because it is entirely and perceptually disclosed to R as a social affordance of T. For example, consider a toddler who, reaching toward a milk bottle far above her, looks toward her parent and starts to cry. BJA is sufficient to explain this sort of imperative interaction because S's referential intention is entirely evident in S's behavior. However, unlike the many animals that coordinate solely in species-typical contexts and ways, humans cooperate flexibly. A great deal of ambiguity can arise when signaling deictically with a multipurpose effector, such as a hand held in an indeterminate pose. Here, because the desired goal is evident in neither the form of the gesture nor the perceptible environment, S's reference to T is a *blank*, which R must somehow fill. R must recognize S's intention, which is underspecified by the occurrently perceived environment but, because it is expressly communicative, is necessarily embedded in the common ground of shared conventions and beliefs.

Consider an example (Kita 2003; Tomasello 2008) in which the very same gesture, produced in the same scenario with the same T, can express entirely different meanings disambiguated only through the common ground. An orator speaking in a crowded conference hall can point to the empty chair in front of him to indicate that a free seat is available (gloss: "sit there"), that it is the only seat left in the hall (gloss: "there is a big audience"), that a certain person is sitting there (gloss: "there she is!"); or that an important guest is absent (gloss: "she's missing"). Declarative pointing can thus target an occurrent T (the empty chair) to refer to an incidental, abstract, or even absent feature X(T). Declarative pointing directs not only the eyes, but also the imagination. By designating and then referencing locations in an arbitrary space, declarative pointing comes to trade in symbols rather than affordances (Kendon 2004). This association is nonnatural to the extent that X(T) can be specified only through the common ground, not from the details of the perceptual environment. Notice that, while common ground information about (X)T conveyed through communicative intention may vary, it must always at least reference itself as the source of the nonnatural meaning X(T). In other words, the beliefs to be shared by the coattenders in order to make sense of pointing must always include the belief that such beliefs exist. If (and only if) this happens, the gesture is implied to have nonnatural meaning and to be a kind of symbolic declaration—it can be interpreted only through reference not to T directly, but to the information associated with T.

We agree with Tomasello that nonnatural meaning is a necessary and sufficient condition for declarative communication, and transforming a perceptual T into a symbol of the *beliefs coattenders share about T* is what specifically distinguishes SJA from BJA. However, Tomasello additionally assumes that nonnatural meaning can be conveyed only when pointing has the *unique function* of sharing attention, hence excluding the possibility that declarative communication could also aim to initiate interaction. We disagree. While a contemplative, noninteractive stance toward T may reinforce declaration by deemphasizing direct affordances of T, it is not indispensable. A communicative act is declarative whenever it references contents from the common ground, regardless of its perlocutionary goal.

What, then, are the cognitive prerequisites of declarative pointing? Two opinions can be found in the literature. According to Tomasello, the common ground must be shared by S and R, and both must be aware that they share it. Drawing on data collected from infants and apes, Tomasello (2008, 189–190, 198) argued that declarative pointing requires coattenders to perform "recursive mindreading"; as implemented by Harris's (1996) model, such metarepresentation explicitly entails that S and R simulate one another's mental simulations (Tomasello 1999). When S points to the empty chair, S is inclined to think that R will correctly interpret S's gesture as an intention to refer to the person who was supposed to sit there; in turn, when R sees S pointing to the chair, R is inclined to think that S knows that R knows that that person is supposed to sit there, and only thus can correctly interpret the symbolic meaning of the gesture. According to this model, JA requires mutual knowledge that is based on a mindreading process that is *internal* (it retrieves private, nonperceptual information via an offline reflective process) and *representational* (it operates on descriptions of one's own mental contents as projected onto a coattender).

However, this is not the only account available. Primatologists remain divided as to whether nonhuman apes can point declaratively—despite general agreement that apes have limited talent for representation. Tomasello (2006) sees no compelling evidence for declarative pointing in apes, and he attributes the absence to the lack of uniquely human simulative devices. Leavens and Racine (2009) respond that declarative pointing has been observed repeatedly among captive apes (Savage-Rumbaugh et al. 1986), and at least once among wild apes (Veà and Sabater-Pi 1998). Even sophisticated pointing behaviors—for example, pointing to absent entities—might not be human specific. While some experiments provoked pointing toward absent entities by prelinguistic children but not by apes (Liszkowski et al. 2009), language-trained apes have reportedly referred to dead individuals (Fouts and Mills 1997), solicited refills of an empty glass (Hoyt 1941), and indicated hidden objects while specifying their type through a symbol language (Menzel 1999). If wild apes don't often point declaratively, Leavens argues, this may be because of differences in developmental enculturation rather than pedigree—in other words, to training, not talent. According to this perspective, all forms of pointing are instrumental and intended to produce social interaction (Leavens 2004). While declarative pointing gestures continue to rely on a common ground, this common ground is not composed of salient facts inferred and projected via mental simulation, *pace* Tomasello, but rather, in accordance with Hutto's (2011) hypothesis, reflects a dispositional knowledge of how to manipulate coattender's beliefs and desires through typically experienced narratives. For Leavens and colleagues, JA requires mutual knowledge derived from a behavior-reading process that is *external* (it directly perceives social affordances in sets of goal-directed intentions engaged by reciprocal attention) and *nonrepresentational* (it enacts appropriate social affordances by means of the R's disposition to respond to S's behavior).

The controversy regarding the cognitive precursors of declarative pointing can thus be characterized as one between representational and dispositional accounts of mutual knowledge. In Tomasello's camp, proponents of a simulationist ToM claim that others' goals are known through introspectively produced models (representational know-that) of inferred covert processes; in Leavens's camp, proponents of embodied cognition and narrative practice claim that others' intentional attitudes are grasped through directly perceived affordances and scriptlike heuristic templates (dispositional know-how). In our opinion, neither approach fully accounts for the properties of human JA, but with appropriate adjustment their complementarity can be demonstrated. The simulationist approach offers *reflectivity* but lacks *reciprocity*. On the one hand, each coattender has access to one's own knowledge state, as is necessary for inference, imagination, and the representation of abstract or absent contents; on the other hand, each coattender's understanding of the other must be achieved by "telementational" methods (Leavens 2011), that is, introspection and projection, because mere reciprocal attention to the outside world is insufficient for cases of nonnatural meaning. The solipsism implicit in telementation would undermine any openness and directness in JA, as Leavens correctly notes: S and R may both have the same representational mapping X(T), but because they each access it privately with no opportunity for overt comparison, they can't truly share it as a joint focus of attention. Conversely, the embodied or interactionist approach offers *reciprocity* but lacks *reflectivity:* MOEA interlinks the coattenders' disposition toward T, but no causal role is assigned to coattenders' explicit modeling of convergent and divergent views. Such an approach is tenable only when S's and R's goals match their established know-how, restricting any declared X(T) to a small library of stereotyped interactions. Tomasello (2008) is right to note that this precludes creative referencing of abstract or absent targets, limiting opportunities for SJA.

Let's return to our example of the lecture hall. Symbolically evoking an absent person by pointing to her empty chair presupposes a knowledge of the arbitrary association between chair and person that is at the same time *open* (occurrently reciprocal, per MOEA) and *nonnatural* (independent of the occurrent physical context or affordances of the chair but instead cuing associations in the common ground). These features are constitutive of declarative pointing as we've defined it, but in contrast to established approaches, they seem to necessitate *both* reciprocity and reflectivity. The current debate in primatology tries to define whether representation is required for declarative pointing, and both Tomasello and Leavens assume representation mediates JA only if each coattender accesses representational mutual knowledge through internal mental devices. We are inclined to reject the assumption, common to both paradigms, that social cognition is either both representational and internal, or not representational at all.

According to Tomasello, SJA is representational in character because S's communicative intention is neither overtly expressed by her pointing finger nor uniquely specified by her referential intention, and is thus made clear only through the context of the common ground. We agree that SJA is representational and that declarative pointing is sufficient to evoke it, but we remain unconvinced that declaration requires preexisting talent for representation. Instead, as we shall develop in section 3, we suspect declarative pointing *produces* representation. According to Leavens, declarative pointing is an elaboration of developmentally enculturated scripts for pointing, rather than a novel phenomenon with its own unique mental foundations. We agree that declarative communication may arise by invoking and abstracting previously experienced narratives related to pointing rather than arising by application of a preexisting internal representational mechanism. We differ from Leavens (2011) in that we believe this cultural–developmental experience implies a substantial transition in social cognition—one that achieves manipulation of sophisticated representations of beliefs through physically embodied media.

We conclude that SJA is neither purely dispositional in character nor predicated upon a preexisting system for mindreading. In the next two sections we will explore this hypothesis, first explaining how pointing is sufficient evidence for BJA, and then describing how declarative pointing is a sufficient scaffold for SJA.

3   Pointing Is Sufficient Evidence for BJA

Pointing is one of the purest means of manipulating JA. In any community that uses pointing, capacity for BJA is assured. Pointing necessarily evokes the core constitutive features of BJA and lacks extraneous features that might muddy its message. We identify three constitutive features of BJA that must be evoked by a coordinating signal. Specifically, a signal must be: (1) *indicative* (it expresses S's intention to deictically reference T, thus sharing attention); (2) *salient* (it makes R aware of S and leaves S able to monitor R, creating reciprocal attention); and (3) *symmetric* (it requires that each coattender, including S, holds equal possibilities of attending to the others' attentional state, hence triggering iterated awareness). By combining these features, pointing coordinates a well-defined and mutually aware community.

It is rare that a signal *necessarily* displays all three of these features at the same time, but hand gestures do. A lieutenant can point out an enemy bivouac, commanding his troopers to silently attack. His order is effective, because he can be confident his troops (whom he can see) can also see where he points, and that he sees them, and that they see one another, coordinating the timing of their response—all while the enemies, sleeping unseen, are none the wiser. The lieutenant's signal achieves its goal because it necessarily coordinates a well-defined and mutually aware community. Other audiovisual signals lack at least one of BJA's three core features. A vocal alarm doesn't need, and is not meant, to be synchronously perceived with its T, so the focus of any shared attention would be either the signal itself (when T is not discernible) or T (when the signal has passed), but rarely both as a coherent intentional unit. Vocal signals are not indicative: They may have referential function but are nondirectional and so specify categories rather than instances (Cheney and Seyfarth 1990). Gaze and heading have indicative function (they refer deictically to some T), but can only effectively communicate S's attention if R's has already been secured through other means (e.g., eye contact, abruptly changed-and-held posture, or emphatic sounds). Worse, they do not easily facilitate symmetry: S cannot generally see her own eyes. More attention-getting bodily gestures (e.g., ground-slapping) are admittedly salient, but since they cannot deictically reference an arbitrary target, they result primarily in dyadic and asymmetric relations (i.e., they draw R's attention to S, not T). While these behaviors may help coordinate BJA, none is sufficient to instantiate it. Certainly, we grant that many animals coordinate BJA through species-typical signal exchange, as when one monkey threatens a foe, looks back toward an ally, and looks back toward the foe while bobbing its head ("appeal-aggression"; Waal 2003). But we argue that pointing gesture does all this and—importantly—achieves *only* this. Pointing is blankly indicative, equally useful in diverse contexts and for different purposes, and this gives it three emergent properties.

(1) *Ostension* (it conveys to R S's attention to T, producing shared attention). An indexical gesture *par excellence*, pointing holds flexible deictic

function able to precisely target any physical object regardless of direction or distance, provided that coattenders and T are all connected by sight-lines. Pointing is more flexible than vocal signals because T is determined by spatial location rather than categorical description; it is more purely indicative than gaze because it is physically removed from the affective context of S's facial expression. Moreover, pointing has an *exclusively* deictic function: whereas postural orienting is at least somewhat action specific and animal vocalizations are associated with stereotyped emotional responses to environmental cues, pointing affords no direct physical consequence and achieves nothing *in itself* but the orienting of R's attention. In this way, the purpose of pointing is entirely (and arbitrarily) ostensive, because it is free from the confound of implicit noncommunicative goals.

(2) *Self-ostension* (it draws attention to S while S attends to R, producing reciprocal attention). Pointing is effective in getting attention because, while the extension of the arm–hand–finger highlights a distal T, it simultaneously highlights the presence of S and emphasizes her communicative intention. Unlike eye contact, hand gestures can be easily displayed to multiple Rs, becoming the T of their shared attention; unlike vocal and olfactory signals, Rs must be immediately linked to S by line-of-sight, thus defining a selected community as the intended audience (e.g., directing allies but not rivals toward a strategic location). The function of pointing is not only ostensive but self-ostensive: S herself becomes *a part* of T, making salient not just the target, T, but the explicit relation, S→T. This is decisive in ontogenesis (Reddy 2005) as the infant needs to become aware of the opportunities afforded by being a *target* of attention (not just an S, but also a T) in order to reciprocate caregiver gaze and, subsequently, to learn how to orient it.

(3) *Symmetry* (both S and Rs can equivalently observe S&R→T, offering both equal access to iterated awareness). Pointing is typically present concurrently in both S's and R's visual fields, unlike most other types of deictic signals. By externalizing the ostensive and deictic signal to a viewable part of one's own body, one's "targetness" becomes perceptually available to R—and also to S. Pointing becomes a concrete, directly perceivable, stand-alone indicator of triadic intent. Whereas mammalian eyes point to one thing at a time, and vocal signals fail to designate precise Ts, pointing can be used to indicate a specific T while simultaneously looking at a specific R (cf. Waal 2003), instantaneously involving R in a triadic relation with S and T. The fact that the R's dual focus of attention (S and T) is mirrored from S's side (R and T) creates the empirical conditions to establish and

iterate symmetric attentional relations. Moreover, because the pointing finger cues S as surely as it cues R, it becomes an externalized public signal, a concrete and observable indicator of the opportunity for collective attention, and more, of the abstract intention of collective action.

As long as S and R are symmetrically situated as two equal coattenders of the pointing gesture, their communication meets Peacocke's criteria of MOEA, because pointing facilitates iterated awareness through a single communicative act. This is a condition not easily achieved by other communicative signals, which require integration of multiple acts and modalities to achieve BJA's three core features. This doesn't mean that pointing is the only signal allowing iteration. It is interesting to note, however, that when (for example) haptic communication establishes JA with blind infants (Bigelow 2003), the hands remain the crucial effector. Here, the task becomes bimanual, for example, putting a toy in the infant's hand while touching his hand with the other hand. In this case, the iteration is enabled because a grasping hand can simultaneously contain and be contained by another hand so that use of two hands can establish a triadic relation.

Furthermore, as we have admitted, BJA can exist without pointing. Many species follow gaze (Shepherd 2010), and social animals coordinate actions to maintain group cohesion, defend against predators, and compete with rival social groups (Shepherd and Cappuccio 2011). However, because pointing accomplishes nothing in itself aside from the manipulation of attentional states, it must be motivated by social goals afforded through JA. Pointing is thus sufficient to establish *at least* BJA, and furthermore, it brings it about through a single enduring, physical, self-contained act. At the same time, pointing can be freely combined with other actions and with nonvisual modalities that emphasize its communicative intentions (e.g., a quizzical eyebrow, a guttural noise, or gaze conveying affective content toward R rather than T; cf. Csibra 2010 and Moore forthcoming). In our view, the most crucial aspect of declarative pointing is that it creates a physically embodied token of the JA of a well-defined community of actors, thus comprising an open and nonnatural representation of their collective state. But first we must digress to explain just what this public representation is and why it matters to SJA.

4   Declarative Pointing Is Sufficient to Scaffold SJA

We have demonstrated that interactions scaffolded by the pointing gesture are sufficient to bring about BJA, in particular by producing MOEA via (I) shared attention, (II) reciprocal attention, and (III) iterated awareness. According to Peacocke's (2005) analysis, these features are sufficient to produce only an "incipient" form of JA, whereas the "full" form additionally requires what he calls "Reflective Social Consciousness": S and R *explicitly* jointly attend to T, becoming reflectively aware of the complex situation in which they are involved. In our terms, SJA doesn't only require the three features of BJA, but also:

(IV) *Reflective social consciousness* that participants share a state characterized by (I)–(IV).

The fact that (IV) refers to itself generates what Peacocke calls a "self-involving situation," that is, a situation U that is constituted by the fact that the agents are aware *both* that they jointly attend to T and that they are in the situation U (ibid., 319). Consider the tale of the Emperor's New Clothes. When the little boy mockingly points to the Emperor on the street, the crowd was already jointly attending to the Emperor's nudity and was mutually aware of one another's attention (as in BJA). What changed when the boy pointed? All parties involved suddenly become *reflectively* aware of their *collective* state of attention—not only that they were all looking at the Emperor, but that they were *all* finding the emperor's state *conspicuous*. Only then does shared embarrassment escalate into public scandal.

This meta-awareness deploys both reciprocity (each coattender's state of awareness depends on and refers to the others') and reflectivity (it implies awareness of one's own state of awareness). It is key that reflective social consciousness is not merely dispositional, as it doesn't only implement an interactional know-how, but also implies at least minimal knowledge that this interactional know-how *exists* and *is mediated by* the representation of S's and R's self-involving state of awareness. Peacocke argues that the reflectivity of SJA "is possible only for beings who have some way of representing attention, mental states, and employing some form of indexical reference to mental states" (ibid., 308). This form of representing attention is minimal, and merely indexical in character; as such, it is different from the representational mutual knowledge as endorsed by simulationists and rejected by the enactivists/interactionists.

In fact, whereas mutual knowledge is representational in that the coattenders' attentional states reciprocally portray one another (R's description of T contains S's description of T *and* S's description of T contains R's description of T), reflective social consciousness is representational in that the coattenders' collective state of attention makes them recognize that they share that very state (*U* contains a jointly attended target, T, and indexically refers to itself, the collective state *U*). In representational mutual knowledge, each attender's representational knowledge about T must fully represent the other's, and so runs into infinite regress; by contrast, in reflective social consciousness, the communal state of attention reveals some information concerning its own structure (ibid., 306–307). Hence, whereas mutual knowledge can be compared to two encyclopedia entries each fully recapitulating the other's recapitulation of their own, *ad infinitum*, reflective social consciousness is like an entry that refers to the encyclopedia that contains it.

So formulated, the fact that SJA is mediated by reflectively produced representation does not lead to a paradox—but it does shift the burden, so that some empirical conditions must simultaneously grant this representation *both* reflectivity and reciprocity. At heart, this problem relates to the foundations of symbolism in JA. Tomasello and Leavens debate whether JA can be mediated by representational processes (simulating, thinking, imagining) or not, but they implicitly assume that representations are *internal* (private, covert, nonperceptual) in character. This certainly *can* be true—but it needn't be. We see no inconsistency in modeling JA as both externally accessible to multiple individuals through occurrent perception (it relies on public, perceptual cues) and representational (the cues in question serve to reflectively access and manipulate private, offline information). We access representations of just this sort whenever we think of immaterial, absent, or abstract contents because of publically available cues. As humans who grow up in a material culture, we routinely traffic in covert ideas and goals crystallized into overt sounds and images.

What we propose is that declarative pointing is the prototypic "external representation" (Kirsh 2010)—a particular instance of extended cognitive processing that doesn't only use the body as an extracranial vehicle of thought (see Clark, this vol.; Wheeler, this vol.), but also makes this thought perceptually available for public consideration, enabling the coattenders to "think about [their own joint] thinking" (Malafouris 2007). Declarative pointing thus represents BJA by indexically referring to it, externalizing it as a communal mental state. This first, minimal, symbolic function of pointing is to openly declare the situation of JA to the participants, bringing about Peacocke's reflective social consciousness, just as the crowd's public awareness is suddenly "awakened" by the little boy's pointing toward the naked emperor. This function makes openly perceptible not only social affordances, but also the explicitly collaborative perspective, and does so independently of context-specific or stereotyped species-typical goals (such as "mob the snake"). The pointing gesture thus makes collaborative intention an object of public knowledge and so renders it available to planning and learning mechanisms. Through the pointing gesture, a prelinguistic and prementalistic human infant can indicate to a novel adult a novel object—or have a novel object indicated to her—and *legitimately expect* they will together do something rewarding with that object, learning to cultivate JA as a useful state. This experience lays a foundation for the second symbolic function of the pointing gesture, which consists in externalizing to an actual perceptual target T the imaginative content X(T) associated with it, making X(T) something that individuals can perceptually attend toward *together*.

The role played by pointing in revealing communicative intent is preeminent for two reasons. First, unlike other indicative gestures such as posture, gaze, and olfactory marking, it is necessarily self-indexical: while pointing's ostensive function expresses S's referential intention, its self-ostensive function can reference its own referential intention S→T and hence solicits the attribution of nonnatural meaning to T. Second, unlike occasionally declarative, and potentially prototypical, gestures such as offering/showing an object, used both by infants (Bates et al. 1975) and apes (Plooij 1978; Russell et al. 1997), pointing operates at distance from T, deemphasizes direct manipulation, and so underlines the noninteractional motivation of the gesture. Various other signals can produce BJA, but their lack of reciprocity, arbitrariness, and explicitly communicative intent make them insufficient to produce explicit awareness of the JA state as such, hindering SJA. Pointing necessarily invites this awareness because it serves only to symmetrically indicate JA.

Because declarative pointing is intrinsically (in Peacocke's terms) "self-involving," reflectivity and reciprocity are both granted: reflectivity, because the ostensive and self-ostensive function of pointing offers to participants an open representation of their intentional JA; reciprocity, because its symmetrical visibility can make both S and R mutually aware that they are representing their direction of attention through the same physical medium. As epitomized by the tale of the naked emperor, declarative pointing necessarily brings about public awareness in which the attended scene indexically refers to itself, incarnating reflective social consciousness in physical form and concretely instantiating SJA. Just as SJA always includes an indication of itself, declarative pointing highlights not merely T but also its own symbolic role in cuing T. Declarative pointing's minimal function includes symbolizing its own capacity to symbolize.

Both apes and humans ritualize gestures to initiate social interactions ("ontogenetic ritualization"; see Tomasello 1999), and, since pointing initiates JA, it is reasonable to assume that under appropriate circumstances, they can ritualize pointing as a symbol of JA. If declaration is implicit to pointing, then why is it rare or absent in wild apes? One reason may be that they don't have the imaginative and perspective-taking skills that allow humans to represent their own bodies as objects of attention, rather than effectors of interaction. A condition for ritualization of pointing in development is that infants expect caregivers' attention (Franco and Butterworth 1996), that is, to share attention to the fact that attention has to be shared. Apes, by contrast, may lack an innate or learned disposition to share imagery and narratives, which may limit opportunities to *notice* self-indexicality and communicative expectations. Another important fact is that most apes lack the intensely (and *arbitrarily*) collaborative child-rearing experience typical of humans, in which parents actively help infants achieve goals that *the infants are not yet aware of*. For example, human infants learn very early to gesticulate to attract and direct the caregiver's attention, to recognize parts of their body as objectual targets of the caregiver's attention and hence to acquire expectations about that attention (Reddy 2005), about its benefits, and about strategies for its manipulation. This self-derived template may then be extended to triadic contexts involving external targets. As we have observed, because human caregivers pervasively attend to their charge's needs, infants learn that manipulation of attention predicts reward across diverse known *and novel* circumstances.

5 Open Knowledge

Our analysis suggests that pointing can arise in communities that possess only BJA abilities, but that once it arises, it instantiates an externalized representation of JA and so scaffolds the acquisition of SJA, which then fully develops only in tandem with other psychological skills (motivational, dispositional, and particularly representational) that promote social cooperation and coordination.

If this hypothesis is correct, declarative pointing is a sufficient condition to bring about SJA, as it both produces and symbolizes JA. But moreover, declarative pointing instantiates the capacity for S and R to share T as a symbol of X(T). Declarative pointing can evoke absent, private contents through publicly and perceptually available targets, hence extending the openness of JA to abstract communications. Let us assume that X is a minimal set of information that allows S and R to map X(T) from T, for example, associating a person of shared interest with the chair in which he or she typically sits. S, who knows X, thinks of the absent person as soon as she sees the empty chair, and wants to share with R this thought. If R knows X and sees S pointing to the chair, R too will think of the absent person. Do S and R both need to know that they both know X? Yes, but only in a minimal sense that does not necessarily involve representational mutual knowledge. A more primitive experience suffices: overtly and jointly representing T (via the pointing gesture) *as a vehicle of occurrent knowledge and common interest*. This "open knowledge" (Peacocke 2005), in our hypothesis, is achieved when pointing is employed as a symbolic medium that serves to make X perceptually salient.

In fact, even if S and R independently imagine, rather than see, the person supposed to sit on the chair, both openly enjoy a public T-centered experience of the person *as if* she were actually in front of their eyes. The symbolic content evoked by pointing has the flavor of direct observation in that it enjoys various features of occurrent perception, such as *passivity*, *spatial localization*, and virtual *autonomy* from the coattenders' beliefs and/ or judgments. R's experience of X(T) would remain anchored to the perceived T even if either S or R believed that X is false, or R judged wrongly the other's particular beliefs about X. Note the difference: Covert contents appreciated through representational mutual knowledge depend entirely on *how* they have been actively inferred and subjectively attributed to the other's mind; on the contrary, those intuitively accessed as open knowledge are passively recognized as objective attributes of T, whether or not they involve inferences about the other's knowledge state. Thus, whereas mutual knowledge refers to a bustle of beliefs and judgments covertly produced by each participant about the others, open knowledge allows participating in a "mutual world," freely accessible by any participant, but virtually independent from them.

Open knowledge is immediate, without nested inferences, because reflective social consciousness grounds in perception the representation of a mutual world: X(T) can be experienced by S and R as open to both if, and only if, their situation of JA to T makes them aware of this very situation. Whereas in BJA only the social affordances related to T are experienced by S and R as openly available to them (knowledge of *how* we can sit on that chair), in SJA the representation of these very social affordances is experienced as equally open and publicly accessible (knowledge *that* somebody usually sits on that chair), provided that S and R jointly attend to T, but regardless of their actual access to each other's minds. This is why every coattender tends to experience (X)T *as if* the others actually know X and know that X is also known by any other coattender.

Open knowledge is more flexible than dispositional mutual knowledge and less demanding than representational mutual knowledge, and occupies an intermediate state between them. Dispositional mutual knowledge requires a repertoire of stereotyped interactive behaviors shared by coattenders: This hinders the declarative character of the communication, because the more communication depends on the immediate interactional context, the more the generality of X(T) is constrained, preventing the necessary reflectivity of SJA. Conversely, representational mutual knowledge requires recursive mindreading between the coattenders, mediated by telementation, which hinders the openness of JA: The more we need to reflect on the other's covert intention, the more we lose the immediacy of symbolic communication and the perceptual-like character of X(T).

Open knowledge is not only phenomenologically distinct from mutual knowledge, it is also less computationally demanding and more robust in creating expectations about its content, as irrelevant recursive mind-reading and behavioral affordances can be bypassed or surrogated to perceptual representation of SJA, provided that both S and R have some at least minimal acquaintance with X and free perceptual access to T. Open knowledge helps explain why infants point "declaratively" even before they gain the ability to adjust their signals to R's attentional state (Franco and Butterworth 1996): For human infant signalers, even the most vague sensitivity to others' attention (which is seemingly innate and shared by all tetrapods; cf. Sewards and Sewards 2002) is sufficient to learn that capturing and manipulating attention is rewarding, because among humans rewards are consistently anchored to associations of the T they have designated (independent of any metarepresentation by S of these associations in R's mind).

6   Conclusion

SJA encompasses three triadic relations that are brought about, directly and simultaneously, by any well-formed act of declarative pointing, that is, any act of pointing that references at the same time the target, T, the self-involving situation, *U*, and the underspecified yet communally relevant referent, X(T).

(1) Perceptual relation (BJA): S and R jointly attend to T, as they directly perceive T and the other's attention toward T;

(2) Self-indexical relation (reflective social consciousness): S and R jointly attend to S's pointing finger and *each of them sees it as a symbol of their U because each of them is aware that they are both aware of attending to it;*
(3) Imaginative relation (open knowledge): Both S and R attend to X(T), and *recognize some X(T) as available for joint contemplation (know-that) as a potential action target or modifier of potential action plans.*

The crucial transition from (1) to (3) is scaffolded by (2): Perceiving the pointing finger as an embodied representation of JA externalizes S's agency from herself and invites the coattenders' identification with a *triadic virtual agent* whose intentions are informed by, but partially dissociated from, coattenders' purely private dispositions. By embodying *U*, pointing comprises a symbolic template for further traffic in externalized representations, and just so, during human development babies learn verbal labels for pointed-out objects. Peacocke (2005) observes that by means of SJA, open knowledge is a precursor, rather than a by-product, of the sophisticated mental devices that usually produce representational mutual knowledge; at the same time, open knowledge is a product, rather than a precondition, of the adaptive interactive skills that constitute dispositional mutual knowledge. Open knowledge provides a representational framework that is more fundamental than representational mutual knowledge but significantly more flexible than dispositional mutual knowledge, in that it manipulates representational contents as well as affordances. In the same way, SJA can arise (in development and evolution) from BJA, for it builds on an open perceptual experience that is immediately and overtly available, but it is not *reducible* to BJA, for it builds on a nonnatural representation of hidden, private knowledge.

If the representational capacity required by reflective social consciousness is reducible neither to specialized adaptive dispositions nor to some preexisting ToM for inference of others' knowledge states, then where does it come from? Hutto's (2008) mimetic ability hypothesis posits the emergence of ToM through symbolic perception, with symbolic perception arising through JA rather than the other way around. In our terms: By symbolically incarnating a communal virtual agent that is only partially dissociated from one's own subjective experience, pointing may provide each coattender with an initial framework for analyzing the covert intentions and beliefs of others. As for the origins of symbolic cognition, Hutto refers to the evolution of a panoply of embodied imaginative-imitative skills: motor memory and reenactments of motor actions, fine-grained imitation of complex motor chains, offline rehearsal of tool-making, perspective shifting, pretended play, and protological and amodal thinking in concrete problem solving. In this account, folk-psychological competence arises not from inferential skills, but from humanity's marked behavioral flexibility, including coordinated behaviors, complex social interactions, and collaborative tool-construction or usage. In other words, the complex imaginative skills necessary to develop hand actions with abstracted or latent goals, such as the construction of tools, are the same skills necessary for gestures to represent abstracted or latent content, as in iconic gesture, ritual dance, or pantomime abilities, which in turn provide embodied support for simple prelinguistic proto-narratives.

An important indicator of how complex behaviors are constructed comes from learning how they are implemented, across species and developmental stages, by actual brains. Mastery of declarative communication surely depends on complex social behaviors related to sharing and modeling experience and on motivation to engage in cooperative or altruistic activities (Tomasello 2009). Such behaviors are unlikely to be independent of the progressive development of motor abstraction (Gallese 2009), metaphorical categorization of social affordances and causal relations (Lakoff and Gallese 2005), or goal emulation and imitation proper (Rizzolatti et al. 2002b). At a more fundamental level, such abilities may depend on the fine intertwinement of observed and enacted brain states including motor intention (Rizzolatti et al. 2002a) and sensory attention (Shepherd et al. 2009), as leveraged by goal-directed motor imagery and remembered and imagined perspective shifts (Currie and Ravenscroft 1997). It will be interesting to see whether and how these proposed neural processes contribute to JA. Our prediction is that hand-centered skillful dispositions, more than specialized inferential mechanisms, play a pivotal role in human social coordination.

References

American Psychiatric Association. 1994. *DSM-IV: Autistic Disorder*. Washington, DC: American Psychiatric Association.

Bates, E., L. Camaioni, and V. Volterra. 1975. Performatives prior to speech. *Merrill-Palmer Quarterly* 21:205–226.

Bigelow, A. E. 2003. The development of joint attention in blind infants. *Development and Psychopathology* 15 (2):259–275.

Brooks, R., and A. N. Meltzoff. 2008. Infant gaze following and pointing predict accelerated vocabulary growth through two years of age: A longitudinal, growth curve modeling study. *Journal of Child Language* 35 (1):207–220.

Campbell, J. 2005. Joint attention and common knowledge. In *Joint Attention: Communication and Other Minds*, 287–297, ed. N. Eilan, C. M. Hoerl, T. McCormack, and J. Roessler. Oxford: Oxford University Press.

Cheney, D. L., and R. M. Seyfarth. 1990. *How Monkeys See the World: Inside the Mind of Another Species.* Chicago: University of Chicago Press.

Csibra, G. 2010. Recognizing communicative intentions in infancy. *Mind & Language* 25 (2):141–168.

Currie, G., and I. Ravenscroft. 1997. Mental simulation and motor imagery. *Philosophy of Science* 64 (1):161–180.

Franco, F., and G. Butterworth. 1996. Pointing and social awareness: Declaring and requesting in the second year. *Child Language* 23:307–336.

Fouts, R., and Mills, S. T. 1997. *Next of Kin.* New York: William Morrow.

Gallese, V. 2009. Motor abstraction: A neuroscientific account of how action goals and intentions are mapped and understood. *Psychological Research* 73 (4): 486–498.

Gallagher, S., and D. Hutto. 2008. Understanding others through primary interaction and narrative practice. In *The Shared Mind: Perspectives on Intersubjectivity*, 17–38, ed. J. Zlatev, T. P. Racine, C. Sinha, and E. Itkonen. Amsterdam: John Benjamins.

Grice, H. P. 1957. Meaning. *Philosophical Review* 66:377–388.

Harris, P. 1996. Desires, beliefs, and language. In *Theories of Theories of Mind*, 200–222, ed. P. Carruthers and P. Smith. Cambridge: Cambridge University Press.

Hoyt, M. 1941. *Toto and I.* New York: Lippincott.

Hutto, D. 2008. *Folk Psychological Narrative: The Sociocultural Basis of Understanding Reasons.* Cambridge, MA: MIT Press.

Hutto, D. 2011. Elementary mind minding, enactivist-style. In *Joint Attention: New Developments in Philosophy, Psychology, and Neuroscience*, 307–341, ed. A. Seemann. Cambridge, MA: MIT Press.

Kendon, A. 2004. *Gesture: Visible Action as Utterance.* Cambridge, MA: Cambridge University Press.

Kirsh, D. 2010. Thinking with external representations. *AI & Society* 25:441–454.

Kita, S., ed. 2003. *Pointing: Where Language, Culture, and Cognition Meet.* Mahwah, NJ: Erlbaum.

Knoblich, G., and N. Sebanz. 2008. Evolving intentions for social interaction: From entrainment to joint action. *Philosophical Transactions of the Royal Society of London* 363 (1499):2021–2031.

Lakoff, G., and V. Gallese. 2005. The brain's concepts: The role of the sensory-motor system in reason and language. *Cognitive Neuropsychology* 22:455–479.

Leavens, D. A. 2004. Manual deixis in apes and humans. *Interaction Studies: Social Behaviour and Communication in Biological and Artificial Systems* 5:387–408.

Leavens, D. A. 2011. Joint attention: Twelve myths. In *Joint Attention: New Developments in Philosophy, Psychology, and Neuroscience*, ed. A. Seemann. Cambridge, MA: MIT Press.

Leavens, D. A., and T. P. Racine. 2009. Joint attention in apes and humans: Are humans unique? *Journal of Consciousness Studies* 16 (6–8):240–267.

Leonard, H. S. 1959. Interrogatives, imperatives, truth, falsity, and lies. *Philosophy of Science* 26 (3):172–186.

Liszkowski, U., M. Carpenter, A. Henning, T. Striano, and M. Tomasello. 2004. Twelve-month-olds point to share attention and interest. *Developmental Science* 7:297–307.

Liszkowski, U., M. Carpenter, T. Striano, and M. Tomasello. 2006. 12- and 18-month-olds point to provide information for others. *Journal of Cognition and Development* 7:173–187.

Liszkowski, U., M. Carpenter, and M. Tomasello. 2007. Pointing out new news, old news, and absent referents at 12 months. *Developmental Science* 10:F1–F7.

Liszkowski, U., M. Schaefer, M. Carpenter, and M. Tomasello. 2009. Prelinguistic infants, but not chimpanzees, communicate about absent entities. *Psychological Science* 20:654–660.

Malafouris, L. 2007. Before and beyond representation: Towards an enactive conception of the palaeolithic image. In *Image and Imagination: A Global History of Figurative Representation*, 289–302, ed. C. Renfrew and I. Morley. Cambridge: The McDonald Institute for Archaeological Research.

McNeill, D., S. Duncan, J. Cole, S. Gallagher, and S. Bertenthal. 2008. Either or both: Growth points from the very beginning. *Interactional Studies* 9 (1):117–132.

Menzel, C. R. 1999. Unprompted recall and reporting of hidden objects by a chimpanzee (*Pan trogolodytes*) after extended delays. *Journal of Comparative Psychology* 113:426–434.

Moll, H., M. Carpenter, and M. Tomasello. 2007. Fourteen-month-olds know what others experience only in joint engagement. *Developmental Science* 10 (6): 826–835.

Moore, R. Forthcoming. Cognising communicative intent.

Mundy, P., J. Block, C. Delgado, Y. Pomares, A. V. van Hecke, and M. V. Parlade. 2007. Individual differences and the development of joint attention in infancy. *Child Development* 78 (3):938–954.

Peacocke, C. 2005. Joint attention: Its nature, reflexivity, and relation to common knowledge. In *Joint Attention: Communication and Other Minds*, 298–324, ed. N. Eilan, C. M. Hoerl, T. McCormack, and J. Roessler. Oxford: Oxford University Press.

Plooij, F. X. 1978. Some basic traits of language in wild chimpanzees? In *Action, Gesture, and Symbol*, 111–131, ed. A. Lock. London: Academic Press.

Povinelli, D. J., J. M. Bering, and S. Giambrone. 2002. Chimpanzees' "pointing": Another error of the argument by analogy? In *Pointing. Where Language, Culture, and Cognition Meet*, 35–68, ed. S. Kita. Mahwah, NJ: Erlbaum.

Reddy, V. 2005. Before the "third element": Understanding attention to self. In *Joint Attention: Communication and Other Minds*, 85–109, ed. N. Eilan, C. M. Hoerl, T. McCormack, and J. Roessler. Oxford: Oxford University Press.

Rizzolatti, G., L. Craighero, and L. Fadiga. 2002a. The mirror neuron system in humans. In *Mirror Neurons and the Evolution of Brain and Language*, ed. M. Stamenov and V. Gallese. Amsterdam: John Benjamins.

Rizzolatti, G., L. Fadiga, L. Fogassi, and V. Gallese. 2002b. From mirror neurons to imitation: Facts and speculations. In *The Imitative Mind: Development, Evolution, and Brain Bases*, ed. A. Meltzoff and W. Prinz. Cambridge, MA: Cambridge University Press.

Russell, C. L., K. A. Bard, and L. B. Adamson. 1997. Social referencing by young chimpanzees. *Journal of Comparative Psychology* 111:185–193.

Ryle, G. [1945] 1971. Knowing how and knowing that. In *Gilbert Ryle: Collected Papers*. vol. 2., 212–225. New York: Barnes and Noble.

Savage-Rumbaugh, E. S., K. MacDonald, R. A. Sevcik, W. D. Hopkins, and E. Rubert. 1986. Spontaneous symbol acquisition and communicative use by pygmy chimpanzees (*Pan paniscus*). *Journal of Experimental Psychology: General* 115:211–235.

Schiffer, S. 1972. *Meaning*. Oxford: Clarendon Press.

Seemann, A. 2007. Joint attention, collective knowledge, and the "we" perspective. *Social Epistemology* 21 (3):217–230.

Seemann, A., ed. 2011. *Joint Attention: New Developments in Philosophy, Psychology, and Neuroscience*. Cambridge, MA: MIT Press.

Sewards, T. V., and M. A. Sewards. 2002. Innate visual object recognition in vertebrates: Some proposed pathways and mechanisms. *Comparative Biochemistry and Physiology, Part A: Molecular & Integrative Physiology* 132 (4):861–891.

Shepherd, S. V., J. T. Klein, R. O. Deaner, and M. L. Platt. 2009. Mirroring of attention by neurons in macaque parietal cortex. *Proceedings of the National Academy of Sciences of the United States of America* 106 (23):9489–9494.

Stephen, S. V. 2010. Following gaze: Gaze-following behavior as a window into social cognition. *Frontiers in Integrative Neuroscience* 4:5.

Shepherd, S. V., and M. Cappuccio. 2011. Sociality, attention, and the mind's eyes. In *Joint Attention: New Developments*, ed. A. Seemann. Cambridge, MA: MIT Press.

Sperber, D. 2000. Metarepresentations in an evolutionary perspective. In *Metarepresentations: A Multidisciplinary Perspective*, 117–137, ed. D. Sperber. Oxford: Oxford University Press.

Sperber, D., and D. Wilson. 1986. *Relevance: Communication and Cognition*. Cambridge, MA: Harvard University Press.

Tomasello, M. 1995. Joint attention as social cognition. In *Joint Attention: Its Origins and Role in Development*, 103–120, ed. C. Moore and P. Dunham. Hillsdale, NJ: Erlbaum.

Tomasello, M. 1999. *The Cultural Origins of Human Cognition*. Cambridge, MA: Harvard University Press.

Tomasello, M. 2006. Why don't apes point? In *Roots of Human Sociality: Culture, Cognition and Interaction*, 506–524, ed. N. J. Enfield and S. C. Levinson. Oxford: Berg.

Tomasello, M. 2008. *Origins of Human Communication*. Cambridge, MA: MIT Press.

Tomasello, M. 2009. *Why We Cooperate*. Cambridge, MA: MIT Press.

Tomasello, M., M. Carpenter, J. Call, M. T. Behne, and H. Moll. 2005. Understanding and sharing intentions: The origins of cultural cognition. *Behavioral and Brain Sciences* 28:675–735.

Veà, J. J., and J. Sabater-Pi. 1998. Spontaneous pointing behaviour in the wild pygmy chimpanzee (*Pan paniscus*). *Folia Primatologica* 69:289–290.

Waal, F. B. de. 2003. Darwin's legacy and the study of primate visual communication. *Annals of the New York Academy of Sciences* 1000 (1):7–31. doi: 10.1196/annals.1280.003.

Wilby, M. 2010. The simplicity of mutual knowledge. *Philosophical Explorations* 13 (2):83–100.

V MANIPULATION AND THE MUNDANE

14 Privileging Exploratory Hands: Prehension, Apprehension, Comprehension

Susan A. J. Stuart

Through our hands we construct our world and through our construction of our world we construct ourselves. We reach with our hands and touch with our hands, and with this reaching and touching we come to understand how things feel and are. It is not an utterable knowledge, yet it is knowing the world in a dynamically engaged, affective, effective way. Through affective feedback our reaching and touching becomes a prehensive grasping which leads, through the enkinesthetic givenness of the agent with its world, to a situated and embodied knowing, and the rudiments of apprehension. With each fresh comprehension a new enkinesthetic enquiry is engendered; with each enquiry we have afresh the anticipatory dynamics of reaching, touching and feeling, with the hand-to-object of world-investigation, the hand-to-body of auto-investigation and investigating the Other.

1 Introduction

It is reputed that Kant once said "The hand is the window to the mind," though a direct quotation that possesses a neat and easy reference is unlikely to be forthcoming. Yet it is possible to follow some of Kant's earlier reasoning[1] to demonstrate how this thought can well be ascribed to him. What's more, we can show how essential it is for his later writing, where, although he avoids any obvious form of ontological commitment to the nature of the human mind and any direct commitment to the physicality of the hands and body, he nonetheless must do so, not merely as a transcendental condition, but because they perform an essential corporeal and affective role in orientating the subject in space and establishing that subject's egocentric point of view. Our hands are the richly sensitive instruments with which we begin the sensory inquiry of lived experience, and it is through their feeling-engagement that we construct our world. I will support this claim in this chapter by appeal to the affective kinesthetic and enkinesthetic engagement with which we are necessarily bound in a relation of felt "withness" with other agents, organisms, and things. As a consequence of constructing our experiential world, we discover ourselves as both transcendental condition—there being a necessary a priori subject of experience—and an antecedent physical entity, with our sensuous embodiment making prehension, apprehension (by which I mean knowing that we know[2]), and ultimately, a broader comprehension possible.

In this discussion, I will use the term "enkinesthesia" to mean the reciprocally affective neuromuscular dynamical flows and muscle tensions that are felt and enfolded between coparticipating agents. We move, touch, and change one another in myriad different ways, with touch having a "phenomenological primacy over the other senses" (Ratcliffe, this vol.). To touch is to awaken to awareness, whether it be direct or indirect touch. Direct touch includes the physical touch of a caress, a pat on the back, a hug, or the rebuff of a pulling away from contact or the strike of a hand. Indirect touch can be achieved through a look where one becomes the object of someone else's attention and experience,[3] for example, in an unspoken admonishment of a look or a wagging finger, a papal blessing that can absolve us of our sins, a friend's wave from a departing train, the touch of eyes that meet across a room, or in the way words and language can alter the way we feel. And, so, it is clear that although there is no distinct organ of touch, our hands and our skin incorporate the other senses, and these inform the proprioceptive, kinesthetic, and, because no touch is ever unidirectional, enkinesthetic senses with their relative feedback. This is especially evident in the intertwining or chiasm of clasped hands, where the touching hand is itself touched, or when I touch my own body in hand-to-body auto-investigation, for then I experience both the touching and the being touched. In that instant I am both subject and object:

This can happen only if my hand, while it is felt from within, is also accessible from without, itself tangible, for my other hand, for example, if it takes its place among the things it touches, is in a sense one of them, opens finally upon a tangible being of which it is also a part. Through this crisscrossing within it of the touching and the tangible, its own movements incorporate themselves into the universe they interrogate. . . . We must habituate ourselves to thinking that every visible is cut out in the tangible, every tactile being in some manner promised to visibility, and that there is encroachment, infringement, not only between the touched and the touching but also between the tangible and the visible. . . . There is double and crossed situating of the visible in the tangible and of the tangible in the visible. (Merleau-Ponty 1968, 133–134)

In sensing our world in the hand-to-object of world-investigation or hand-to-body of Other-investigation, our hands make contact in a way that our eyes and ears do not: "in touch we directly appeal to the tactile properties of our own bodies in investigating the self-same tactile properties of other bodies" (O'Shaughnessy 1989, 38). Yet, like the other senses, touch is inextricable from a dynamic moving, sensing, and acting living body.

I will develop my argument by using Kant's 1768 essay "Concerning the Ultimate Ground of the Differentiation of Directions in Space" in conjunction with Woelert's recent article on "Kant's hands" (Woelert 2007) to draw out why a failure in functional vertical symmetry establishes our physical spatiality and provides us with a situated perspective on a world of other agents, organisms, and things.[4] In addition I will re-present Woelert's claim that Kant's arguments, in this early period, anticipate claims for the constitutive force of the body, yet these are claims that are more usually associated with the phenomenology of the early twentieth century. This emphasis on the body is important because an agent's action constitutes an affectively laden interrogation of its world, that is, a nonpropositional somatosensory questioning of how our world is for us now and how we anticipate it will continue to be, and this is facilitated by our kinesthetic and enkinesthetic engagement. What we discover in this fundamental action is not, at first, an utterable knowing—it is not associated with speech or formal conceptualization; it is preconceptual. Reaching and touching constitute a prehensive grasping[5] that forms the essential background to a situated and embodied knowing, and this provides the rudiments of apprehension. With each fresh knowing, or embodied comprehension, a new enkinesthetic enquiry is engendered. With each inquiry we have afresh the anticipatory dynamics of reaching, touching, and feeling, with the hand-to-object of world-investigation, the hand-to-body of auto-investigation, and the hand-in-hand of investigating the Other. In the second half of the chapter I will turn to the body and specifically to the hands, arguing that all the senses but, with greatest intensity, the hands and touch open new horizons of sensuous opportunity, proximity, and closeness to others: agents, organisms, and things.

2   Enantiomorphic Hands and Their Constitutive Role in Our Experience

By means of a geometrical proof, Kant demonstrates that our hands play a crucial role in proving the existence of an absolute space that exists around us and through which we move. Though his proof is geometrical in spirit, its consequence is phenomenological, and most interestingly phenomenological because, of all the possibilities within our functional vertical symmetry, ears, nostrils, eyes, arms, legs, and so on, he chooses the hands, hands that feel and grasp and bring us explicitly into a prenoetic epistemic phenomenology.

If we permit ourselves a momentary consideration of this issue, we can begin to make sense of some of the reasons why Kant may have chosen the hands to exemplify the notion of irreconcilable spatial opposition and our physical being in the world. For one thing, our hands, but not our nostrils, eyes and ears, or tongue, are laid out before us and can become the focus of these other senses. For another, in their dexterous, and sometimes not so dexterous, fingering and handling engagement, our hands express our predominantly prepropositional questions about how our world continues to be—searching, fumbling, reaching, touching, grasping, lifting, holding, caressing, stroking, or feeling the size, shape, texture, tension, density, softness, and pliability of that with which we are in immediate encounter. Through this swiftly habituated affective tactual perception, our hands become the means for a great deal of our skillful engagement with our world. Yet, there is a problem. As Frank R. Wilson states in his prologue: "Our lives are so full of commonplace experience in which the hands are so skillfully and silently involved that we rarely consider how dependent upon them we actually are" (1998, 3). They are our sensuous root in and route into our world; yet as a result of their success they become, in many ways, invisible.[6] From as early as eleven weeks, the fetus stretches, reaches, and touches the local world within which it is situated.

Human foetuses tentatively touch the placenta, umbilicus and the uterine wall with their hands at 11 weeks. They make jaw movements and swallow amniotic fluid, expressing pleasure or disapproval at tastes injected into it by sucking and smiling or grimacing with disgust. Complex movements of trunk, arms and legs position the body, and may react to the mother's body movements and the contractions of the muscles of her uterus (Lecanuet et al., 1995; Piontelli, 2002; Trevarthen et al., 2006). (Trevarthen and Reddy 2007, 52)

When asked "where are you" we often gesture vaguely to our chest, the site of our felt breathing and heartbeat, or to somewhere behind our eyes, even though we only sometimes consciously direct their gaze. We rarely open our hands out in front of our body and say "This is where I am." We can look down at our chest, but our eyes, ears, and nostrils are invisible to us unless we perceive them in a mirror or touch them with our hands.

Our arms and legs may raise us up and even propel us forward into the space around us, yet it is our hands that disclose our world to us, sensually eliminating the apparent barrier between us and our world.[7] They express explicitly our intertwining with our world where we oscillate between experiencing as perceiving object and being the subject of perception: "When I press my two hands together, it is not a matter of two sensations felt together as one perceives two objects placed side by side, but of an ambiguous set-up in which both hands can alternate the rôles of 'touching' and being 'touched'" (Merleau-Ponty 1962, 93; see also Depraz, this vol.). These points, I will maintain, were, in some form, apparent to Kant.

Let's return to Kant's proof. His intention is to prove that "absolute space, independently of the existence of all matter and as itself the ultimate foundation of the possibility of the compound character of matter, has a reality of its own" (Kant 2002, 366; emphasis in original), and the basis for his claim is that it is only with reference to the sides of our body, and more specifically our hands, that we can even begin to make sense of geographical orientation (*Gegend/Gegenden*, according to Walford 2001) and the notion we have of distinct places (*Lagen*). His formal argument states that

If two figures drawn on a plane surface are equal and similar, then they will coincide with each other. But the situation is often entirely different when one is dealing with corporeal extension [bodies]. . . . They can be exactly equal and similar, and yet still be so different in themselves that the limits of the one cannot also be the limits of the other. . . . The most common and clearest example is furnished by the limbs of the human body, which are symmetrically arranged relative to the vertical plane of the body. The right hand is similar and equal to the left hand. And if one looks at one of them on its own, examining the proportion and the position of its parts to each other, and scrutinising the magnitude of the whole, then a complete description of the one must apply in all respects to the other, as well. . . . I shall call a body which is exactly equal and similar to another, but which cannot be enclosed in the same limits as that other, its incongruent counterpart. (Kant 2002, 369–370)

His point is clear: The intrinsic descriptions of the relations of the parts of a hand are equivalent whether we refer to the left or the right hand, but the hands themselves are mirror images of one another. They are enantiomophs, which evade "integration into a homogenous and purely self-referential form of space" (Woelert 2007, 139), within which they could dissolve and disappear from the physical world. It is the fact that they do not and cannot dissolve and disappear that is crucially important. Our hands are equal and similar in the relation of their parts, but their distinct orientations demonstrate that they are topologically nonidentical. As Hanna says:

Incongruent counterparts are perceivable mirror-reflected spatial duplicates that share all the same monadic properties, have exactly the same shape and size, and correspond point-for-point, but are in different places and cannot be made to coincide by rigid translation within the same global orientable space (an orientable space is a space with intrinsic directions). Even more briefly put, incongruent counterparts are enantiomorphs. Enantiomorphs are qualitatively identical but topologically non-identical. (Hanna 2008, 53–54)

It is their topological nonidentity, emphasized by their inherent directionality, that forces them apart into different places within a space that is ontologically, but—though it seems unnecessary to say it—not epistemically or phenomenologically, independent of us. Kant continues:

Imagine that the first created thing was a human hand. That [hand] would have to be either a right hand or a left hand. The action of the creative cause in producing the one would have of necessity to be different from the action of the creative cause producing the counterpart.

Suppose that one were to adopt the concept entertained by many modern philosophers, especially German philosophers, according to which space simply consists in the external relation of the parts of matter which exist alongside each other. . . . [Since] there is no difference in the relation of the parts of the hand to each other, and that is so whether it be a right hand or a left hand; it would therefore follow that the hand would be completely indeterminate in respect of such a property. In other words, the hand would fit equally well on either side of the human body; but that is impossible. (Kant 2002, 371)

This impossibility, because there exist failures in our functional vertical symmetry, makes it possible to establish both our physical spatiality as corporeal bodies and our individually situated perspectives on our world. Additionally, and crucially, it makes possible our ability to align or orient ourselves in space, first in relation to the directionality of our hands in their touch and in their grasp, and then in their motile relation to other objects, agents, people, and our own body.[8] As Hanna says, Kant's argument thus demonstrates "that the actual space of perceivable material bodies is intrinsically directional (i.e. orientable) and egocentrically-centered" (Hanna 2008, 55). To take a concrete example, we understand the orientation of a map and the direction shown by a compass needle in relation to how we, with our hands, are holding the instruments and, thus, in orientation to our corporeally extended body. This is not to deny other salient cues (or clues), like the orientation of the written script, the positioning of the cardinal points, and so on, but recognizing them and grasping them as meaningful comes much later in a mediated conceptualization of what is, at first, sensuously and immediately to hand. This two-phase understanding is not lost on Kant, for he says, "Even our judgements relating to the cardinal points of the compass are, in so far as they are determined in relation to the sides of our body, subject to the concept that we have of directions in general" (Kant 2002, 367). Fundamentally we possess a "referential anchoring in the oriented structure of the human body" (Woelert 2007, 142), and this is understood with reference to "universal space as a unity" (Kant 2002, 365).[9] Thus Woelert continues: "It is only via our body that we can distinguish between these Gegenden[10] in space, and it is only in doing so that we (as situated in space through our body) are able to actually grasp space as a whole in the first place. . . . space is of direct practical use only when we can relate it to the primordial spatiality of our body" (Woelert 2007, 142).

Key to this passage are the notions of "via our body" and "grasp"; the former because of its emphasis on the processual physicality of movement, sensing, orientation, and location, and the latter because, although it is used as a metaphor for understanding, it appeals to our original prelinguistic manual prehension of our world and how our sensory engagement comes to have meaning for us.[11] It is within this intricate sensory, manual, somatic, affective interplay—the kind we associate primarily with the reaching, touching, groping, and grasping of young children and mistakenly consider secondary in the language-using organism—that we embody our pragmatic concernful relations (Merleau-Ponty 1962; Heidegger 1968; Gendlin 1992). This is merely to say, as others have said before, that it is from the cycle of perception and action that meaning for the organism emerges. (See, e.g., Lakoff and Johnson 1980; von Uexküll 1982; Johnson 1987; Hodges 2007; Hodges and Baron 2007; Steffensen and Hodges 2010.) "Every action, therefore, that consists of perception and operation imprints its meaning on the meaningless object and thereby makes it into a subject-related meaning-carrier in the respective Umwelt (subjective universe)" (von Uexküll [1940] 1982, 31). In this respect, the moving, feeling body has a constitutive role, for every action is affectively replete with concernful, values-realizing exploration within the immanent habitus of the agent.

3   Hands as the Orienting Structures of Self-Referential Anchoring

Referential anchoring begins with the body but is phenomenally most striking through the orientating structures disclosed in the unfolding of our hands in space before us and around us. As our hands open out with their palms facing up, we feel, as it were, from the inside, and see, as it were, from the outside, ourselves, oriented against background phenomena. In the dark, my groping with my right hand does not feel like groping with my left; they have a distinct phenomenal orientation that forms within us, passively and involuntarily, and constructs a distinct phenomenal-somatic alignment with our world. When I grasp the apple in front of me with my right hand, my thumb is on the left; when I grasp it with my left hand the position of my thumb is reversed, and the same is true if one hand is grasping the other; they each possess a unique orientation or directionality. So it is that each hand, in its affective-anticipatory, enkinesthetic engagement, has a distinctive feel that brings with it, implicitly and involuntarily, the direction of my body. If we think again of the map, there are lots of cues for its orientation in relation to us, but primarily these are phenomenal and distinctly left and right; I could not have my hands crossed, or have the map behind me or to one side, without the phenomenal experience being entirely distinct and, what's more, distinguishable at a somatosensory kinesthetic, and thus, enkinesthetic level.

This deeply felt engagement takes place amid the multifarious modes of prenoetic self-givenness,[12] but in the movement of our hands this givenness has a distinctive flavor; it is one in which we experience simultaneously the interiority of "auto-affection" (Henry 1963/1973) alongside the putative exteriority of the visual sensation. This is not a characteristic shared by any other sense, because of the phenomenological primacy of touch. In the plenisentient activity of the senses and, thus, in our experience more broadly, our hands are inextricably entangled. Their entanglement possesses a "presentational immediacy," whether it is of one hand touching or grasping the other or the "perceptive mode in which there is clear, distinct consciousness of the 'extensive' relations of the world" (Whitehead 1978, 61); it brings about the collapse of the artificial interior-exterior sensory division and is suggestive of an altogether more radical interiority (Henry 1963/1973). The synthesis of such a radical interiority is characterized by its immanence, its direct nonduality, through which the senses, including the enkinesthetic sense and the proprioceptive and vestibular systems, constitute the agent's "sense-full" givenness. In this context, "sense-full" implies replete, that is, full of the senses (plenisentient), but it also implies coherent in the sense that "the reactions of an organism are not edifices constructed from elementary movements, but gestures gifted with an internal unity" (Merleau-Ponty 1963, quoted in Baldwin 2004, 51), and full of meaning in its pragmatic values-realizing, concernful enaction.[13]

Concernful enaction emphasizes agential, sensory, and "corporeal anticipation" (Sennett 2008), described enkinesthetically, because of its inherent affective community and reciprocity, as the organism's anticipatory affective dynamics (Stuart 2010, 2011, 2012); and such dynamics make possible the formation of kinesthetic memories, melodies, and habits. "With the development of motor skills the individual impulses are synthesized and combined into integral kinesthetic structures or kinetic melodies" (Luria 1973, 176) which possess an "internal articulation and as a kinetic melody gifted with a meaning . . . [they carry within themselves] . . . an immanent intelligibility" (Merleau-Ponty 1963, quoted in Baldwin 2004, 51). As Merleau-Ponty says, "habit has its abode neither in thought nor in the objective body, but in the body as mediator of a world" (Merleau-Ponty 1962, 168); and there is no more palpable sensory mediator than our hands and our skin. It is through our hands that we first develop kinesthetic habits, such as grasping a rattle or taking hold of our foot, which bring with them expectations and anticipations of how our world will continue to be.

In a prenoetic sense, we speak of the lived and knowing body, as opposed to the "Cartesian corpse" (Leder 1992)[14] of our practices as skillful, and of the way in which habitual actions can be characterized by the readiness-to-hand of absorbed coping, possessing a "practical non-thetic intentionality [that] has nothing in common with a cogitatio (or a noesis) consciously orientated towards a cogitatum (a noema), . . . [but is] rooted in a posture, a way of bearing the body (a hexis), a durable way of being of the durable modified body which is engendered and perpetuated, while constantly changing . . . in a twofold relationship, structured and structuring, to the environment" (Bourdieu 2000, 143–144). By "twofold" Bourdieu is not endorsing a dichotomy; rather, he is emphasizing the processual dynamic structural coupling that exists for organisms and their worlds. As a recursively dynamic coupling it must be enkinesthetic, having the capacity to move and be moved, to shape and be shaped, and comprising no well-defined boundaries between agents, actions, and objects.

If one may consider the environment of a system as a structurally plastic system, the system and its environment must then be located in the intricate history of their structural [enkinesthetic] transformations, where each one selects the trajectory of the other one. (Varela 1989; quoted in Bitbol and Luisi 2004, 102)

Cotterill (2001) comments on anticipation when he speaks of the difference between probing-by-movement associated with simple organisms like *Eschericia coli* and *Euglena gracilis* and probing-by-proxy associated with organisms that have both a central nervous system and muscle spindles. In both cases, movement is involved, but in the latter it can be as a covert modification of existing reflexes, making possible the anticipation of potential actions, something that is crucial because "understanding is not concerned with grasping a fact but with apprehending a possibility of being" (Ricoeur 1981, 56). It is only in this way that we can prepare for fulfilled anticipations, and for contingency and surprise.

We are not conscious of something unless [we] are covertly setting up the muscular movements associated with its perception, and such movements will have had to be learned from prior experience. The mechanism thus mediates influence on a currently proceeding (or currently planned) muscular act. That influence stems from motivation-triggered anticipation of the act's outcome. (Cotterill 2001, 7)

In the preceding paragraph in his text, there is even greater emphasis on the anticipatory mechanism in relation to the prehension and possible apprehension of the intentional trajectory of the Other's action and movement: "The clue, I strongly feel, lies in the word anticipation. If we are not actively attending to something in our environment, our nervous system will not be able to predict the likely continuation of that something's present trend" (ibid.).

Let's bring this discussion back to Kant and the crucial constitutive role in experience played by the hands and the body. We have already seen that it is through our hands that we begin the richly sensitive felt inquiry that composes our lived experience, and synchronous with this inquiry is the composition or construction of a spatially extended experiential world. With their nonreductive nature, Kant has shown that our qualitatively identical but topographically nonidentical hands play an essential corporeal role in establishing an external world and orientating us in space; but he is also aware of the affective nature of this experience, that the phenomenal subject is not an abstract yet logically necessary transcendental condition, but a sensuous physical entity.

In an intriguing paper written in 1766, Kant claims that "I am just as immediately in the tips of my fingers, as in my head" (Kant [1766] 1900, 149), for, he argues, "where I sense, there I am":

No experience teaches me to believe some parts of my sensation to be removed from myself, to shut up my Ego into a microscopically small place in my brain from whence it may move the levers of my body-machine, and cause me to be thereby affected. . . . *My soul is as a whole in my whole body, and wholly in each part.* (Ibid.; emphasis in the original)

In these claims, Kant brings together the notions of a transcendental egocentric point of view with an embodied, phenomenologically rich subject of experience. In other words, it is a geometrically borne, physically extended, enkinesthetically embedded subject that feels its way, searching and fumbling, reaching and touching, grasping, lifting, holding, caressing, and stroking, toward a prenoetic prehensive comprehension. In this unceasing process of investigation and manipulation the agent experiences—from moment to moment—felt epiphanies, when the immanence of our engagement is disclosed[15] and reveals an ever-changing range of new action possibilities. Through this perpetual exploratory sensuous entanglement, we can know how to go on without knowing reflexively and propositionally that it is we who know how to go on. This is not to suggest an absence of egocentricity, for it is clear from Kant's writing that a first-person perspective, as a nomologically necessary transcendental subject, is still present. What we need to understand from his writing is that the egocentric point of view exists but is known only because of the agent's primitive and pervasive sensuous phenomenology:

I hold, [that] this ability of a conscious living organism like us to have a single point of view is grounded in egocentrically centered embodiment, and a primitive bodily awareness that includes proprioception (the sense of one's own body parts and limbs), kinesthesia (the sense of bodily movement), the sense of orientation and balance, bodily pleasures and pains, tickles and itches, the feeling of pressure, the feeling of temperature, the feelings of vitality or lethargy, and so-on. (Hanna 2008, 59)

It is with, and only with, my body that I am able to exist in relation to things in the world, and it is with my hands that I am able to directly manipulate and bring about change in my world, and yet at the same time it is through my bodily sensing and touching engagement that I can experience, most readily, the immanence of the Other in my enkinesthetically attuned absorbed coping. Whitehead presents this forcefully when he says that "the 'withness' of the body is an ever-present, though elusive element in pure perceptions of presentational immediacy" (Whitehead 1978, 474–475). The affectively rich, or plenisentient, givenness of the body is a necessary accompaniment to my experience, for the "primordial locality of the lived human body becomes something that I am . . . passively and habitually instituting the field of our subjectivity" (Woelert 2007, 147). Though Kant doesn't refer to it in these terms, it is still the field of subjectivity, the "withness" of the body, and the inherent directionality of felt engagement, to which he refers in this passage: "Considering the things which exist outside ourselves: it is only in so far as they stand in relation to ourselves that we have any cognition of them by means of the senses at all. It is not therefore surprising that the ultimate ground on the basis of which we form our concept of directions in space, derives from the relation of these intersecting planes to our bodies" (Kant 1992, 366).

There is an enkinesthetic emphasis on the field of subjectivity that is necessarily composed of the sensuous activity of affective "being-with" other agents and objects in the world: "The human body invokes a correction according to which space . . . is conceived not as a passive representation but as an active process essentially structured by the subject's directed and sensuous activity themselves" (Woelert 2007, 146). Thus we can understand what it means when we say that it is through our hands that we construct our world, that is, our field of subjectivity, and through our construction of our field of subjectivity that we construct ourselves.

4   Conclusion

Let's consider the place of our hands in the world; they don't emanate from nowhere, they belong in orientation and alignment with the rest of the body, but they are extended from the core of the body and reach out to touch and manipulate what they perceive. They move in relation to one another, frequently acting in conjunction, but sometimes not, and in this way they cannot be thought passive. Their activity in our world is located in the intricate history of our structural, enkinesthetically felt coupling. They encounter perturbations, feel difference, and engender change. It is swiftly apparent just how different they are from the eyes or other sense organs, though it is the eyes that are usually taken to be our predominant sense. While it is true that we can touch someone with a look or affect someone with a gaze, and it is true that all our sensory activity contributes to the formation of our subjective field, it is with our hands that we most directly grasp and engage with our sensory engagement. Through the eyes alone we have managed to imagine ourselves a disembodied ego with a narrow to invisible field of subjectivity. But this cannot be done if we accept our hands and our skin as our fundamental or primordial means for affective, effective engagement with our world. Where the hands are recognized as the body's animate instruments of direct touch and change, the field of subjectivity cannot consist only of the pure ego or of transcendental structures alone. Historically, we seem to "have ignored the more inarticulate background of feeling in which those distinct elements are imbedded" (Cory 1933, 31), and if now we "go back to the facts" and accept Kant's arguments for the constitution of an absolute and phenomenal space, we will be compelled to accept both that "the transcendental functioning of space implies the transcendental functioning of the human body" (Woelert 2007, 145) and that our field of subjectivity is wide, sensuously rich, and constant only in its capacity for change.

If the hand is the window to the mind, it is only because the hand can grasp its world, prehending and comprehending it in ways that make the organism more effective in its engagement, and to do this it needs a moving, feeling body that, in its affective enkinesthetic activity, can anticipate and enact the hand-to-object of world-investigation, the hand-to-body of auto-investigation, and the investigation of the Other. It is only then that the hand can construct its world and the mind can be revealed.

Acknowledgments

The author would like to acknowledge many very enjoyable conversations with Norman Gray and Alex South about geometry, tentacles, position, directionality, and the fivefold rotational symmetry of the starfish.

Notes

1. The "earlier" period referred to here is Kant's "precritical" period, and is associated with the period 1755–1770. It's rather a shame to use the term "precritical" as Shell's marvelous book, *The Embodiment of Reason Kant on Spirit, Generation, and Community* (1996), attests. Kant's "critical" period ranges from 1781 to 1796, in which, perhaps most significantly, the *Prolegomena* and the three *Critiques* were published.

2. The action of grasping with the intellect, including forming an idea or conception.

3. For an interesting elaboration of how we can be affected by the look of another, see ch. 1 of Part 3 of *Being and Nothingness* by Sartre ([1945] 1956).

4. The term "functional vertical symmetry" is used deliberately to emphasize the necessity of movement and action in speaking about alignment and orientation, for enantiomorphism alone cannot explain our egocentric orientation. Our hands might very well not have been enantiomorphic—it is, after all, only a contingent feature of our bodily structure—yet we would still have and would be able to speak about a right and a left side; and right and left can only be demonstrated and comprehended in the functioning of our hands, or arms, legs, shoulders, and so on.

5. The term "grasping" is deliberately ambiguous, taken to mean both taking hold with our hands and understanding.

6. One might consider in this context the phrase that has become a byword for familiarity: "I know X like the back of my hand"—though when asked to describe the back of our hand we find ourselves at a complete loss.

7. As Stuart puts it:

The boundaries which seem to separate us from our worlds open us up to those worlds and reveal to us our inseparability from them. Those boundaries which can appear, at first, rigid and fixed are often malleable and semi-permeable. We need think only of the skin with its surfaces within surfaces (Hoffmeyer 2008, 17–38), the biological membranes of stratum corneum, epidermis, dermis and subcutaneous tissue, and our sense receptors and nerves; then there are the hairs that respond to temperature, which can stand erect if we are suddenly fearful, and which can be brushed by a sleeve or touched gently by a breeze; and then there are the non-biological membranes of clothes with their textures and degrees of translucency, and our personal and social boundaries which vary in relation to our moods and emotions, our confidence, our company, our feeling of well-being and health, and so on. Our natural assumption is to see the boundary of the body as the limit of our experiential world, but it is precisely its semi-permeable nature its breach which provides us with the possibility of experience in the first place. The skin, overrun with an abundance of receptors—sixty kilometers of nerve fibers, fifteen kilometers of veins, with millions of sense receptors for pain, temperature, pressure and touch (Hoffmeyer 2008, 18)—opens us up to the world and discloses it through our inescapable engagement with it, and then, of course, the skin is supplemented by the plenisentience of visual, proprioceptive, kinesthetic, auditory, gustatory, and olfactory senses which open us up in their own way, are affected by change or motion within our world and which, with internal feedback, can bring about affective change within themselves. (Stuart 2010, 306)

8. More will be said about this in the next section; for now our emphasis is on orientation and directionality, and establishing a corporeally based perspective.

9. The greater context is given as: "The direction [*Gegend*], however, in which this order of parts is directed [*wohin diese Ordnung der Teile gerichtet ist*], refers to the space outside the thing. To be specific: it refers not to places [*Lagen*] in the space—for that would be the same thing as regarding the position of the parts of the thing in question in an external relation [*in einem äusseren Verhältnis*]—but rather to universal space as a unity [*auf den allgemeinen Raum als eine Einheit*] of which every extension must be regarded as a part" (Kant 2002, 365–366). Walford (2001) presents a compelling case for Kant's argument making sense only if *Gegend* is translated as "direction" or *Richtungen*, as Kirchmann (1873) does, and not as "region" as other commentators have done. (Walford's own recommendation for where this "error" can be seen most clearly is in van Cleve and Frederick's [1991] collection of essays.)

10. Woelert translates *Gegenden* as "regions." Whether we use "regions" or "orientations" in this particular quotation makes no material difference to the import of the present essay.

11. Kant is notable among his fellows for claiming that "There can be no doubt that all our knowledge begins with experience" (Kant 1929, B1 41) without

dissolving into skepticism as Hume (1978) and Berkeley ([1710] 1913) did before him. Berkeley's famous dictum *esse est percipi (aut percipere)*—to be is to be perceived (or to perceive)—follows from a marvelously clear claim he makes in Principles:

> It is indeed an opinion strangely prevailing amongst men, that houses, mountains, rivers, and in a word all sensible objects, have an existence, natural or real, distinct from their being perceived by the understanding. But, with how great an assurance and acquiescence soever this principle may be entertained in the world, yet whoever shall find in his heart to call it in question may, if I mistake not, perceive it to involve a manifest contradiction. For, what are the fore-mentioned objects but the things we perceive by sense? and what do we perceive besides our own ideas or sensations? and is it not plainly repugnant that any one of these, or any combination of them, should exist unperceived? (Berkeley [1710] 1913, 31)

It is the imagination, as "metaphysical glue," in Hume's writing that gives us our natural tendency to believe in the existence of real and persistent mind-independent objects. Otherwise, "what Hume discovered is that rational analysis of our ideas shows that, except for mathematics, our ideas are (inadequate) representations of the reality they purport to be about" (Livingston 1984, 194).

The central intention of the first Critique, from which the quotation is taken, is to determine the necessary preconditions for the very possibility of experience and, as mentioned earlier, Kant makes no explicit ontological commitment to the nature of such transcendental conditions. (This is by no means a criticism of Kant; over two hundred years later, our position on the categorial structures for experience and the integration of the senses is still some distance from resolution.) However, the claim for experience as fundamental is not made lightly, and it becomes clear to the reader that the transcendental conditions include not only space and time as a priori forms of intuition, the a priori categories of the understanding, and the transcendental unity of apperception, but also the existence of a spatially extended physical world that we come to know through our affective corporeal engagement and which orientates us and establishes our point of view.

But Kant's triumph in this respect is relatively short-lived; all too soon again we cognitivize and disembody the human mind. Thought is given a preeminence over feeling and bodies become the otherwise invisible vehicles of content. Under this conception, our referential anchor shifts from the body to the relative latecomer, the disembodied pronoun, "I," which, through its use as some sort of communicative prosthesis, claims to refer to the (entirely illusory) "self" that is neither animated physical agent, nor historically, socially, and morally formed person (Stuart 2006). Cognitivism and computationalism are two of the greatest culprits in this area, for they present the mind as symbolic, representational, and reducible to a set of physical states and processes that are fully explicable through scientific experiment and analysis. These theories emphasize the "mental," in this case the propositional and disembodied, at the expense of the physical, and continue to forge a spurious dichotomy, while claiming not to be dualist. It is most perplexing.

12. The givenness and ownership of an agent's experience results from the agent's neuromuscular dynamics—without movement, there can be no affect; without affective movement, there is no agential givenness. Affective movement is always with, and within, the agent's world, so givenness is at its core necessarily enkinesthetic.

13. "Sense-full" is a term coined by O'Connor (2007) when she speaks of the bodily intentionality manifested in the performance of a skilled action. She says: "Thus, in virtue of bodily intentionality, the particular techniques become 'sense-full'" (190). In addition, it is very satisfyingly redolent of "hand-full" and full hands.

14. Feuerbach is an excellent nineteenth-century example of such writing: "Only the sensuous is clear and certain. Hence, 'the secret of immediate knowledge is sensuousness'" (Feuerbach 1843/1972, GW IX: 321; PPF 55). Whereas the old philosophy started by saying, "I am an abstract and merely a thinking being to whose essence the body does not belong," the new philosophy, on the other hand, begins by saying, "I am a real, sensuous being and indeed, in its totality is my ego, my essence [*Wesen*] itself" (ibid., GW IX: 320; PPF 54). "When sensuousness begins all doubts and quarrels cease. The secret of immediate knowledge is sensuousness" (ibid., §38).

15. One might also think of this in terms of Heidegger's (1968) concepts of "de-distancing," "disclosedness," and "attunement," but that must await another opportunity.

References

Baldwin, T. 2004. *Maurice Merleau-Ponty: Basic Writings*. London: Routledge.

Berkeley, G. [1710] 1913. *A Treatise Concerning the Principles of Human Knowledge*. Reprint edition. Chicago: Open Court.

Bourdieu, P. 2000. *Pascalian Meditations*. Trans. R. Nice. Cambridge: Polity Press.

Cichetti, D., and D. J. Cohen. 2006. *Developmental Psychopathology*, vol. 2. New York: John Wiley.

Cory, D. 1933. Dr. Whitehead on Perception. *Journal of Philosophy* 30 (7):29–43.

Cotterill, R. M. J. 2001. Evolution, cognition, and consciousness. *Journal of Consciousness Studies* 8 (2):3–17.

Feuerbach, L. [1843] 1972. *Principles of the Philosophy of the Future (Part III)*. Trans. Z. Hanfi. New York: Doubleday.

Feuerbach, L. 1981. *Gesammelte Werke*. Ed. Werner Schuffenhauer. Berlin: Akademie.

Gendlin, E. T. 1992. The primacy of the body, not the primacy of perception. *Man and World* 25 (3–4):341–353.

Hanna, R. 2008. Kantian non-conceptualism. *Philosophical Studies* 137 (1):41–64.

Heidegger, M. 1968. *Being and Time*. Trans. J. Macquarrie and E. Robinson. New York: Harper & Row.

Henry, M. 1963/1973. *L'Essence de la manifestation/The Essence of Manifestation*. The Hague: Nijhoff.

Hodges, B. H. 2007. Values define fields: The intentional dynamics of driving, carrying, leading, negotiating, and conversing. *Ecological Psychology* 19 (2):153–178.

Hodges, B. H., and R. M. Baron. 2007. Values as constraints on affordances: Perceiving and acting properly. *Journal for the Theory of Social Behaviour* 22 (3):263–294.

Hoffmeyer, J. 2008. *Biosemiotics: An Examination into the Signs of Life and the Life of Signs*. Trans. J. Hoffmeyer and D. Favareau. Scranton: University of Scranton Press.

Hume, D. 1978. *A Treatise of Human Nature*. Ed. L. A. Selby-Bigge. (First published 1739, translation 1888). Oxford: Clarendon Press.

Johnson, M. 1987. *The Body in the Mind: The Bodily Basis of Meaning, Imagination, and Reason*. Chicago: University of Chicago Press.

Kant, I. [1766] 1900. *Dreams of a Spirit-Seer Illustrated by Dreams of Metaphysics*. Ed. and trans. E. F. Goerwitz, with an introduction and notes by F. Sewall. London:Swan Sonnenschein. http://ia600204.us.archive.org/0/items/dreamsofspiritse00kant/dreamsofspiritse00kant.pdf.

Kant, I. 1992. Concerning the ultimate ground of the diffentiation of direction in space (1768). Trans. D. Walford. In *The Cambridge Edition of the Works of Immanuel Kant: Theoretical Philosophy, 1755–1770*, 365–372, ed. D. Walford and R. Meerbote. Cambridge: Cambridge University Press.

Kant, I. 1929. *The Critique of Pure Reason*. Trans. N. Kemp Smith. New York: Macmillan. (A edition 1781; B edition 1787.)

Kant, I. 2002. Concerning the ultimate ground of the differentiation of directions in space. In *The Cambridge Edition of the Works of Immanuel Kant: Theoretical Philosophy, 1755–1770*, 365–372, ed. D. Walford and R. Meerbote. Cambridge: Cambridge University Press.

Kirchmann, J. H. von. 1873. *Erläuterungen zu Kants kleinern Schriften über Logik und Metaphysik (Philosophische Bibliothek 58)*. Berlin: L. Heimann.

Lakoff, G., and M. Johnson. 1980. *Metaphors We Live By*. Chicago: University of Chicago Press.

Lecanuet, J.-P., W. P. Fifer, N. A. Krasnegor, and W. P. Smotherman, eds. 1995. *Fetal Development: A Psychobiological Perspective*. Hillsdale, NJ: Erlbaum.

Leder, D. 1992. *The Body in Medical Thought and Pra*ctice. Dordrecht: Kluwer.

Livingston, D. 1984. Review of *The Sceptical Realism* of David Hume. *Human Studies* 10 (2):193–200.

Luria, A. R. 1973. *The Working Brain: An Introduction to Neuropsychology*. Trans. B. Haigh. London: Allen Lane.

Merleau-Ponty, M. 1962. *Phenomenology of Perception*. Trans. C. Smith. London: Routledge & Kegan Paul.

Merleau-Ponty, M. 1963. *The Structure of Behavior*. Trans. A. Fisher. Boston: Beacon Press.

Merleau-Ponty, M. 1968. *The Visible and the Invisible*. Evanston, IL: Northwestern University Press.

O'Connor, E. 2007. Embodied knowledge in glassblowing: The experience of meaning and the struggle towards proficiency. *Sociological Review* 55 (Suppl. s1):126–141.

O'Shaughnessy, B. 1989. The sense of touch. *Australasian Journal of Philosophy* 67 (1):37–58.

Piontelli, A. 2002. *Twins: From Fetus to Child*. London: Routledge.

Ricoeur, P. 1981. *Hermeneutics and the Human Sciences*. Cambridge: Cambridge University Press.

Sartre, J.-P. [1945] 1956. *Being and Nothingness: An Essay in Phenomenological Ontology*. Trans. H. Barnes. New York: Philosophical Library.

Sennett, R. 2008. *The Craftsman*. New Haven: Yale University Press.

Shell, S. M. 1996. *The Embodiment of Reason Kant on Spirit, Generation, and Community*. Chicago: University of Chicago Press.

Steffensen, S. V., and B. H. Hodges. 2010. The ecology of values-realizing in dialogical and social systems. Paper given at the symposium "Expression, Engagement, Embodiment: The Ecology of Situation Transcendence," University of Glasgow, February 2010.

Stuart, S. A. J. 2006. Extended body, extended mind: The self as prosthesis. In *Screen Consciousness: Mind, Cinema and World*. Amsterdam: Rodopi.

Stuart, S. A. J. 2010. Enkinaesthesia, biosemiotics, and the ethiosphere. In *Signifying Bodies: Biosemiosis, Interaction and Health*, 305–330, The Faculty of Philosophy, University of Braga, Braga.

Stuart, S. A. J. 2011. Enkinaesthesia: The fundamental challenge for machine consciousness. *International Journal of Machine Consciousness* 3 (1):145–162.

Stuart, S. A. J. 2012. Enkinaesthesia: The essential sensuous background for co-agency. In *Knowing without Thinking: Mind, Action, Cognition, and the Phenomenon of the Background*, 167–186, ed. Z. Radman. Basingstoke: Palgrave Macmillan.

Trevarthen, C., K. J. Aitken, M. Vandekerckhove, J. Delafieldt-Butt, and E. Nagy. 2006. Collaborative regulations of vitality in early childhood: Stress in intimate relationships and post natal psychopathology. In *Developmental Psychopathology*, vol. 2, ed. D. Cichetti and D. J. Cohen. New York: John Wiley.

Trevarthen, C., and V. Reddy. 2007. Consciousness in infants. In *A Companion to Consciousness*, 41–57, ed. M. Velman and S. Schneider. Oxford: Blackwell.

van Cleve, J., and R. E. Frederick, eds. 1991. *The Philosophy of Right and Left: Incongruent Counterparts and the Nature of Space*. Dordrecht: Kluwer Academic.

Varela, F. 1989. Reflections on the circulation of concepts between a biology of cognition and systemic family therapy. *Family Process* 28:15–24.

von Uexküll, J. [1940] 1982. The theory of meaning. *Semiotica* 42 (1):25–87.

Walford, D. 2001. Towards an interpretation of Kant's 1768 *Gegenden im Raume* essay. *Kant-Studien* 92 (4):407–439.

Whitehead, A. N. 1978. *Process and Reality: An Essay in Cosmology*. Ed. D. R. Griffin and D. W. Sherburne. New York: Free Press.

Wilson, F. R. 1998. *The Hand: How Its Use Shapes the Brain, Language, and Human Culture*. New York: Vintage Books, Random House.

Woelert, P. 2007. Kant's hands, spatial orientation, and the Copernican turn. *Continental Philosophy Review* 40:139–150.

15 The Enculturated Hand

Richard Menary

1 Introduction

Sometimes we think with our hands. Is that an outrageous claim? What would make it less outrageous? Perhaps showing that mental concepts like intentionality are perfectly compatible with the claim would be a start. To establish this compatibility is the aim of the first section of this chapter. The compatability can be established by showing that a minimal intentional directedness can be found in the biological world. Animals are often intentionally directed at an object for some biologically specifiable end. This establishes the idea that an organism's body can be intentionally directed without also needing a contentful representation to guide it.

We then need a broad framework for understanding cognition in which the claim that we think with our hands makes sense. The framework I provide is one in which cognition is understood in terms of integration and enculturation. This is the aim of the second section. While the first section establishes that bodies can be intentionally directed, it does so by showing that intentional directedness is mediated by the environment. The second section elaborates on this by showing that human intentionality is no different, except that the human environment is a culturally and socially constructed niche. Human intentionality is culturally mediated; both the body and the brain are enculturated.

I then go on to provide some of the conditions required for encultura-tion, in particular phenotypic and neural plasticity, in the third section. The conditions include the triple inheritance model of niche construction (Laland, Feldman, and Odling-Smee 2003) and the plasticity of the human phenotype, in particular its ability to learn culturally transmitted skills and practices.

The fourth and fifth sections roll out a discussion of cognition as the iterative process of manipulating informational structures in a cultural niche—a kind of cultural practice, a cognitive practice (Menary 2007), and the consequent transformation of neural circuits and functions. The resulting biocultural framework for cognition is quite compatible with the claim that the hand is not just an organ for physically changing the world; it is also an organ for thinking.

2   Minimal Intentional Directedness

Is there continuity between biological intentional directedness and high-level contentful thoughts directed at the world? The answer to this question is yes. However, there is an intermediary between noncontentful but cognitive responsiveness and contentful thought. That intermediary is the sociocultural complexity of the human niche. The biological and the cultural merge to form the peculiarly human form of cognition; but human cognition is nevertheless continuous with the rest of the biological world. Call this the continuity thesis (Menary 2009). Let us start by identifying biological intentionality.

The humble sponge attaches itself to a rock and allows the vortices of the surrounding water to aid in the pumping function of its flagella (Vogel 1981). The solution may be an elegant adaptation to its local niche, but this is not an instance of intentional directedness. Contrast this process of the humble but elegant sponge with the process of phonotaxis in female crickets (*acheta domesticus*). The female cricket can locate and move toward the location of male cricket songs (Webb 1994). She does so by the activation of two dedicated interneurones, each of which is connected to one of her ears. The strongest activation determines the direction in which she will fly. If the interneuron connected to her left ear is more strongly activated than that connected to her right ear, then she flies to the left. There is an exquisite coupling between the iterated song of the male cricket and the activation threshold and decay time of the female's dedicated interneurones. Intuitively, we might think that the female cricket is intentionally directed at the male cricket's song. Can we provide some criteria of intentional directedness that would help to make our evaluation more theoretically robust?

To do so, I need to demonstrate that there is a theoretically robust sense of intentionality that is compatible with complex intentionally directed actions of organisms such as the female cricket. Intentionality as we receive it from Brentano and the Scholastics is derived from the Latin word intentio (or in + tendere)—"to stretch or extend (toward)." The primary sense of intention for the Scholastics is an intentional act; Aquinas analyzes it as an act of will:

Intention, as the very word denotes, means to tend to something. Now both the action of the mover and the movement of the thing moved tend to something. But that the movement of the thing moved tends to anything is due to the action of the mover. Consequently intention belongs first and principally to that which moves to the end; hence we say that an architect or anyone who is in authority, by his command moves others to that which he intends. Now the will moves all the other powers of the soul to the end. Therefore it is evident that intention, properly speaking, is an act of the will. (Aquinas 1997, 272)

The teleological notion of moving toward an end is the core of the concept of the intentional act. When we lose the Scholastic trappings, we are left with the bare bones definition of intentional directedness as the organism moves or tends toward an object for some naturally specifiable end.[1] Biological intentional directedness has the virtue of giving an account of how the intentional relation becomes established. For example, a trait of an animal might have the function of detecting prey; it is intentionally directed at an object for the end of prey detection. Furthermore, the intentional directedness toward the object is end directed, where that end is to be acted upon by catching the prey, and the trait is coordinated with other traits to achieve that outcome. Unlike the Aristotelian–Scholastic tradition, here we don't think that the end functions as a final cause; the traits evolved under the pressures of natural selection. But we do think that the traits can be end directed, that is, intentionally directed. We also expect that the traits will sometimes be intentionally directed at an object when there is no object present; perhaps, for example, the animal is highly attuned to movements that indicate the presence of a predator, upon the detection of which the animal flees (despite the fact that movements are not always caused by predators).

Let us return to acheta domesticus. Given our analysis of organisms that are intentionally directed at an object for some naturally specifiable end, we can more clearly see why the female cricket is intentionally directed at the male cricket's iterated song patterns. This analysis is really a triangulation analysis of intentionality. In other places (Menary 2007, 2009), I have traced the triangulation analysis back to the work of C. S. Peirce. Ruth Millikan is a more recent exponent of the analysis, in terms of producer–consumer relations (Millikan 1984, 1993). The core features of a triangulation account are as follows:

a. The vehicle has certain intrinsic or relational properties that make it salient to a consumer.
b. The vehicle's intentional function is established by the exploitation of the salient properties of the vehicle by a consumer.

c. The intentional relation is established only when the intentional function is recruited for some further end, such as the detection of food.

In our cricket case, the triangulation is produced by the coordination of
the male cricket song with the female cricket interneurone activation and
decay patterns. The male cricket song is the vehicle with salient properties,
and the interneurone is the consumer of those properties, establishing
its intentional function. The iterated song is exploited by a consumer that
recruits the vehicle to the production of some end, in this case directing
the female toward the male.

In Millikan's terms, the producer mechanism has the proper function
of producing the song for the consumer mechanism, the female's interneurones. The producer and consumer mechanisms can function properly
only if they are both present and coordinating: this is their "normal condition" (Millikan 1993).

Intentionality is connected to other traits, behaviors, and the environment of the organism. Intentionality is not a property of mental acts, as
it were, bracketed off from all other traits of the organism and from the
complex environment (including complex social and group environments)
in which the organism is situated. The organism is intentionally directed
at an object through its body and actions; this intentional directedness is
not a matter of conscious volition or symbolic representations, but rather
is mediated by the environment.

If such a view of minimal intentional directedness is correct, then there
is no bar to considering the body, and therefore the hand, as intentionally directed at objects for some end. In the next section, I turn to the
integrationist and enculturated framework for cognition. The aim is to
provide the theoretical basis for how our hands become intentionally
directed through processes of bodily enculturation.

3   Cognitive Integration and Enculturated Cognition

Cognitive integration is a position I have developed over recent years
(Menary 2006, 2007, 2009, 2010a,b) that takes cognitive systems to be
integrated wholes with interacting parts, but where those parts can include
neural, bodily, and environmental components. One way to understand
integration is to focus on the coordination dynamics of interacting components of a system. The components here might be processes, or they
might be structures. The global behaviors of a system are a product of the
coordinations between system components (which may themselves be

complex systems). The system is constituted by its components and their interactions, and its successful functioning requires all its proper parts to be in good working order. We often find this kind of relationship in nature, where organisms become deeply integrated with parts of their environment, such that they become part of the phenotype (they fall under selective pressure). Nature is no stranger to the kind of hybrid, integrated system that is at the core of cognitive integration (Turner 2000).

The interesting thing about dynamical work on cognition is that the interacting components of cognitive systems are sometimes located spatially outside the central nervous system of an organism. Because the system components coordinate with one another to produce the global behavior of the system, it does not matter that some of the components are not located within the skin of the organism.

The relationship of cognitive integration to enculturation rests in the acquisition of cultural practices that are cognitive in nature. What I mean by this is that the practices are aimed at augmenting our cognitive capacities, allowing us to complete cognitive tasks, sometimes in ways that our unenculturated brains will not allow. These practices are what I call cognitive practices (Menary 2007, 2010a). Cognitive practices can be thought of as bodily manipulations of informational structures in public space (Menary 2007; Rowlands 2010). According to cognitive integration, processes and informational structures can be system components even though they are not within the skin of the organism. Cognitive practices are culturally endowed bodily manipulations of informational structures. They also intentionally direct the body, and in some cases the hands, to salient properties of informational structures. The environment mediates the intentional directedness of the body and hand, because cognitive practices are cultural practices and as such exist as patterns of action across cultural groups (Roepstorff, Niewöhner, and Beck 2010).

The practices are normative: There are right and wrong ways to do them, and they are often encoded as rules or procedures to be followed (especially for the neonate or the novice). However, once they are internalized, they are enacted without the need for reference to those rules and procedures. The body and the hand are intentionally directed without the need for conscious reference to rules, or explicit representations of what to do.

The main thesis of enculturated cognition is that there are cognitive processes best performed by bodily manipulating information structures. The main extension of our cognitive processes is by processes that are part of a cultural practice: This is the position adopted by cognitive integration (Menary 2007). Many of these practices involve artifacts such as tools, writing systems, number systems, and other kinds of representational systems. These are not simply static vehicles that have contents, but are active components embedded in dynamical patterns of cultural practices. These practices originate in the world, and the practices that govern them are also in the world. These artifacts and practices are products of cultural evolution, evolving over faster timescales than biological evolution. Writing systems, for example, are only thousands of years old; consequently, there is no gene for writing. This is important: Cognitive capacities for reading and writing, mathematics, and other culturally recent forms of cognition could not be biological adaptations. The timescales for their evolution are too short. It follows that these recent cognitive capacities must be acquired through learning and training. The cortical circuits with which we are endowed through evolution are transformed to perform these new culturally specified cognitive functions, even though they evolved to perform different functions.

The brain's plasticity for learning and the development of new cortical functions is remarkable and results in new cognitive wholes that did not exist before. It is in this way that cultural practices and representations can get under the skin and transform the processing and representational structure of cortical circuitry (Ansari 2008; Dehaene 1997, 2009; Menary 2010a,b). Cognition enculturated results in extended cognitive capacities and the transformation of the brain to function according to the rules and representations of writing systems, mathematical systems, and so on.

In these cases, when the processing routines cross from the world into the brain, our cognitive abilities become enculturated: We get to be readers and writers, mathematicians, and so on by a process of transforming existing cognitive abilities to perform new, cultural functions.

I have already alluded to the relationship between neural plasticity and a structured cognitive and developmental environment. Learning to read and write is a wonderful, cognitive example of the brain's plasticity. Recent work in cognitive neuroscience (Dehaene 1997; Dehaene and Cohen 2007) gives us a very clear picture of how the plasticity of the brain in learning allows for the redeployment of neural circuitry for functions that were not specified by biological evolution. In other words, those neural circuits acquire new culturally specified functions, functions that have only existed for thousands, not millions, of years. However, the functions of the new interconnected circuits are dependent on the cultural practices that determine those functions. Consequently, not only is the brain an open system, but it is also functionally unbounded: Its functions can be extended beyond those we are endowed with by evolution. As Dehaene himself puts it:

"Writing created the conditions for a proper 'cultural revolution' by radically extending our cognitive abilities" (Dehaene 2009, 307). The hand is complicit in enculturating the brain: It gathers the physical traces of cultural systems of representation to the brain, it grips the pen or stylus, and it begins the process of physical feedback for the body and brain to learn which movements produce the right inscriptions of phonetic sounds. The enculturated hand is the writing hand, the hand that plots a graph or writes out a form of inference. Through the hand's new role as a representation maker, I can be intentionally directed at phenomena that are simply not present to my field of vision. My memories can be structured and stored and made available for future reference, as, for example, an autobiographical narrative. This is a remarkable achievement; but how did the conditions for enculturation come about?

4   The Conditions for Enculturation: Phenotypic Plasticity and Neural Plasticity

Humans are behaviorally and developmentally plastic (Sterelny 2003). They can perform a wide variety of actions and learn an almost endless array of knowledge and skills. How is this so, when our most genetically similar relatives are not nearly so developmentally and behaviorally plastic?

In evolutionary terms, humans are capable of developing a wide range of skills that allow them to cope with a wide variety of environments (and their contingencies). For example, even where skills are (broadly) of the same type, such as hunting, they will vary to cope with the differences of local environment—think of the differences between Aboriginal hunters in the Pilbara, hunter-gatherers in the Central American rainforests, and Inuit seal-hunters (Sterelny 2003, 167).

One interesting reason for this is that development (plasticity) is extended in humans relative to other species. Humans take a long time to learn how to walk and talk, and much, much longer to read and write. Other primates have much faster developmental timescales. While this might make humans more dependent on their caregivers for longer, it also allows them to refine skills and produce a greater array of them before entering adulthood.

Why should we be like this, especially if, as Tooby and Cosmides (1992) think, we are already exquisitely cognitively adapted to our ancestors' Pleistocene environments? Sterelny's argument is that rather than thinking of humans as adapted for Pleistocene hunting and gathering environments, we should think of humans as developmentally plastic as an adaptive response to the variability and contingency of the local environment. Rather than being born with "a confederation of hundreds or thousands of functionally dedicated computers (often called modules) designed to solve adaptive problems" (Tooby and Cosmides 1995, xiii), we are born to be phenotypically and developmentally plastic, and our brains exhibit high degrees of plasticity. But Cosmides and Tooby will claim that "*it is in principle impossible for a human psychology that contained nothing but domain-general mechanisms to have evolved, because such a system cannot consistently behave adaptively*" (1994, 90; italics in original). How can a phenotypic plasticity account of human cognitive evolution be defended against the massive modularity thesis?

The first answer is to deny that genetic inheritance is the only kind of inheritance available to humans. Recent work on niche construction (Laland, Feldman, and Odling-Smee 2003) has shown that organisms inherit the niches that were constructed, or at least physically influenced, by the previous generation. Over time this can have profound influences on the phenotype:

When organisms adapt their environment rather than adapting to their environment, they often establish a feedback loop that results in evolutionary cascades. Thus the establishment of life in burrows selects for further evolutionary changes. In some cases these changes are extreme: living in mounds influences every facet of termite morphology, physiology and behaviour (Turner 2000). (Sterelny, 2005, 22)

A third line of inheritance is cultural inheritance; knowledge, skills, and artifacts are passed on to the next generation, but learning environments and learning techniques are also passed on, so that the next generation can acquire and be transformed by the inherited cultural products. This last point is important for our purposes, because developmentally plastic creatures like us need scaffolded learning environments in which to develop. So Tooby and Cosmides's worry that a system without specialized modules could not be consistently adaptive turns out not to be a worry if the environments in which we develop our cognitive capacities are highly structured and we are scaffolded by them in development.

How, though, are we capable of acquiring these new cultural capacities in development? The reason is neural plasticity. Rather than the process of synaptogenesis or lesion-induced plasticity, the kind of plasticity I will discuss here is what I call learning-dependent plasticity (see Menary, forthcoming). Learning-dependent plasticity—L-plas for short—can result in both structural and functional changes in the brain. Structurally, L-plas can result in new connections between existing cortical circuits. Functionally, L-plas can result in new representational capacities and new computational capacities.

For example, when we learn mathematics and learn to read and write, the representational capacities of our neural circuitry are transformed (from approximate quantities to precise numbers). Neural circuitry is redeployed from an evolutionarily older function to a new cultural function—in this case, recognizing Arabic numerals in the temporal lobe. The brain's capacity to perform mathematical computations is transformed by linking together what were previously independent neural circuits into a system for mathematical cognition. The transformation results in new connections between the frontal lobe for number word recognition and association, the temporal lobe for the visual recognition of number form, and the parietal lobe for the approximate recognition of magnitudes across both left and right hemispheres (Dehaene 1997).

The deeply transformative power of our learning histories in the cognitive niche is one that reformats the representational capacities of the brain in terms of public representational systems. We internalize public representational systems in the way Dehaene suggests, but we also learn techniques for manipulating inscriptions in public space (Menary 2010a). In learning the manipulative techniques, the first transformation is one of sensorimotor abilities for creating and manipulating inscriptions; we learn algorithms, such as the partial products algorithm, and this is an example of the application of cognitive practices (as outlined above). This is something we learn to do on the page and in the context of a learning environment, in public space, before we do it in our heads. Our capacities to think have been transformed, but in this instance they are capacities to manipulate inscriptions in public space. This is a way of showing that the transformation of our cognitive capacities has publicly recognizable features. This should not be a surprise, given that the cognitive niche is socially and culturally constructed.

That we have evolved to be phenotypically and developmentally plastic is in no small part due to the plasticity of our brains. Our developmentally plastic brains exhibit learning plasticity when they are coupled to a highly scaffolded learning environment; the brain is profoundly transformed, and consequently, we are cognitively transformed in a deeply profound way. The hand is again complicit in this transformation; but let us now delve more deeply into the ways in which we bodily manipulate public information structures with our hands, before returning to the process of transformation in the final section.

5 Manipulating the Niche

We create, maintain, and manipulate cognitive niches (Sterelny 2010). We do so with our bodies, but primarily with our hands. We group objects together into classes; we store information in easily accessible places to use in future bouts of thinking; we update the information, delete it, reorder it, reformat it—and we do so always with our hands. We write or type, we push or pull; the hand is ever present in our construction of cognitive niches. These practices are located within the normative, social, and semantic structure of the cognitive niche. Cognitive practices are bodily engagements with the niche that are regulated by norms. The norms and social systems and structures of the environment regulate the coordination in two ways: first by transformation of our basic cognitive capacities and second by restructuring the informational and physical structure of the environment.

Let us return to integrated cognition to make sense of bodily manipulations as cognitive processes. Cognitive integration is the coordination of bodily processes of the organism with salient features of the environment, often created or maintained by the organism. A coordinated process allows the organism to perform cognitive tasks that it otherwise would be unable to do; or allows it to perform tasks in a way that is distinctively different and is an improvement upon the way that the organism performs those tasks via neural processes alone. Integrationists think that some cognitive processes turn out to be coordinated. It is important to note that integrationists are not committed to the view that artifacts and tools are themselves cognitive or mental, nor to the view that a simple causal interaction between two states X and Y makes X part of Y (see Adams and Aizawa 2008).

Rather, we should think of integration as the outcome of the process of enculturation through scaffolded learning, as discussed in the last section. This is a quite different position from that of the extended mind as it is usually construed. Such a position is usually taken to be committed to the idea that cognition is first in the head and then gets extended into the world. The brain recruits artifacts into its problem-solving routines (Clark 2008), and as long as the organism is causally coupled to the artifact in the right way (with a fair dosage of trust and glue), they together constitute an extended cognitive system. By contrast, enculturated cognition takes the transformation of existing cortical circuits to be part of our developmental pathway, and the functional integration of practices of manipulating public information structures with enculturated neural functions results in impressive new cognitive wholes. The "outside" and the "inside" are not so functionally different after this process of enculturation, because culture gets under the skin during development.

What, though, is this process of internalization like? I turn now to an example. There are neural forms of bodily manipulations, but we should not think that the development of motor programs (or body schemas) for manipulating informational structures needs to be a suite of subpersonal states that represent and thereby control the manipulations themselves. Rather, we should think of body schemas as being subpersonal processes that dynamically govern posture and movement in a close to automatic way (Gallagher 2005, 26). Repertoires of body schemas function together as motor programs (Gallagher 2005). Some motor programs are learned, such as riding a bike and writing, and some are innate, such as swallowing.

Body schemas can be selectively tuned to the environment, and such tuning is developed through learning, habituation, practice, and training. A physical instance of this would be learning how to catch a ball in a game like baseball or cricket. In the first instance, one learns how to catch the ball, and then one trains to perfect the technique; one might do so by imitating the movements of a teacher, or by explicit direction by the teacher. The learned movements are practiced until one is habituated to make the right moves appropriate to the situation. Learning and training require us to learn the right movements and to apply them in the right ways, to do so in conformity with the rules of the game.

The cultural and social sedimentation of our body schemas fundamentally shapes the ways that we move and act. In this process of development, body schemas become intentionally directed at the world—one's hands become intentionally directed at the flight of the ball (Rowlands 2006), but without the timescales that allow for slow deliberation before action.

There is also a clear and fundamental sense in which body schemas and motor programs become highly integrated with the environment. In these cases, the body schema incorporates parts of the environment—such as the hammer in the carpenter's hand (Gallagher 2005, 37). A clear example of this is the work by Maravita and Iriki (2004) on training Japanese macaques in tool use. The macaques were trained to hold and use a rake to pull food placed on a table toward them. Maravita and Iriki (2004) report that several changes to the receptive field of bimodal neurons were observed in macaques who were trained (enculturated) to use the rake. These changes in the properties of bimodal neurons occurred only after "active, intentional usage of the tool, not its mere grasping by the hand" (Maravita and

Iriki 2004, 81). The visual receptive fields of these bimodal neurons now extend to include the hand and entire tool: The tool is now incorporated into the body schema. The hand-tool must also be intentionally directed, aiming at the salient object (food), for some further end—to eat.

Body schemas do not just initiate behavior; they are fully integrated with the environment. They are constrained by the environment because they often require its perceptual navigation and the manipulation of environmental objects. Therefore, the body shapes itself to meet the environment, to hold a glass in hand or grip a pencil between the fingers and thumb for writing. Body schemas are attuned to environmental affordances for action (Gibson 1979): The glass affords drinking and the pencil writing. Again, the intentional directedness of the hand is mediated by the environment.

It is in the fluid manipulation of objects in the environment and in fluent skilled activities that we are most likely to find the unconscious integration of the body schema with the environment. Experienced drivers will understand the nature of this integration, which involves the seamless coordination of body and car, sometimes to such an extent that one cannot recall the details of the journey when the destination is reached (Gallagher 2005). Although the body image involves conscious experience of our own bodies and that experience is of a bounded body, the body schema has no such boundary; it directs our primary embodied engagements with the world, and it is because of this that we feel ourselves to be both in and part of the world. Furthermore, it is constitutive of our first cognitive engagements with the world, our perceptual navigation, our imitation of others, and our manipulation of the environment.

Let us turn to a cognitive example, the bodily manipulation of numerical inscriptions. What are the properties of numerical inscriptions such that they can be manipulated as a practice? The inscriptions of mathematical notations are structured in terms of a spatial arrangement on a page, monitor, or some other physical space. Some inscriptions have a complex structure with parts that are organized into a whole, with spatial relationships between these parts. The inscriptions are manipulated in terms of this structure. The inscriptions might be transformed, composed, or combined in various ways according to the relevant norms of the system. The last is important, because atomic tokens can be combined into more complex tokens. A simple way of understanding manipulative norms is that they are the result of these combinatory rules of the system.

For example, the numbers 2, 3, and 6 can be composed into the more complex string 236. Mathematical notations, as well as letters, words, and sentences, can be arbitrarily complex, according to a recursive rule system. Recursivity allows for a potentially infinite variety of strings and allows for a potentially infinite variety of meaningful expressions.

Since recursively recombinable inscriptions are meaningful, it follows that they are potentially representationally salient for some consumer. For example, a graph that depicts change in some domain is significant for a consumer who can interpret the salient properties of the inscription for some end. Inscriptions are, therefore, interpretable, but only according to some background norms of interpretation and meaning that are part of the public system of representation.

Manipulative norms and interpretative norms apply to inscriptions of a public representational system and are never simply dependent on an individual. Indeed, it is the individual who must come to be transformed by being part of the community of representational system users. Mastery is gained through immersion in cognitive practices, a sustained period of familiarity with explicit norms, and fluent real-time enaction of those practices.

In the mathematical case, there are norms for manipulating mathematical inscriptions, and those inscriptions are interpreted as having some kind of significance for some purpose. For example, we are taught manipulative norms for writing down the numbers in this fashion to complete a long multiplication:

343

822

The inscription of the notation allows us to perform the simpler set of multiplications starting with 2 times 3 (the partial products algorithm). Those who are well trained in performing the procedure will be able to fluidly run through the series of operations. They will have well-trained body schemas. Experts will not need to think about what they are doing (in simple cases at least), nor will they need to interpret (consciously) the inscriptions. Novices, by contrast, will require conscious rehearsal of the manipulative norms for successful completion of the task.

The hand must glide the pen or pencil across the page (or the fingers across the keyboard), enacting the norms embedded in the cognitive practices of mathematics. Novices learn in a painstaking fashion, but the expert produces a fluid display of practiced activity. The transformation from novice to expert is a profound one; and it is this process of transformation to which I now turn.

6 Cultural Transformations

We are not born with the ability to perform complex mathematical calculations. At best, we have an inherited ability to approximately discriminate between small sets in terms of the quantity of their members. The inherited cortical functions for approximation are sometimes collectively referred to as the "number sense" (Dehaene 1997). To be able to do long multiplication or algebra, our basic neural circuitry needs to be redeployed to perform new, culturally specified functions. This redeployment results in a massive transformation of our cognitive capacities. In this section, I outline the process of transformation from an inherited ability to approximately discriminate quantities to the learned ability to perform exact calculations using mathematical notations and norms for their manipulation.

There is strong evidence to suggest that we have a basic analogical and nonlinguistic capacity to recognize quantity and number (numerosity, or the number sense). There appears to be an ancient evolutionary capacity to discriminate cardinality, that is, to determine in an approximate way the quantity of membership of sets, known as subitizing. It is obvious how this capacity, only for very small sets, would be beneficial for activities such as foraging, hunting, and so on. Neural populations that code for number are distributed in the intraparietal sulci (IPS; Dehaene and Cohen 2007).

A growing number of studies show that both animals and humans possess a rudimentary numerical competence, which is an evolutionary endowment. For example, red-backed salamanders have been shown to choose the larger of two groups of live prey (Uller et al. 2003). Single-neuron activation studies in rhesus monkeys (Nieder, Diester, and Tudusciuc 2006) have discovered that individual neurons respond to changes in number when presented visually (DeCruz 2008). These neurons are also located in the intraparietal sulci, indicating a cross-species similarity. The neurons peak at the presentation of a specific quantity, but then decrease as the numbers presented differ from the original. So a neuron that peaks at the presentation of two items responds less to three or four items. Therefore, the ancient capacity for numerosity is an approximate function, not a discrete one (DeCruz 2008). As Dehaene puts it: "This commonality [of function in the IPS] suggests that humans and macaques have a comparable labeled line coding mechanism for the representation of numerosities" (2009, 191). Since our inherited cortical circuitry endows us only with these approximate functions for numerosity, it appears to be a mystery how we developed the abilities to perform discrete mathematical functions.

It becomes less of a mystery if we do not assume that we rely only on the cortical circuits responsible for numerosity. To be able to perform exact calculations, we have to acquire number words, recognize numerical inscriptions, and associate the number words with those inscriptions: "Exact calculation is language dependent, whereas approximation relies on non-verbal visuospatial cerebral networks" (Dehaene et al. 1999, 970). This means that we have to invent and develop a public system for representing quantities and their relations. Since we don't first have these representational systems in the head, they must have developed as part of ongoing cultural practices, ways of keeping track of things in the world.

It is clear, then, that we make fast, intuitive judgments of numerosity and that we also acquire symbolic representations of numerals. But what is the importance of a public mathematical symbol system? Its importance lies in the ability to perform computations that cannot be performed by ancient neural functions for numerosity. For example, the neural circuits responsible for numerosity cannot perform functions like the square root of 54, and yet this is simply represented in terms of public mathematical symbols (DeCruz 2008).

The inherited systems for numerosity are evolutionary endowments; we can be reasonably sure of this because they are constant across individuals and cultures and they are shared with other species (macaques). The numerosity systems are "quick and dirty"; they are approximate and continuous, not discrete and digital. By contrast, discrete mathematical operations exhibit cultural and individual variation; there is a big difference between Roman numerals and Arabic numerals. The discrete operations are subject to verbal instruction (they actually depend on language); one must learn to count, whereas one does not learn to subitize. Mathematics depends on cultural norms of how to reason (mathematical norms). The ability to perform exact calculations of mathematics depends on the public system of representation and its governing norms. We learn the interpretative norms and manipulative norms as a part of a pattern of practices within a mathematics community, and these practices transform what we can do. They are constitutive of our exact calculative abilities; without them we would be at a loss. Mathematical practices get under our skins by transforming the way that our existing neural circuitry functions.

Thus, our mathematical practices are an example of cognitive transformation. As Dehaene puts it: "I emphasise that cultural reconversion or 'neuronal recycling' transforms what was initially a useful function in our evolutionary past into another function which is currently more useful within the present cultural context" (2004, 152). The relationship between the evolutionarily earlier system and the recent development of public mathematical systems, norms, and symbols is one of the "invasion" of the cortical territories that are dedicated to evolutionarily older functions by novel cultural articts (representations, norms). The transformation results in new connections between the frontal lobe for number-word recognition and association, the temporal lobe for the visual recognition of number form, and the parietal lobe for the approximate recognition of magnitudes across both left and right hemispheres (Dehaene 1997).

The deeply transformative power of our learning histories in the cognitive niche is one that reformats the representational capacities of the brain in terms of public representational systems. We internalize public representational systems in the way Dehaene suggests, but we also learn techniques for manipulating inscriptions in public space. In learning the manipulative techniques, the first transformation is one of sensorimotor abilities for creating and manipulating inscriptions; the second transformation is of our ability to represent abstract mathematical relationships as bequeathed to us by our cognitive niche. Without our hands we cannot write, we cannot draw, and therefore we cannot bequeath these cultural products to future generations.

Development in the cognitive niche results in an integrated cognitive system. The integrated cognitive system has been through the dual component process of transformation where one gains mastery over the symbol system in public space, which leads to a transformation of our limited cognitive capacities such that we can complete cognitive tasks by manipulating symbols in public space or by manipulating symbols in neural space, but most importantly by a combination of both sets of resources. This transformation results in new body schemas for action and new cortical functions for reading, writing, and mathematics.

7 Conclusion

Our journey has taken us from intentionality through niche construction to developmental and neural plasticity, on to the transformation of body schemas and neural circuitry, and finally to the norm-guided manipulation of public inscriptions of representational systems. The enculturated hand is the intentionally directed tool by which we manipulate those inscriptions. Our hands are one of the means by which our brains are transformed in development. Thus the hand is an organ of thought par excellence.

Note

1. The account given here is quite compatible with recent work in biological approaches to intentionality; see, e.g., Millikan 1993, Godfrey-Smith 1996, Hutto 2008, Sterelny 2003, and Menary 2007, 2009.

References

Adams, A., and K. Aizawa. 2008. *Defending the Bounds of Cognition.* Oxford: Blackwell.

Aquinas, T. 1997. *Basic Writings of Saint Thomas Aquinas: Man and the Conduct of Life,* vol. 2. Ed. A. C. Pegis. Indianapolis: Hackett.

Ansari, D. 2008. Effects of development and enculturation on number representation in the brain. *Nature Reviews: Neuroscience* 9 (4):278–291.

Clark, A. 2008. *Supersizing the Mind: Embodiment, Action, and Cognitive Extension.* Oxford: Oxford University Press.

Cosmides, L., and J. Tooby. 1994. Origins of domain-specificity: The evolution of functional organization. In *Mapping the Mind: Domain-Specificity in Cognition and Culture,* ed. L. Hirschfeld and S. Gelman. New York: Cambridge University Press.

De Cruz, H. 2008. An extended mind perspective on natural number representation. *Philosophical Psychology* 21 (4):475–490.

Dehaene, S. 1997. *The Number Sense: How the Mind Creates Mathematics.* London: Penguin.

Dehaene, S. 2004. Evolution of human cortical circuits for reading and arithmetic: The "neuronal recycling" hypothesis. In *From Monkey Brain to Human Brain,* ed. S. Dehaene, J. R. Duhamel, M. Hauser, and G. Rizzolatti. Cambridge, MA: MIT Press.

Dehaene, S. 2009. *Reading in the Brain: The Science and Evolution of a Human Invention.* New York: Viking.

Dehaene, S., and L. Cohen. 2007. Cultural recycling of cortical maps. *Neuron* 56 (2):384–398.

Dehaene, S., E. Spelke, P. Pinel, R. Stanescu, and S. Tsivkin. 1999. Sources of mathematical thinking: Behavioral and brain-imaging evidence. *Science* 284 (5416): 970–974.

Gallagher, S. 2005. *How the Body Shapes the Mind.* Oxford: Oxford University Press.

Gibson, J. J. 1979. *The Ecological Approach to Visual Perception.* Boston: Houghton Mifflin.

Godfrey-Smith, P. 1996. *Complexity and the Function of Mind in Nature*. Cambridge: Cambridge University Press.

Hutto, D. 2008. *Folk Psychological Narratives: The Sociocultural Basis of Understanding Reasons*. Cambridge, MA: MIT Press.

Laland, K., M. Feldman, and J. Odling-Smee. 2003. *Niche Construction: The Neglected Process in Evolution*. Princeton: Princeton University Press.

Maravita, A., and A. Iriki. 2004. Tools for the body schema. *Trends in Cognitive Sciences* 8 (2):79–86.

Menary, R. 2006. Attacking the bounds of cognition. *Philosophical Psychology* 19 (3):329–344.

Menary, R. 2007. *Cognitive Integration: Mind and Cognition Unbounded*. Basingstoke: Palgrave Macmillan.

Menary, R. 2009. Intentionality, cognitive integration, and the continuity thesis. *Topoi* 28 (1):31–43.

Menary, R. 2010a. Dimensions of mind. *Phenomenology and the Cognitive Sciences* 9 (4):561–578.

Menary, R. 2010b. Cognitive integration and the extended mind. In *The Extended Mind*, ed. R. Menary. Cambridge, MA: MIT Press.

Menary, R. Forthcoming. Neuronal recycling and neuronal plasticity. *Mind & Language*.

Millikan, R. 1984. *Language, Thought, and Other Biological Categories*. Cambridge, MA: MIT Press.

Millikan, R. 1993. *White Queen Psychology and Other Essays for Alice*. Cambridge, MA: MIT Press.

Nieder, A., I. Diester, and O. Tudusciuc. 2006. Temporal and spatial enumeration processes in the primate parietal cortex. *Science* 313 (5792):1432–1435.

Roepstorff, A., J. Niewöhner, and S. Beck. 2010. Enculturing brains through patterned practices. *Neural Networks* 23 (8–9):1051–1059.

Rowlands, M. 2006. *Body Language: Representation in Action*. Cambridge, MA: MIT Press.

Rowlands, M. 2010. *The New Science of the Mind: From Extended Mind to Embodied Phenomenology*. Cambridge, MA: MIT Press.

Sterelny, K. 2003. *Thought in a Hostile World: The Evolution of Human Cognition*. Oxford: Blackwell.

Sterelny, K. 2005. Made for each other: Organisms and their environment. *Biology and Philosophy* 20:21–36.

Sterelny, K. 2010. Minds: Extended or scaffolded? *Phenomenology and the Cognitive Sciences* 9 (4):465–481.

Tooby, J., and L. Cosmides. 1992. The psychological foundations of culture. In *The Adapted Mind: Evolutionary Psychology and the Generation of Culture*, ed. J. Barkow, L. Cosmides, and J. Tooby. New York: Oxford University Press.

Tooby, J., and L. Cosmides. 1995. Mapping the evolved functional organization of mind and brain. In *The Cognitive Neurosciences*, ed. M. Gazzaniga. Cambridge, MA: MIT Press.

Turner, J. S. 2000. *The Extended Organism: The Physiology of Animal-Built Structures*. Cambridge, MA: Harvard University Press.

Uller, C., R. Jaeger, G. Guidry, and C. Martin. 2003. Salamanders (*Plethodon cinereus*) go for more: Rudiments of number in an amphibian. *Animal Cognition* 6:105–112.

Vogel, S. 1981. Behaviour and the physical world of an animal. In *Perspectives in Ethology*, vol. 4, ed. P. Bateson and P. Klopfer. New York: Plenum Press.

Webb, B. 1994. Robotic experiments in cricket phonotaxis. In *From Animals to Animats 3: Proceedings of the Third International Conference on the Simulation of Adaptive Behavior*, 45–54, ed. D. Cliff, P. Husbands, J.-A. Meyer, and S. Wilson. Cambridge, MA: MIT Press.

16 On Displacement of Agency: The Mind Handmade

Zdravko Radman

If habit is neither a form of knowledge nor an involuntary action, what then is it? It is knowledge in the hands, which is forthcoming only when bodily effort is made, and cannot be formulated in detachment from that effort.

—Maurice Merleau-Ponty

1 Introduction: Illusions of Intellectualism

A widespread understanding of agency—both commonsense and philosophical—presupposes a *centralized knower*, an agent in the head that knows pretty much exactly the organism's next move, initiates purpose of action, and knows how to achieve the organism's goal. This view claims to intuit a conscious supervisor, one that knows what the effector system is about to realize in action. This is basically the position of intellectualism (whose roots, needless to say, go all the way back to the philosophy of antiquity), which, roughly, claims that whatever actions an organism performs in the world are the result of a previous mental activity with propositional content.

According to what Gilbert Ryle calls the "intellectualist legend," "whenever an agent does anything intelligently, his act is preceded and steered by another internal act of considering a regulative proposition appropriate to his practical problem" (1949, 31). Or, as Bengson and Moffett put it: "according to intellectualism, an individual's act is an exercise of a state of Intelligence only when the individual has a distinct mental state of engaging some propositional content" (2012, 11). One can interpret this as saying that to engage in action one must contemplate some proposition. Anti-intellectualists clearly disagree with this claim (cf. Ryle 1949; Stanley and Williamson 2001; Noë 2005) and argue that intellectualism fails to do justice to a more variegated understanding of agency and a more encompassing conception of mind.

One of the major implications of intellectualism is that to be a human agent, as a rule, presupposes an attentive planer, who is always clear about his intended next moves and is in control of their execution. What I call "intellectualist illusion" refers to just this hegemony of conscious thought and deliberation. Yet to be critical of intellectualism does not imply that a disintellectualized agent can only be conceived in terms of blind embodiment and naïve coping; it does not imply that, in other words, the only alternative to the dominance of the thinking mind is a submission to the ignorant body. To assume that it does imply this generates the scheme according to which the two views remain opposed in an exclusive relation, with a strict "division of labor" between them, and it also implies that there is a temporal distribution according to which, for example, every motor action is preceded by a goal initiated in the conscious mind of a purposeful thinker.

Theorists from both camps seem to adhere to this exclusiveness—an adherence that makes them insensitive to the fact that from the very possibility of discriminating between the propositional and practical it need not follow that there must be a sharp dividing line between them (though see Snowdon 2004; also comments by Noë 2005); for these thinkers it is probably also inconceivable that sometimes one form of knowledge can converge with another (e.g., that what counts as a proposition at t_1 can be acquired by a skill at t_2). Accordingly, discrimination between "knowing that" and "knowing how" is useful, but its strict application may turn out to be problematic or even impossible. The imposed demarcation either does not exist in reality or is not nearly as firm as one might want it to be. The exclusions and polarizations we create in our theories may not exist as such in the phenomena we attempt to describe and analyze. In other words, the mere possibility of discrimination and characterization does not imply that we will actually find such an organization in the phenomena we investigate.[1]

I am convinced that the best way to demonstrate that the core assumption of intellectualists finds little or no support in current empirical studies and theories is to focus on the hand—its skills and art of manipulation. The hand belongs most intimately to "us" and at the same time is an organ well suited to participate in worldly affairs, in which manual behavior is largely independent of preconceived propositional plans. (I also think that it would be useful to coin the term "handing" to refer to all the various forms of performance, from simple sensorimotor reactions to more complex cultural creations, since standard English lacks a verb that would account for everything the hand is capable of doing.[2])

In this chapter, I will attempt to provide arguments for the assumption that our hands are empowered to deal competently with natural, social, and cultural circumstances without engagement of a conscious thinker, while at the same time recognizing that such an embodied faculty is never blind, naïve, or uneducated, and that neither can our motor abilities be understood in such a way. Much of the motivation for this chapter is to try to show not only that there exists something like practical or, more concretely, manual knowledge, but also that it is generated in a semiautonomous way, apart from contemplation and deliberation.

2 From Affordances to Actions

Ever since J. J. Gibson (1977, 1979), with his "ecological approach," has infected theorists with the idea of "affordances," this concept has rightly become a matter of consideration (e.g., embodied philosophy often takes the concept as a necessary to an organism's embeddedness in the environment), but it seems to me that it has been merited with more weight than it deserves. Understanding perception as a "direct pickup" of invariant properties has been criticized, for instance, by Jerry Fodor and Zenon Pylyshyn, who conclude that in order for perception to be "direct" and "unmediated" visual percepts need only be inferred from the "properties of the light that are directly detected" (1981, 141) and "from the properties of the environment that are . . . 'specified' by the sample of the light" (ibid., 142). Yet, as we now know, there is no "information in the light" (ibid., 165), and the role of optics in visual perception is negligible compared to the enactive competence that an agent has to display in order to make input become relevant and meaningful. In addition, "directness" that has no memory, no possibility of or need for learning, no sense of history or culture, could hardly afford anything like our knowledge of the world.

First of all, not everything to which an organism is exposed can possibly matter, or be an affordance (if that were the case, we would soon collapse under the burden of data we would have to process). In other words, there must be some (pre)selection made by the organism on the basis of some relevance or interest that potential affordances have for it.

In any of the conceivable cases, the initiation of what makes things affordances starts, as Chris Frith would say, "on the inside" (Frith 2007, 126); this, however, needs to be complemented with actions, conventionally conceived as being directed "outside." In other words, affordances do not speak for themselves, nor can they, for that reason, be a decisive element in coping. Preconditions have to be fulfilled in order for something to count as an affordance for the organism at all, and that cannot happen to a passive being. In short, nothing can be an affordance if there is no agent capable of recognizing any object as such (i.e., as affordable). Neither are objects possible affordances to all agents in the same way. Consequently, criteria for affordability are to be sought more in the experience and competence of the agent than in the "given" quality of objects or their direct representation in the features of light.

For example, what a razor affords depends on whether it is in the hand of a male adult, a small baby, or a killer. Further, pills are medicine in the hands of a responsible and informed person but can suddenly become dangerous if not kept "out of reach of children." Affordances mutate instantly according to modes of manipulation (it is not only with new technical devices that we have to read the instructions for usage; generally, we have to learn how to deal with objects and tools or have some self-educated experience with them). Objects change their status (e.g., from useful tool to weapon) according to the user's intentions and possible modes of handling them.

Virtually all objects are cognitively innocent, or epistemologically nonexistent, until an agent relates to them. If they are ignored (i.e., subjectively inhibited), they afford nothing. They have to matter for the agent in some way if they are to be discovered as affordances. It is quite important that an organism learns *not to react to insignificant stimuli* and to ignore most of what objects may potentially afford in the surroundings if attended to, but which are irrelevant for the present behavior. Our perception of our environment is constantly shifting with changing modes of how to cope with it; it undergoes cognitive transformation according to the agent's attitudes, and also according to the slogan "different actors—different affordances."

For instance, the role of the pen in writing is negligible, as is that of the keyboard in piano-playing. What a keyboard affords lies in the hands of the pianist, and what the creative hand does is least dependent on what the pen affords. Similarly, what a kitchen affords depends on how good of a cook you are, and what a cockpit affords depends on whether you are a qualified pilot or not.

Things that surround us are "invisible," innocent, and helpless until we attend to and treat them by "handing" them. They are used, misused, and change their status according to agent's intentions (e.g., a hammer is a tool maid for nailing, but it can also be turned into a weapon). For example, liquid—until recently, considered a fairly benign element—is strictly prohibited in the personal luggage of airport passengers, because in the hands of terrorists it might turn into a means of destruction. In the hands of a pervert, the primary function of any object may be betrayed. You can read the book or you can burn it; you can light a cigarette or set fire to a house. Yet, no matter how motivated humans are for destruction, we are fortunately also creative beings. In artistic hands, the alteration of the primary function of objects is undertaken to create artifacts that enrich our living world with cultural values.

Regardless of intentions, hands, more than anything else, inform us about "other bodies" (an aspect that should philosophically be no less relevant than the problem of "other minds"). Could we not then also, alongside the concept of intersubjectivity, introduce the term of *interobjectivity*[3] (meaning both our own, active corporeal exchanges as well as a sense—a "mirroring"—of other bodies' mutual interactions) to refer to an elementary way of human coping in the *world of objects* (biological, natural, and artificial)? With interobjectivity in the (theorist's) mind, we are better equipped to appreciate Merleau-Ponty's view that "my body is a thing among things" (1964b, 164), that "things and my body are made of the same stuff" (ibid.), or that "things have an internal equivalent in me" (ibid.).

From what has been said thus far, we can conclude that the world is not *given* (in any philosophically relevant sense) but *graspable* (understood not only in terms of the mere mechanics of motor behavior). A further, by no means trivial, consequence is that *actions are underdetermined by the physical features of surroundings, by affordances and appearances* (and certainly by light patterns on photoreceptors). Believing also that "merely motor" is a poor way of characterizing what hands do for us in the world, I proceed in the following sections to outline some core notions (manual perception, practical cognition, manual guesswork, etc.) that may help us gain a more profound understanding of the nature of manual behavior and how it relates to the world.

3   Manual Perception

There is probably no better way of demonstrating of the way that humans deal with the world unconsciously than by attending to the role of the hand in its participatory relation to the natural, social, and cultural milieus. This section is thus meant to provide some (selective rather than extensive) evidence of manual intervention in the world that is characterized by promptness, effortlessness, and adaptability. This type of manual navigation enables a cognitive organism to cope with his or her surroundings without recourse to reflection or any centralized control. As Sean D. Kelly rightly observes: "many of our most basic ways of relating intentionality to the world are precisely *not in the form of having thoughts about it*" (2000, 163; emphasis added). Yet it should not be taken for granted that the only alternative to "thoughtlessness" is either an automatic sensory reception by the passive bodily apparatus or a bodily ignorance. Consider the following example: "If you see a stick in the water, it looks curved, but if you *know* it is a stick, you *know* that you can grasp it the same way you would grasp a straight object. A stick in the water and out of the water look different, but they afford the same actions" (Prinz 2009, 429; emphases added). The example is simple but the evidence far reaching: The hand is obviously not deceived by appearances (but for another perspective, see Gallagher, this vol.). The conclusion is far from trivial: Vision, though important, is not an exclusive guide to action. The hand, it seems, knows its way around its environment without recourse to our dominant cognitive source—vision (however, cf. Mattens, this vol.). The example shows that the agent is simply not fooled by appearances.[4]

Chris Frith (2007, ch. 3) analyzes another optical illusion, originally designed by Roelofs (1935) and experimentally applied by Bridgeman et al. (1997; see also, e.g., Glover and Dixon 2001). What is experimentally measured is precision of hand movements corresponding to the visual stimulus. Briefly, an observer is placed in front of a screen that shows a spot in a frame. When the frame moves to the right, it creates the visual illusion that it is the spot that moved and not the frame. However, when the observer is asked to reach for the spot, he touches it at its real position and not where his visual data suggest it is. No matter how intimately interconnected the sensory and the motor are, and no matter how strongly vision dominates cognition, the hand is, in this particular situation, not deceived by the eye. The appropriate movement is applied as if there is a bodily know-how that functions efficiently and also apart from surveillance of the eyesight. "So your hand 'knows'"—says Frith—"that the target has not moved even though you think that it has" (2007, 69). Other experiments provide support for the claim that "I am unaware of what precisely my hand is doing" (ibid., 64). Perception and grasp, thinking and movement need not go hand in hand, though it does not follow that they do not interact.

A mismatch between perceptual judgment and motor act has also been observed by Wong and Mack (1981), but it is particularly Goodale et al. (1986), Goodale et al. (1991), Goodale and Milner (1992) who provide sound evidence that the visual and motor systems may work independently. Further, they claim that not only is there a dissociation between perception and action but also that these are processed in separate neural pathways (Goodale et al. 1994), that there are double neural routes (dorsal and ventral) that do not work in accord with each other (Goodale et al. 2005), and also that there exist "two parallel visual systems—each constructing its own vision of reality" (Goodale and Milner 2006, 663).

Although the "two-system hypothesis"[5] is from the contemporary perspective not uncontroversial, its consequences offer a ground for a profound assumption: What rules our actions is not determined by conscious processes. What (conscious) perception tells the agent may prove not to be relevant, and what is achieved under volition may be a tiny fraction of what an organism does, or may be an illusion that we only slowly realize as such.

As Salvatore Aglioti et al. succinctly put it: "what we think we see may not always be what guides our actions" (1995, 684).[6] Or, in a more recent version: "what we see is not necessarily what is in charge of our actions" (Goodale and Milner 2006, 663). Further, in slightly different phrasing: "what you see is not always what you get" (ibid.).

The phenomenon of "blindsight," first analyzed and baptized by Lawrence Weiskrantz and his collaborators (see Weiskrantz et al. 1974; Weiskrantz 1986, 2006), also deals with the dissociation of the visual and the motor. What is so amazing in these experiments is that subjects—patients suffering damage to the primary visual cortex (due to which there is no visual stimulation in that part of the brain)—are still capable of properly reacting to what was going on in their visual field. Claiming that they have no visual (i.e., no conscious or phenomenal) experience, their own explanation of the successful performing is that they were making "guesses." (I will return to the idea of guesswork in section 4.)

I assume that it is not only in blindsight that the hand knows what to do in action autonomously from visual guidance. It seems there is less in optics that informs the doer than we are habitually prone to admit.[7] For one thing, we know that photoreceptors cannot determine what should be ignored and what should be attended to, that is, they cannot discriminate irrelevant information from that which matters in some sense for the organism. To put it bluntly, what is projected on the retina does not bear instructions on what to do, when to react, and how to proceed.

Generally, what we see is dominated more by the function and possible use of objects (as well as virtually our entire experiential history of them) than by their appearance and physical features alone. Knowledge of the former is what influences our attitude toward objects and our way of handling them. Accordingly, a coin and a plate that look elliptical are treated by the hands as round objects and a box that is optically shortened in depth is handled as a cube.

Vision is a useful aid to manual performance, but it can neither guide nor fully anticipate what actions are possible. Vision does help in orienting ourselves in the environment, but there is a huge leap from seeing to doing and manipulating. What can be done, and what cannot (or should not) be done, in what way and how—all these are ultimately constrained by possibilities of manipulation that are not determined by what is sensorily given and often cannot be decided in reflection. Let us refer to this capacity as to *manual perception*. It not only combines the (multi)sensory and motor, but also incorporates the ability to discriminate objects in the world according to modes of possible actions upon them. This sort of competence alters our experience of our surroundings according to our exposure to "other bodies." The mere presence of objects transforms the field of the subject's perception and action. The neuroscientific perspective reveals exactly what I mean by manual perception. We learn from Chris Frith that "there are neurons in the parietal cortex of monkeys (and presumably in humans too) that become active when the monkey sees and object near its hand. It doesn't matter where the hand happens to be. The neurons will become active when anything comes near the hand" (Frith 2007, 62; see also Graziano et al. 2004). Thus, the environment changes according to manual matrices that the organism implicitly imposes on its living space.

Just as there is no naked eye, there is also no naïve body and no innocent hand. For instance, how a hairdresser sees hair depends on whether he has a comb or a pair of scissors in his hand. Much in the same way, how a painter sees a landscape depends on whether he is holding a pen or a paintbrush. His modes of depiction are defined by the sort of tools he has at his disposal. In other words, what hair or landscape afford, or rather what shapes manual perception, lies literally in the hands of a *coiffure* and painter and the tools they use.[8] Similarly, our manual perception of food (and, indeed, taste) differs according to whether we have chopsticks or a knife and fork in our hands. Atsushi Iriki et al. (1996) provide more specific support for such general claims. Though the investigation of the research group focuses on the use of tools and how they reshape neural maps in monkeys (but see Maravita and Iriki, 2004; also Baccarini and Maravita, this vol.), their observations are also recognized as relevant by philosophers of mind. Noë, for instance, says: "tool users undergo cortical reorganization as they acquire new tool-using skills" (2005, 284). Learning from experience with tools is so profound that after just a little practice a tool feels to us "as direct as if it were part of our body" and as if "our body is extended out into the rest of the physical world whenever we use tools" (Frith 2007, 63; see also, e.g., Johnson Frei 2003; Cardinali et al. 2009). (For more on how arm-schemas seem to lengthen as a result of human tool use, see also Maravita and Baccarini, this vol.)

What I think is essential about manual perception is that it is not limited to the confines of the physical body nor by the tools one uses; that is, it is not restricted to sensations, and neither does its scope stop at the outer skin of the body nor at the body's reach as extended through tools.[9] What Norman Doidge says generally of mental perceptual experience is also applicable to this specific form of it, namely, it takes place "not on the skin surface but *in the world*" (Doidge 2007, 11; emphasis added). Yet it is important to stress that the world of humans is both natural and cultural, found and fabricated, ready-made and invented. In Merleau-Ponty's words: "Everything is both manufactured and natural in man" (1962, 220), or, in Jerome Bruner's terms: "*Logos* and *praxis* are culturally inseparable" (1990, 81). We increasingly decipher the cognitive world in terms of manual coping capacitated to provide a cultured "view" on what our mind shapes as "reality."

"The hand manipulates, the hand knows and the hand communicates," says Raymond Tallis (2003, 31). It is "an organ of exploration and cognition in its own right: it interrogates objects not merely as a preliminary to acting upon them; it interrogates them to determine what they are and whether it is worth acting upon them" (ibid., 28).[10] In this way, for an experienced cognitive organism space exists as an *arena of possible actions*. One relates to it through matrices of motor possibilities. Just as a glimpse at a chessboard (for chess-players at least) is never neutral but pregnant with possible moves of the figures on the board, so is the entire environment shaped in terms of potential moves and manipulations; realized mostly as "involuntary movements," these possibilities "leave innumerable ideas in our mind" (James 1890, 488), and in our memory.

Merleau-Ponty, for instance, says: "When I sit at my typewriter, a *motor space opens up beneath my hands*, in which I am about to 'play' what I have read" (1962, 167; emphasis added). As already noted above, the mere presence of objects recognized in terms of manual perception transforms the space into an environment shaped by motor significations and inspires much of our involuntary motor behavior. Eugene Gendlin talks about *a space of behavior possibilities*. "Objects are clusters of behavior possibilities. Many possible behaviors come with any object. The objects exist not just in locations but *in* the space of behavior possibilities. *That is the behaviour space in which we act and perceive"* (Gendlin 2012, 154–155; emphasis in original). On this approach, we begin to understand situations in terms of what we have "at hand" prior to any manifest purposefulness or deliberation. Space becomes "hand centered" (see Holmes, this vol.). The implication of "seeing by the hand's eye" is that it is possible to say: Not only do we see in order to act, but we also act in order to see.

Manual perception, unlike visual perception, is not concerned merely with what is the case, but is rather tuned to figuring out how to do things. Napier points out that the hand is "principal vehicle of motor activity"; as such it "has advantages over the eye because it can observe the environment by means of touch, and having observed it, it can immediately proceed to *do* something about it" (1980, 8; emphasis added). In a similar vein, Tallis clarifies: "While reaching in animals is largely 'happening,' human fingering and manipulation are 'doing'"; the human world, then, appears increasingly as "the product of doing" (Tallis 2003, 69). Contemporary neuroscience too has already realized that "the shape of our brain maps changes depending upon what we *do* over the course of our lives" (Doidge 2007, 49; emphasis added).

Thus, what we do with our hands shapes the way things are seen, remembered, imagined, or generally kept in mind. Erich Harth emphasizes the primacy of the manual when he says: "Images are facsimiles, incomplete and fanciful copies of reality that we *fashion with our hands for our eyes to look at*" (1993, 79; emphasis added). Similarly, Merleau-Ponty says of the eye that "it is *that which* has been moved by some impact of the world, which it then restores to the visible through the offices of an *agile hand*" (1964b, 165; second emphasis added). We can thus see that visual inferences and judgments rely on the experience of the hands; that it is not optics alone that helps agents orient themselves in the world, but that rather it is a sort of practice that provides clues for acting.

Manual perception is to be viewed as an instance of "operative reason" (Merleau-Ponty 1962, 57). What it facilitates includes, but also exceeds, the sensory or tactile; it is a dimension of "practical understanding" (Bruner 1990, 74). Manual perception, just like visual or auditory perception, helps discriminate the relevant from the accidental, the "meaningful" from the "invisible," the "doable" from the "unhandy." From a practical point of view, it recognizes the organism's surroundings as touchable, graspable, gestured, movable, heavy, light, hart, soft, hot, shaky, slippery, rough, wet, dry, filled, empty, warm, cold, smooth, round, edgy, thin, palpable, breakable, and so on, and the objects in it as ones that can be cut, closed, fixed, opened, pushed, pressed, pulled, raised, dropped, hit, squeezed, passed over, split, brought together, stapled, handed, and the like. "We perceive objects with the ways we could behave with them, for example hold them, or push them, eat them, sit on them," says Gendlin (2012, 156). Strictly speaking, the eye can only "know" that the door is in a certain position; but to see it as "open" is to have an experience of hand intervention that brings the door to the position opposite to "closed." The "open" is not information read from the retina; it is a result of practical understanding.

There is something like a vocabulary of movement that the body has internalized—a language of manual embodiment according to which the environment *means* something to the organism just as other forms of perception do in their own way. (See also, in this respect, Natalie Depraz's chapter, this vol.) Manual perception presupposes a specific sort of understanding analogous to other forms of perception. Just as we implicitly and instantly understand, for example, mirror images (e.g., that there is nothing "inside" the mirror but that it reflects what is in front of it) and passport photos (e.g., that the miniature two-dimensional image of a head with no body represents a whole person), so too we "know" or "understand" how far we have to stretch our arm to reach a glass, how firmly we have to grasp it so that we do not drop it, and how gently we have to hold it so that it does not break. (Just as we are surrounded by familiar sounds in our perceptual field even when we do not attend to them, so do we attune to our surroundings in a sensorimotor way even when the conscious self is dormant.)

Now, though I am sympathetic to the view that "much of meaning is based on action" and also to the view that "meaning can only be determined by utilizing the motor system" (Glenberg 2007, 363), I am inclined to take a step further and assume that the *motor is meaningful in its own way*. So it is not only that semantics *draws on* the motor; I conceive of embodiment as possessing its own language of performative possibilities. These performative possibilities are captured and recognized by the "pupil" of the manual eye that perceives things that are not conveyed by vision or by other sensory modes. We then "muscle read" just as we "mind read."

The conclusion to this discussion of manual perception takes the form of the following hypotheses: The eye provides us with hints—scarce signals, fragmentary data—that the cognitive organism has to decipher not only in visual terms but also by consulting other instruments of practical understanding; optical stimuli often bypass conscious vision; mere light does not provide sufficient instructions to guide action; and manual perception enables the shift from what things look like to how can they become objects of "handing."

We may thus conclude that *hands are not faithful to the conscious mind* (and we should be tremendously grateful that this is the case). Marc Jeannerod is straightforward on this point: "There is a great advantage of not being aware while you are doing something" (quoted in Gallagher 2008, 62). And, as he further explains: "In the conscious movement, the velocity is slower and accuracy is poorer" (ibid., 63). In sum, "both the automaticity and the *lack of consciousness* of the motor process are essential attributes for behavior" (ibid.; emphasis added).

Neither do our hands conform to appearances. They do not always trust what the eyes see and are not obedient servants to deliberation and volition. In a way, hands have a sort of "reason" of their own and often act independently of data provided by light and the eyes. Possessing such know-how, they can "see" in the dark, under the table, and behind the back, and can provide an intimate physical bond to our partner,[11] among other things.

Thus, the habit of thought that favors the role of visual perception in action has a neglected but highly important counterpart: It is manual experience that significantly informs seeing. It therefore turns out that *observation is largely derivation from deeds.*

4 Manual Guesswork and Manual Intelligence

Treating the manual as something exclusively sensorimotor is like treating letters as something purely graphic or visual. That is, what letters generally do exceeds their physical (visual or tactual) nature. It is (except perhaps for graphic designers) irrelevant how they look or what their formal features are; what matters (for the reader) is what and how they mean, that is, what are they *about.*[12] For instance, when you turn a book's page while reading, you typically recall not what the previous page looked like graphically but what the text was about. Analogously, manual behavior is "inscribed" in the motor "scripture," which is *about* things and situations in the world, that is, it takes place beyond the boundaries of the body, and even beyond that of the tools. These boundaries can be extended as far as to include the "world." "We must therefore recognize that what is designated by the terms 'glance,' '*hand*,' and in general 'body' is a system of systems devoted to *inspection of a world*" (Merleau-Ponty 1964a, 67; emphases added).[13]

The world as inspected by the "sight" of the manual appears not as one of light but one of space (peripersonal and beyond), invested with significations known to the agent by means of practical cognition. Merleau-Ponty speaks of "immanent significance"[14] and Mark Wrathall of "motor significance."[15] And because "every sensation is already pregnant with a meaning" (Merleau-Ponty 1962, 346), the space in which we perform is never empty. Consequently, the motor aspects of our experience do not happen in vacuum. Movements, as I have said, are *motivated* (Radman 2008),[16] and when the agent enters the scene it immediately appears as patterned by possible behavior. In this sense, encountering empty space is like "hearing" silence; neither the former nor the latter is ever free from "filling in"—a capacity that an experienced agent permanently practices by inhabiting the void with potential happenings. Thus there is a readiness for movement before anything has even been moved.

Understanding the world "in muscular terms" (Gombrich 1972, 36) turns it into an *arena of possible moves and manipulations*. The perceptual field (in which not everything can count as relevant and what does matter can be presented in numerous ways) is charted in terms of possible manual behavior motivated to find the best possible way of *adapting to the most likely scenario* of what is the case in the surrounding world, and to choose the most appropriate mode of action. Indeed, only a doer is capable of possessing *"knowledge in the hands*, which is forthcoming only when bodily effort is made, and cannot be formulated in detachment from that effort" (Merleau-Ponty 1962, 166; emphasis added). Or, as C. I. Lewis put it: *"It is only because we are active beings that our world is bigger than the content of our actual experience"* (1929, 140; emphasis in original). As active and participatory beings, we permanently practice a "muscular" and "manual knowing" that is nothing like the abstract "aboutness" that theorists have in mind, but whose primary function is rather *adaptation to mundane situations*.

Yet, no matter how counterintuitive it might appear, this sort of adaptation is neither direct nor straightforward, and neither can it be adequately theoretically explored in elusive intentionalist terms. Since I doubt that the somewhat phantom concept of "aboutness" can do the job, a more challenging question would be: What is this "aboutness" about? In other words, how are the objects to which our mental states refer further related to our mundane circumstances?

If, as neuroscientist Walter Freeman claims, "all that our brains can know has been synthesized within themselves, in the form of *hypotheses about the world*" (1999a, 121; emphasis added), then we could conclude that what our neural/mental mechanism basically does is provide us with the *most plausible scenario of reality*—the best bet, so to speak, about what the organism implicitly judges as being most likely the case in the world. If brain's basic function is to prepare us for the "next step" in acting, so that we are not puzzled by the new situations that we must permanently embrace, then there must be also a mechanism cognitively equipped to enable such projections.

Because the motor is no exception to this general mode of the brain's functioning, it can be referred to as an implicit "motor guesswork." One way to illustrate this is to recall of C. I. Lewis's "if-then" proposition (1929, ch. 5), even if only in some loose way. For instance, if I stretch my hand with sufficient energy then I can fetch my suitcase from the turning belt of the baggage claim; if I am not cautious at the table full of wine glasses I might break some; if I don't wear an oven mitt I might burn myself when taking a hot pan from the oven; if I don't push the heavy glass door with enough power it won't open; if I press the toothbrush more than necessary I might hurt my gums; if I don't handle the knife with caution I will cut myself; and so on. In a similar vein, Tallis claims that "motor programmes include not only 'commands' and 'calculations,' but also 'if-then' and other logical operations" (2003, 65). However, the sort of if-then motor "reasoning" need not be based in logic, and neither is its exercise propositional, as it "sounds" in language. A further theoretical refinement of the if-then operation is discussed by Gendlin, who says: "If we kick the ball we can no longer pick it up and throw it. . . . If we boil the eggs, we can't then fry them. *Each behavior is a change of the cluster of implicit* 'cans.' If we do *this* we can no longer do *that*, or not in the same way as before. On the other hand, after each behavior we can do some that we couldn't do before" (2012, 157; emphases in original).

Motor and, above all, manual guesswork make us approach the world, and enables us to navigate it, by casting a net of the probable onto what will be picked out as our actuality. Equipped with corporeal know-how that enables us to feel at home in relating to an environment that is changing all the time, agents can anticipate behavior with skillful effortlessness. It is this capacity that allows us to perform with ease such actions as singing a familiar song or calculating a simple algebraic operation. We don't have to think about the sounds of the melody we sing; and we don't have to literally think of the numbers with which we operate; we just sing and calculate.

Such "just doing" (see also Radman 2012) can be seen as an instance of a broader competence to which some authors refer as to "manual intelligence (MI)."[17] However, it is one thing to say that human manual activity facilitates the growth of intelligence, as Tallis (2003) seems to imply,[18] and quite another to claim that the manual, and the body in general, possesses a know-how or "intelligence" of its own, meaning that the hand's role in human coping in the world is authentic and to a high degree independent of other (e.g., sensory) sources.

If intelligence is to be taken as a mode of adaptation (see Plotkin 2003, ch. 2), then MI too must be seen as a way of adapting to our ever-changing mundane circumstances. Similarly, Horst Bredekamp's usage of "*manuelles Denken*" (2009; quoted also by Gallagher, this vol.) brings this aspect to the foreground. Maycock et al. also emphasize the cognitive dimension of MI: "Given the richness of human manual interaction, we argue for the consideration of the wider field of 'manual intelligence' as a perspective for manual action research that brings the cognitive nature of human manual skills to the foreground" (2010, 1). They are also convinced that a "deeper understanding of manual intelligence will constitute a major step towards our understanding of general cognition" (ibid.). For Steven Pinker as well, the hand is a "pilot of intelligence" (1997, 194). He explains: "Hands are levers of influence on the world that makes intelligence worth having. Precision hands and precision intelligence co-evolved in the human lineage, and the fossil record shows that hands led the way" (ibid.).

But "manual thinking," or MI, is not merely an appealing quasi-literary syntagma (as it may appear to some); it fits well with growing scientific evidence that speaks in favor of "intimate interactions between motor skills and higher cognitive functions" (Olivier et al. 2007, 644). There is evidently a "considerable overlap that exists between the cortical areas controlling skilled hand movements and those involved in higher cognitive functions, such as language and number processing" (ibid. 644). Even more explicit is Jacques Paillard, who says that "even if the mind is not reducible to movement, movement plays an important explanatory role in our cognitive life. Cognitive systems are not separable from motor systems" (quoted in Gallagher 2008, 73). Wolpert et al. go even further and claim that motility is primary and dominant, and speak of "motor chauvinism": "From the motor chauvinist's point of view the entire purpose of the human brain is to produce movement. Movement is the only way we have of interacting with the world. All communication, including speech, sign language, gesture and writing, is mediated via the motor system" (2001, 493).

The "smart hand," then, knows what to do and how to do it before intelligence engages in action, or, as Merleau-Ponty would say: "My movement is not a decision made by my mind" (1964b, 162). It is in similar way that I read Jeannerod: "Action *execution* is the ultimate test of whether the represented anticipation of its goal was adequate or not" (2006, 168; emphasis added). In other words, an organism learns about the effects of its own action after that action is accomplished; hands take the lead, and many an action is initiated, and even accomplished, before volition plays any role.

Given what we have called the "intellectualist illusion," Jesse Prinz and Andy Clark come to a far-reaching conclusion that "different kinds of vehicles can fill the thinking role" (2004, 58). Accordingly, "it is reasonable to think of practical abilities as a kind of knowledge" (Noë 2005, 285). Reasoning, then, comes not only from words[19] but also from moves and gestures.

It can be said that the mundane constellation of our environment changes according to our guesses and expectations of our manipulative "thought." And the latter is irreducible; having things in sight, in words, or in reflection, and having them as objects of practical concern, are different things (even though there is lively interaction among them). Indeed, as we well know, not everything thinkable is doable, and it is mostly the hand that decides the latter. Indeed, very little of what emerges in the mind is manageable by hand or apt for manipulation; also, some things are such that only as manufactured can they be presented to the mind. It can thus be said that the manual possesses imagination "in the fingers" that produces marvels the thinking mind could never fully preconceive and words can hardly express. It could well be that "under the fingers" products emerge that stimulate thought and nourish imagination in a way that thought and imagination, by their own means alone, could never achieve. To say that is to imply that hands possess their own know-how and also a degree of autonomy that allows for action that is not under the surveillance of the conscious "self." As such, our hands can be emancipated from actuality and "see" ahead of the presently given. Phenomenology already knows the phenomenon; Merleau-Ponty, for instance, says: "It is through my body that I go to the world, and tactile experience appears 'ahead' of me" (1962, 369). When we further read that "perception anticipates, goes ahead of itself" (Merleau-Ponty 1964b, 36), then we have to be aware that this holds also for manual perception. Interestingly, this view is in perfect accord with contemporary neuroscientific observations, such as the following: "Not only does it [brain] elaborate on the hard facts which the senses report, it also *computes ahead and anticipates* much of what is about to happen" (Harth 1982, 164–165; emphasis added).

It is this notion of being "ahead" that I am inclined to see as characteristic also of manual behavior. It provides a possible reading of the general phenomenological credo of being "always already" in the world and specifically being "ahead of oneself" (Heidegger 1962; for the latter chiefly sec. 41) by means of corporeal—dominantly manual—navigation that charts the path for conscious processes to follow.

5 Making Comes before Matching

Hands, of course, are not only for pointing, touching, reaching, and grasping (though these reveal a lot of human manual capacities) and the scope and impact of their work in shaping the physiognomy of the world has to be viewed and judged within a larger picture that should also include the cultural (an aspect already addressed above in section 4).

A great deal of contemporary research focuses on subtle aspects of sensorimotor performance, delivering us valuable new knowledge of the underlying mechanisms, but I believe that we should take a step further and see how these mechanisms contribute to "making" or, in different terms, *how our works mold our worlds*.[20] One domain where this is most authentically demonstrated is that of art and aesthetic experience. "Art involves seeing; the seeing is meaningful and structured by its meaning, the meaning is conveyed, as it is achieved, only in the *making*" (O'Malley, 1970, 53). Or, in a more compact form: "Art . . . involves seeing the meaning in *making*" (ibid.; emphasis added).

Very much along the same line of thought is Nelson Goodman's claim that "The eye comes always ancient to its work. . . . Not only how but what it sees is regulated by need and prejudice. It selects, rejects, organizes, discriminates, associates, classifies, analyzes, constructs. It does not so much mirror as *take and make*; and what it *takes and makes* it sees not bare, as items without attributes" (1976, 7–8; emphases added). Or even more explicitly: "having a mental image may be construed in terms of ability to perform certain activities" (Goodman and Elgin 1988, 91).

"Making comes before matching"—a dictum formulated by the historian of art Ernst H. Gombrich (in his 1960 work and others)—is a perfect slogan to label a form of interaction even more basic than the one he is referring to (he primarily applies it to pictorial representation); in my usage, it is taken to account for all forms of transformation that recreate the "given" as the "taken"[21] and remold the natural into the cultural (see also Menary, this vol., and Cole, this vol.).

Inspiration and imagination may be set in motion by unintended moves and spontaneous manipulations. "Trivial" manual interventions may trigger both commonsense behavior and artistic endeavors. "In fashioning . . . sketches, paintings, or sculptures we allow an often nebulous idea to initiate the creative process, which is further driven by the product that emerges *under our hands.*" (Harth 1993, 109; emphasis added). Clearly, the eye can "take and make" only with the recourse to the embodied (and enhanded) agent (or maker), who becomes what he is by means of the mechanisms that constitute him. Recognizing the meaning and impact such a mechanism-in-action has in constructing what emerges in our experience as "reality" throws new light on the cognitive merit of that very mechanism: It increasingly itself gains attributes of agency. The idea is nicely expressed by Tallis when he speaks of "*the mystery of awakening to agency out of mechanism*" (2003, 12)—or, in another formulation, of "our wakening out of our bodies and their mechanisms to ourselves as agents using our bodies for definite purposes" (ibid., 122).

Therefore, contrary to what our commonsense beliefs suggest, we are not the commanders of most of what "handing" facilitates, and even less can we influence the effects of the doing on the doers. We are full-blooded makers, but we are not masters of what the making does to the mind. In our actions we change the current constellation of our surroundings; but doing in turn transforms the mental world of the doer, and that is the process on which the "self" has virtually no or little influence. In what we do, we permanently pattern our mental world in a way we cannot consciously monitor, modulate, or control. After the motto "learning transforms the learning device itself,"[22] we can say that making transforms the makers themselves. We thus increasingly come to perceive ourselves as *products of the process.* And, if I am correct, the consequence of such an assumption is that *we are not only actors but also targets of our own actions.* The lesson from the primacy of making may thus be that we are *both subjects and objects of our own doing.*

6   Aiming Away from the Centralized Agency

From very simple manual operations, such as cutting bread, to more complex forms of "handing," such as conducting an orchestra, hands are not in our conscious focus; they are always directed toward accomplishing deeds that exceed the merely motor and involved in manipulations that accomplish tasks within the larger scale of human motivations, interests, desires, and so on, which, however, are not fully preconceived in the conscious mind. Because hands are always *about* doing things, they themselves remain outside of conscious awareness. It is this engaged aboutness that erases hands' presence in consciousness; or rather, it is thanks to their absence from the focus of awareness that they can perform so effortlessly and skillfully. In this sense, hands remain "invisible," as the background does,[23] and often do their job best just when this is the case.[24] Manipulating a knife and handling a baton are both forms of manual behavior that are *about* performance as exemplified in cutting and conducting, forms of which, mostly, do not originate in deliberate strategies. In other words, to understand "handing" we must expand the range of observation far beyond the sensory and the motor. When we do so, we realize that the hand itself dissolves from conscious attention. The true mystery of manual wisdom is thus contained in the challenging phenomenon that the organism can best perform complex tasks when mind is not consciously aware of those tasks.

"The point is," as Alva Noë says, "that we don't need to think about our bodies or pay attention to them to act" (Noë 2009, 77). He further says: "You don't need to locate your hands before you put them to use in reaching for something, and as a general rule you don't need to pay attention to your body parts (hands, fingers, whatever) in order to use them effectively" (ibid.).

And it is not only in reaching and grasping that hands act as more or less sovereign instruments whose performance allows adjustments not attended to in awareness. In more demanding forms of manipulation, such as painting, their performance also has a relatively high degree of autonomy, the results of which an artist learns after her exposure to and interaction with the medium. A painter knows how to proceed in her enterprise *after* she has received a response provoked by the manual intervention onto the pictorial medium, not necessarily as a result of the clear-cut, mental (artistic) plan. What emerges on the canvas is as much a result of strictures and potentialities of the medium that the hand meets as it is an outcome of the painter's "artistic vision." A picture in the mind and a picture on the canvas are so different that it is only the common term that affiliates them. It is the hand that grants existence to the latter, and converts imagination into embodied (painted) image.

Understanding *action as aiming*, from the perspective outlined here, allows for a conception of agency without preconceived purposefulness. From the fact that an action, upon its completion, appears goal-oriented to an observer it does not follow that it was purposeful from the point of its initiation. For aiming can be initiated prior to any purpose or goal

manifested in consciousness. Most of what the hands do can be characterized as *participation without purposefulness.*

This must be the sort of conclusion that Walter Freeman has also come
to when he, for instance, talks about "action 'flow' without awareness"
(1999b, 156) and action "without recourse to agency" (ibid.), which then
drives him to assume that what happens is a sort of a "disappearance of
agency" (ibid., 159). Freeman's intent is clear and well founded, but his
phrasing is unfortunate. Agency clearly does not disappear if we realize
that there is no *thinking* subject in the role of a doer and no conscious
controller attentive to performances of the body; it simply has to be found
in other instances, that is, at more basic or elementary levels of embodiment. The "disappearance of agency" seems inadequate, as it implies that
only a "centralized knower," with the power of governing its own behavior
in an "intellectual" manner, can be considered an agent. I have attempted
in this chapter to justify the attribution of agency to an organism even if
it is not engaged in conscious processes and deliberation, a situation that
holds for most of manual behavior. I have tried to make us more sensitive
to the idea (and perhaps to prepare us for a new scientific paradigm) that
an agent is capable of competently coping with the world before or apart
from conscious intervention in it. I thus think that it is more appropriate
to conceive of this sort of shift, from a thoughtful and deliberative agent
to embodied participator guided by autonomous mental processes, as one
of *displacement.*

Hands are coupled to such a "displaced" agency that largely functions
as an autonomously working system. In this sense hands can be seen as
operating, more than we are prone to admit, free from the supervision of
the conscious and volitional "self." The "I" mostly only recognizes the
results of actions already accomplished; "goals" emerge in awareness as
actions completed; "purposes" are recognized when the *results* of manipulation become visible. However, the "self" *witnesses what has been done and
thereby creates the illusion that it authored the deeds.* Yet most of what we
do is exercised in "displaced" manner—or, as William James would say,
"unforeseen by the agent" (1890, 486)—that is, as a result of skilled action
unaccompanied by the effort of a cerebral knower, and with the same ease
with which we breathe and digest.[25]

It must be that we are uncomfortable with the idea that each of us is a
child of own work; instead, we seem to prefer a self-image characterized
by parental authority that rules over our courses of action. However, the
"authority" is not where our habit of thought localizes it; it is mostly *away
from the conscious and thought-centered "self,"* facilitating our actions in a

prompt and effortless way and guiding us competently by the compass of corporeal know-how.

7 Conclusion

Aiming to extrapolate and emphasize the most specific trait of humans, cultural history takes the human to be a "social animal," an "animal rationale," an "animal symbolicum," a cultural creature, a political being, "*Homo ludens*," and so on. But there are equally philosophical reasons to define the human as a "manual being," for the role of the hand in shaping of our mental world is unique and profound, and it deserves to be listed among the distinctive features of the type of animal we are (for indeed, we are the only species privileged in this way).

What follows from the above discussion is that hands are not faithful to the purposive mind, nor are they obedient servants to what objects afford us; that vision provides hints for deciphering surroundings but does not prescribe modes of action upon it; that the ways of the hand bypass conscious agency, or, in other words, that manual enaction enables competent coping with the world without recourse to reflection. This is a tremendous mercy to the organism: it need not engage in contemplation at each and every instance of its behavior, and yet it can adapt to mundane situations with skillful ease.

The hands belong to a participatory being that engages in worldly affairs—natural, social, and cultural—without having always to consciously deliberate. An agent with rich manual experience does not only react to what mind experiences as "reality," but rather *guesses and anticipates what might be the most likely situation in the world* and prepares the entire organism for how to deal with it. Thanks to bodily know-how, and specifically to manual intelligence, a cognitive organism is tuned to cope with the world without the aegis of the so-called higher cognitive processes.

One conclusion we can draw is that the living world of humans is less depicted in purely visual terms than it is dominated by practical experience and understanding. It is this practical cognition, mostly given the impact of manual know-how, that enables an agent to read its surroundings in terms of blueprints of its own actions, past and possible. In this sense it can be said that the landscape of behavioral possibilities bears the handwriting of manual coping based on remembered and imagined experience. This amazing feature makes us feel at home wherever we move, and whatever we meet, until the familiar is challenged by the new and unexpected.

Hands act ahead of deliberation and create the very conditions for the organism to feel at home in the *present situation, which is cognitively inherited from the anticipated future.* Having such a horizon of the potential always ready "at hand" makes us (on an experiential level) feel that we approach future with the certainty of knowing it before it actually happens, and (on a theoretical level) provides us with a possible interpretative key to the phenomenological credo of "always already" being in the world (whereby there are enough philosophical reasons to want to convert *being*-in-the-world into *acting*-in-the-world). And yet aiming oneself in the world can never fully allow one to anticipate what can, and what cannot, be manually achieved, and in what ways. There are "reasons" known only to the body and knowledge uniquely "in the hands," of which the conscious mind is often only a late eyewitness. This affirms the body as corporeal connoisseur on its own terms and the hands as the ultimate arbiter on doability.

Acknowledgments

I want to thank the Croatian Ministry of Science, Education, and Sport for providing support to my research project Embodied Mind and Intentional Act. Many thanks also to Raphael Bene, Andrew Bremner, and Harry Farmer.

Notes

1. A typical illusion of this sort is taking mathematics—a useful *means* of analysis of mental processes—and then attributing its algorithms to the "language of thought" (Fodor 1975), or using computers as explanatory tools and then claiming that the mind is nothing but a program and the brain nothing but hardware.

2. Interesting enough, Natalie Depraz (this vol.) also makes use of the term.

3. It is closely associated with Merleau-Ponty's "intercorporeity" (1962; see also Gallagher's and Ratcliffe's comments, this vol.).

4. Compare my discussion of the same example in the context of background knowledge (Radman 2012).

5. A brief exposition of it can be found in Goodale and Milner 2006.

6. See also Hutto's comment (this vol.) on Milner and Goodale's experiment; also those by Andy Clark (2000).

7. For an interesting case of human infants whose reaching is guided without visual feedback of their hands, see Bremner and Cowie, this volume.

8. How the use of tools alters brain functions in monkeys has been extensively researched by Atsushi Iriki and his research group at the Laboratory for Symbolic Cognitive Development in Tokyo. Among other things, they have shown that the monkey's brain grows as it learns to use tools. For a survey of the research see, e.g., Maravita and Iriki 2004.

9. Compare Merleau-Ponty as he observes: "Once the stick has become a familiar instrument, the world of feeble things recedes and now begins, not at the outer skin of the hand, but at the end of the stick" (1962, 175–176).

10. Tallis also draws attention to the fact that the word *manu* is etymologically connected to the Sanskrit verb that means "to think," which he takes as a further support for his assumption that "prehension and apprehension are, deeply, one" (2003, 16, n. 16).

11. More than that, our hands are irreplaceable in the erotic experience that unfolds for us the sensual world, which is unknowable through any other perspective. (For more on this aspect of the hand, see, e.g., Depraz, this vol.)

12. "Aboutness" is a term coined by R. A. Fairthorne (1969), but its philosophical usage (basically as a synonym for intentionality) is due to John Searle (esp. 1983). This is a huge area in the philosophy of mind, but unfortunately we do not have the space to discuss its important aspects here.

13. In a similar way, Freeman talks about "exploratory manipulations" (1999a, 36).

14. "To experience a structure is not to receive it into oneself passively: it is to live it, to take it up, assume it and discover its *immanent significance*" (Merleau-Ponty 1962, 301; emphasis added).

15. "The objects and situations that we encounter in the world thus act on us through an ambiguous and indeterminate *motor significance*" (Wrathall 2005, 115; emphasis added).

16. For John Searle, motivation counts among features of "reasons for action." They are, as he says, "essentially *motivational* in the sense that they must be capable of motivating an action" (2001, 141).

17. When I coined the term MI for my own use I was not aware that Helge Ritter (e.g., Ritter et al. 2011) and his research group had already been operating with the term. (Thanks to Rolf Pfeifer for drawing my attention to that fact.) The term is now used also by other authors.

18. This is, namely, what Tallis claims: "manipulative activity enhances intelligence; intelligence permits more complex manipulative activity; this in turn enhances intelligence" (2003, 35). And also: "There seem to be at least two aspects of human manual activity that make it particularly apt to be the driver for increased *individual and collective intelligence*. The first relates to its manipulative function; the second to its cognitive or exploratory function" (ibid., emphasis added).

19. In a straightforward formulation, Merleau-Ponty says: "thought is not an effect of language" (1962, 223).

20. The Goodmanian phrasing is here intentional. See for instance his (1975) essay, "Words, Works, Worlds."

21. As Dewey suggests: "The history of the theory of knowledge or epistemology would have been very different if instead of the word 'data' or 'givens,' it had happened to start with calling the qualities in question 'takens'" ([1929] 1960, 178).

22. La Jolla group with Sejnowski; quoted by Gazzaniga (1998, 14).

23. "As the precondition of Intentionality, the Background is as invisible to Intentionality as the eye which sees is invisible to itself" (Searle 1983, 157). For more on the background, see Radman 2012.

24. Imagine a pianist who suddenly shifts her attention to the very mechanics of her finger movements; that would most likely paralyze her playing. When the mechanics of the fingerwork is brought to the forefront, the interpretation of the music piece is endangered. Similarly, a surgeon who focuses on the very workings of his fingers, and not on what they are trying to accomplish, would also likely be paralyzed.

25. Automaticity has been extensively studied and elaborated notably by John Bargh and his research group (see, e.g., Bargh 1994; Bargh and Chartrand 1999; Hassin et al. 2005). On a specific meaning of automaticity in relation to the hand see, e.g., Pisella et al. 2000.

References

Aglioti, S., J. F. X. DeSouza, and M. A. Goodale. 1995. Size-contrast illusions deceive the eye but not the hand. *Current Biology* 5 (6):679–685.

Bargh, J. A. 1994. The four horsemen of automaticity: Awareness, intention, efficiency, and control in social cognition. In *Handbook of Social Cognition*, vol. 1, ed. R. S. Wyer and T. K. Srull. Hillsdale, NJ: Erlbaum.

Bargh, J. A., and T. L. Chartrand. 1999. The unbearable automaticity of being. *American Psychologist* 54 (7):462–479.

Bengson, J., and M. Moffett. 2012. Two conceptions of mind and action: Knowing how and the philosophical theory of intelligence. Introduction to *Knowing How: Essays on Knowledge, Mind, and Action*, ed. J. Bengson and M. Moffett. Oxford: Oxford University Press.

Bredekamp, H. 2009. *Galilei der Künstler: Der Mond. Die Sonne. Die Hand.* Berlin: Akademie Verlag.

Bridgeman, B., S. Perry, and S. Anand. 1997. Interaction of cognitive and sensorimotor maps of visual space. *Perception & Psychophysics* 59 (3):1368–1370.

Bruner, J. 1990. *Acts of Meaning*. Cambridge, MA: Harvard University Press.

Cardinali, L., F. Frassinetti, C. Brozzoli, C. Urquizar, A. C. Roy, and A. Farné. 2009. Tool-use induces morphological updating of the body schema. *Current Biology* 19 (12):R478–R479.

Clark, A. 2000. Visual awareness and visuomotor action. In *Reclaiming Cognition: The Primacy of Action, Intention, and Emotion*, 1–18, ed. R. Núñez and W. Freeman. Thorverton: Imprint Academic.

Dewey, J. [1929] 1960. *The Quest for Certainty: A Study in the Relation of Knowledge and Action*. New York: Capricorn Books.

Doidge, N. 2007. *The Brain That Changes Itself: Stories of the Personal Triumph from the Frontiers of Brain Sciences*. London: Penguin.

Fairthorne, R. A. 1969. Content analysis, specification, and control. *Annual Review of Information Science & Technology* 4:73–109.

Fodor, J. A. 1975. *Language of Thought*. Cambridge, MA: Harvard University Press.

Fodor, J. A., and Z. W. Pylyshyn. 1981. How direct is visual perception? Some reflections on Gibson's "ecological approach." *Cognition* 9:139–196.

Freeman, W. J. 1999a. *How Brains Make Up Their Minds*. London: Phoenix.

Freeman, W. J. 1999b. Consciousness, intentionality, and causality. *Journal of Consciousness Studies* 6 (11–12):143–173.

Frith, C. 2007. *Making Up the Mind: How the Brain Creates Our Mental World*. Malden, MA: Blackwell.

Gallagher, S. 2008. *Brainstorming: Views and Interviews on the Mind*. Charlottesville: Imprint Academic.

Gazzaniga, M. S. 1998. *The Mind's Past*. Berkeley: University of California Press.

Gendlin, E. T. 2012. Implicit precision. In *Knowing without Thinking: Mind, Action, Cognition, and the Phenomenon of the Background*, 141–166, ed. Z. Radman. Basingstoke: Palgrave Macmillan.

Gibson, J. J. 1977. The theory of affordances. In *Perceiving, Acting, and Knowing: Toward an Ecological Psychology*, ed. R. Shaw and J. Bransford. Hillsdale, NJ: Erlbaum.

Gibson, J. J. 1979. *The Ecological Approach to Visual Perception*. Boston: Houghton Mifflin.

Glenberg, A. M. 2007. Language and action: Creating sensible combinations of ideas. In *The Oxford Handbook of Psycholinguistics*, 361–370, ed. M. G. Gaskell. Oxford: Oxford University Press.

Glover, S., and P. Dixon. 2001. Motor adaptation to an optical illusion. *Experimental Brain Research* 137:254–258.

Gombrich, E. H. 1960. *Art and Illusion: A Study in the Psychology of Pictorial Representation*. Princeton, NJ: Princeton University Press.

Gombrich, E. H. 1972. The mask and the face: The perception of physiognomic likeness in life and in art. In *Art, Perception, and Reality*, ed. E. H. Gombrich, J. Hochberg, and M. Black. Baltimore: John Hopkins University Press.

Goodale, M. A., G. Kroliczak, and D. A. Westwood. 2005. Dual routes to action: Contributions to the dorsal and ventral streams to adaptive behavior. *Progress in Brain Research* 149:269–283.

Goodale, M. A., J. P. Meenan, H. H. Bülthoff, D. A. Nicolle, K. J. Murphey, and C. I. Racicot. 1994. Separate neural pathways for the visual analysis of object shape in perception and prehension. *Current Biology* 4 (7):604–610.

Goodale, M. A., and A. D. Milner. 1992. Separate visual pathways for perception and action. *Trends in Neurosciences* 15:20–25.

Goodale, M. A., and A. D. Milner. 2006. One brain—two visual systems. *Psychologist* 19 (11):660–663.

Goodale, M. A., A. D. Milner, L. S. Jacobson, and D. P. Carey. 1991. A neurological dissociation between perceiving objects and grasping them. *Nature* 349:154–156.

Goodale, M. A., D. Pelisson, and C. Prablanc. 1986. Large adjustments in visual guided reaching do not depend on vision of the hand or perception of target displacement. *Nature* 320:748–750.

Goodman, N. 1975. Words, works, worlds. *Erkenntnis* 9 (1):57–73. Reprinted in *Ways of Worldmaking*. Indianapolis: Hackett, 1978.

Goodman, N. 1976. *Languages of Art*. Indianapolis: Hackett.

Goodman, N., and C. Z. Elgin. 1988. *Reconceptions in Philosophy and Other Arts and Sciences*. London: Routledge.

Graziano, M. S. A., C. G. Gross, C. S. R. Taylor, and T. Moore. 2004. A system of multimodal areas in the primate brain. In *Crossmodal Space and Crossmodal Attention*, 51–67, ed. C. Spence and J. Driver. Oxford: Oxford University Press.

Harth, E. 1982. *Windows on the Mind: Reflections on the Physical Basis of Consciousness*. New York: William Morrow.

Harth, E. 1993. *The Creative Loop: How the Brain Makes a Mind*. Reading, MA: Addison-Wesley.

Hassin, R. R., J. S. Uleman, and J. A. Bargh, eds. 2005. *The New Unconscious*. New York: Oxford University Press.

Heidegger, M. 1962. *Being and Time*. Trans. J. Macquairre and E. Robinson. New York: Harper & Row.

Iriki, A., M. Tanaka, and Y. Iwamura. 1996. Coding of modified body schema during tool use by macaque postcentral neurons. *Neuroreport* 7:2325–2330.

James, W. 1890. *The Principles of Psychology*, vol. 2. New York: Dover.

Jeannerod, M. 2006. *Motor Cognition: What Actions Tell the Self*. Oxford: Oxford University Press.

Johnson Frei, S. H. 2003. What's so special about human tool-use? *Neuron* 39: 201–204.

Kelly, S. D. 2000. Grasping at straws: Motor intentionality and the cognitive science of skillful action. In *Heidegger, Coping, and Cognitive Science: Essays in Honor of Hubert Dreyfus*, vol. 2, ed. M. Wrathall and J. Malpas. Cambridge, MA: MIT Press.

Lewis, C. I. 1929. *Mind and the World Order: Outline of a Theory of Knowledge*. New York: Dover.

Maravita, A., and A. Iriki. 2004. Tools for the body (schema). *Trends in Cognitive Sciences* 8 (2):79–86.

Maycock, J., D. Dornbusch, C. Elbrechter, R. Haschke, T. Schack, and H. Ritter. 2010. Approaching manual intelligence. *Künstliche Intelligenz* 24: 287–294.

Merleau-Ponty, M. 1964a. *Signs*. Evanston, IL: Northwestern University Press.

Merleau-Ponty, M. 1964b. *The Primacy of Perception and Other Essays on Phenomenological Psychology, the Philosophy of Art, History and Politics*. Evanston, IL: Northwestern University Press.

Merleau-Ponty, M. 1962. *Phenomenology of Perception*. Trans. C. Smith. London: Routledge.

Napier, J. 1980. *Hands*. (Revised by R. H. Tuttle.) Princeton, NJ: Princeton University Press.

Noë, A. 2005. Against intellectualism. *Analysis* 65 (4):278–290.

Noë, A. 2009. *Out of Our Heads: Why You Are Not Your Brain, and Other Lessons from the Biology of Consciousness*. New York: Hill & Wang.

Olivier, E., M. Davare, M. Andres, and L. Fadiga. 2007. Precision grasping in humans: From motor control to cognition. *Current Opinion in Neurobiology* 17:644–648.

O'Malley, J. B. 1970. Seeing—meaning—making. *The Human Context* 11 (1):48–68.

Pinker, S. 1997. *How the Mind Works*. New York: W. W. Norton.

Pisella, L., H. Grea, C. Tilikete, A. Vighetto, M. Desmurget, G. Rode, D. Boisson, and Y. Rosetti. 2000. An "automatic pilot" for the hand in human posterior parietal cortex: Toward reinterpreting optic ataxia. *Nature Neuroscience* 3 (7):729–736.

Plotkin, H. 2003. *The Imagined World Made Real: Towards a Natural Science of Culture*. New Brunswick, NJ: Rutgers University Press.

Prinz, J. J. 2009. Is consciousness embodied? In *The Cambridge Handbook of Situated Cognition*, 419–436, ed. P. Robbins and M. Aydede. New York: Cambridge University Press.

Prinz, J. J., and A. Clark. 2004. Putting concepts to work: Some thoughts for the twenty-first century. *Mind & Language* 19 (1):57–69.

Radman, Z. 2008. The motivated movement. Paper presented at the conference "Kinaesthesia and Motion." University of Tampere, October 2–5.

Radman, Z. 2012. The background: A tool of potentiality. In *Knowing without Thinking: Mind, Action, Cognition, and the Phenomenon of the Background*, 224–242, ed. Z. Radman. Basingstoke: Palgrave Macmillan.

Ritter, H., R. Haschke, F. Roetling, and J. J. Steil. 2011. Manual intelligence as a Rosetta stone for robot cognition. *Springer Tracts in Advanced Robotics* 66:135–146.

Roelofs, C. 1935. Optische localisation. *Archiv für Augenheilkunde* 109:395–415.

Ryle, G. 1949. *The Concept of Mind*. New York: Harper.

Searle, J. R. 1983. *Intentionality: An Essay in the Philosophy of Mind*. Cambridge: Cambridge University Press.

Searle, J. R. 2001. *Rationality in Action*. Cambridge, MA: MIT Press.

Snowdon, P. 2004. Knowing how and knowing that: A distinction reconsidered. *Proceedings of the Aristotelian Society* 104 (1):1–29.

Stanley, J., and T. Williamson. 2001. Knowing how. *Journal of Philosophy* 97:414–444.

Tallis, R. 2003. *The Hand: A Philosophical Inquiry into Human Being*. Edinburgh: Edinburgh University Press.

Weiskrantz, L. 1986. *Blindsight: A Case Study and Implications*. Oxford: Oxford University Press.

Weiskrantz, L. 2006. *Blindsight: A Case Study Spanning 35 Years and New Developments*. Oxford: Oxford University Press.

Weiskrantz, L., E. K. Warrington, M. D. Sanders, and J. Marshall. 1974. Visual capacity in the hemianopic field following a restricted occipital ablation. *Brain* 97: 709–728.

Wolpert, D. M., Z. Grahramani, and J. R. Flanagan. 2001. Perspectives and problems in motor learning. *Trends in Cognitive Sciences* 5:487–494.

Wong, E. and A. Mack. 1981. Saccadic programming and perceived location. *Acta Psychologica* 48:123–131.

Wrathall, M. 2005. Motives, reasons, and causes. In *The Cambridge Companion to Merleau-Ponty*, ed. T. Carman and M. B. N. Hansen. Cambridge: Cambridge University Press.

VI TOMORROW'S HANDS

17   A Critical Review of Classical Computational Approaches to Cognitive Robotics: Case Study for Theories of Cognition?

Etienne B. Roesch

The human body is the best picture of the human soul.
—Ludwig Wittgenstein

The behavior of any system is not merely the outcome of an internal control structure (such as the central nervous system). A system's behavior is also affected by the ecological niche in which the system is physically embedded, by its morphology (the shape of its body and limbs, as well as the type and placement of sensors and effectors), and by the material properties of the elements composing the morphology.
—Rolf Pfeifer and Josh Bongard

1   Introduction

Exploring the environment through touch, pointing, grasping, carrying, or manipulating objects is something that we, humans and other animals, perform effortlessly. As I write these words, my fingers touch the keyboard; each finger is assigned a set of keys and, without any visual feedback, I type one letter at a time to compose words and sentences in an orchestrated and precise order. I am not the slightest bit conscious of the particular movements that each of my muscles elicits. My mastery of the keyboard is a simple feat compared to that of the pianist who is playing the piece by Chopin that I'm listening to in the background. Everyday actions of this sort are of the utmost importance for a wide range of activities. Not surprisingly, a number of projects in the field of engineering, medicine, and prosthetics attempt to produce artificial devices that resemble the human hand. This work has employed extremely varied strategies, the success of which, however, remains debatable.

In the next section, I introduce selected attempts at designing functional humanoid robots. The differences between these designs are most notable when looking at the robotic hands produced. Next, I describe some of the issues that arise when building control architectures for robotic devices. I describe the steps involved and point out some of the shortcuts that roboticists often take. I demonstrate that the process of building an artificial device implies making radical decisions that constrain what can ultimately be done and, more importantly, bias what can be learned from this enterprise. Finally, I take a critical perspective over this body of work and discuss the consequences for the way we conceptualize our own body and view our own interactions with the world.

2 What Does a Robotic Hand Look Like?

A wide range of robotic devices have been developed whose aim is to function as a human hand, the nature of which may surprise readers who are not familiar with science fiction movies. First, not all robotic hands are attached to a body; most devices sit on a table, or are attached to an arm of some kind, which may itself be mounted on the floor. In addition, most robotic hands do not have five fingers—if they have fingers at all. They tend to have a functional extremity, however, that is used to perform a (small) number of specific tasks—and nothing else. Most robotic hands achieve a rather low number of degrees of freedom (DOFs), ten or so, that is, the total number of movements (e.g., rotations, displacements) that actuators (motors and the like) can produce. To give an example, a human hand has about 40 muscles, which provide 23 degrees of freedom in the hand and wrist. It might not sound like a lot, but it provides us with the ability to produce extremely complex movements in three-dimensional space. While these degrees of freedom provide us, humans, with great flexibility, we are not half as dexterous and precise as our artificial counterparts: the advertised specifications for industrial robotic hands (see fig. 17.1), for instance, state that movements in three-dimensional space will be reliably reproducible with a margin of error that ranges between 0.02 and 0.1 millimeters. For devices that can carry payloads exceeding a ton, that's quite a feat, one that we will never achieve.

Industrial robots have a narrow purpose: They are designed to perform a few particular tasks, and to do them well. Humanoid robots, on the other hand, aim to emulate our human hands, which are remarkably adaptable and used for an infinite number of tasks. The motivation for creating a humanoid robot with hands as functional as ours is therefore easy to understand, as is the extreme difficulty in actually doing so.

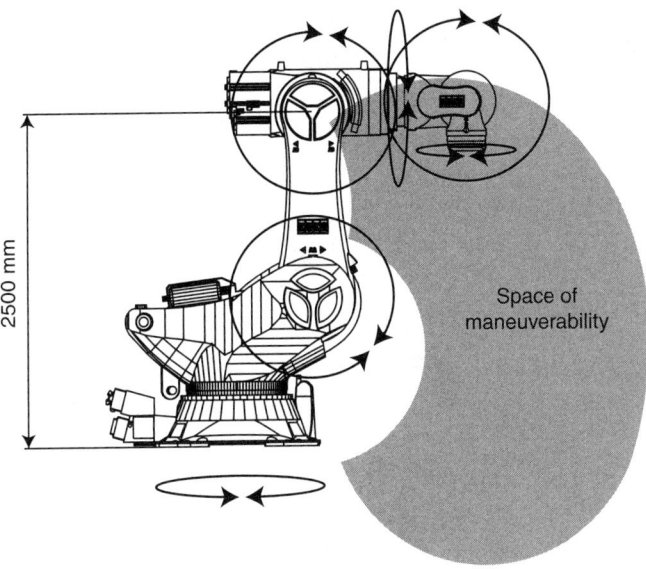

Figure 17.1
The KR 1000/1300 Titan PA by KUKA Roboter GmbH is a 4690 Kg palletizing robotic hand, which is typically mounted on the floor, and can manipulate a payload of up 1300 kilograms. It has 6 degrees of freedom, and can reproduce any movement with a margin of error less than 0.1 millimeters. Lighter weight devices typically reach margins of error below 0.02 millimeters. See http://www.kuka-robotics.com for details.

The first humanoid robots were produced in the 1970s by a team of four groups from the School of Science and Engineering at the University of Waseda, Japan. Each group had a specialty, and, over the years, they produced a series of humanoid robots—the WABOTs. These robots were composed of a heavy metal body, two legs, and two arms. Later versions also had a head with one camera (an eye) and a set of microphones (ears). Some robots could walk up a ramp and climb a few steps. Of special relevance is the WABOT-2 (1984), which was famous for playing the keyboard (fig. 17.2). Its fingers were actuated by rigid tendons, which were linked to a large number of motors. The hand, fingers, and wrist were not very flexible, but were workable enough to push the keys of the keyboard, as ordered by its computer program.

Since this early work, engineers have miniaturized many of the robotic components. Yet, today's humanoid robots function no more like humans

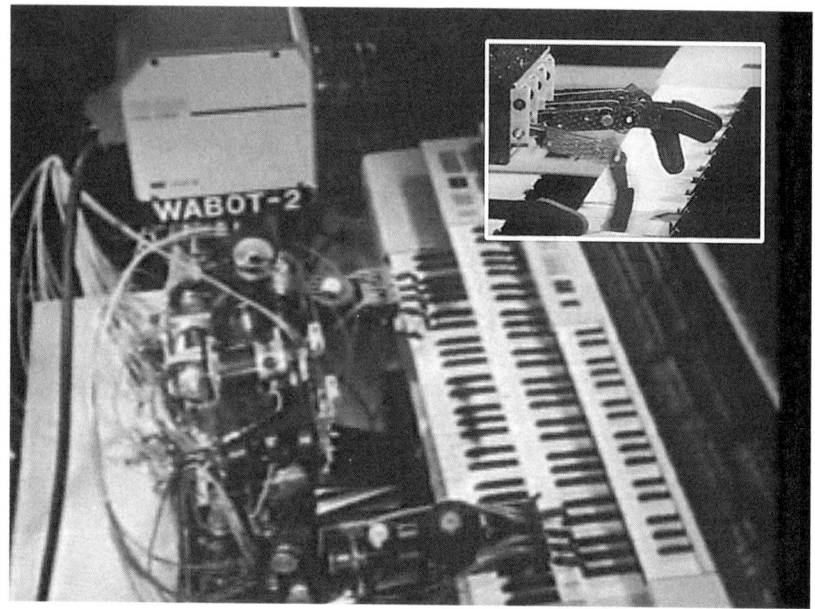

Figure 17.2
In the 1970s, researchers from the School of Science and Engineering at the University of Waseda, Japan, produced a series of humanoid robots, the WABOT. WABOT-2 (1984) was famous for being able to play the keyboard.

than does a can with two arms, even though robots in research labs these days are not too far from C-3PO from the *Star Wars* series. In 2005, for instance, the European Commission funded the RobotCub project, a five-year-long attempt at building a humanoid robot the size of a three year old, called the iCub. The project ended in 2010 with the delivery of a number of iCubs to research groups around the world. The robot is 104 centimeters tall, and it has 53 actuated degrees of freedom that permit the movement of the head, the torso, arms, hands, and legs. It also has a pair of cameras in place of eyes, and microphones in place of ears. Facial expressions are simulated with LEDs on a plastic mask. At the present time, the iCub also ships with 108 tactile sensors based on capacitive technology, on each hand: 48 in each palm and 12 in each fingertip.

The iCub robotic platform is constructed and distributed under the Open Source banner; thus, anyone can access the blueprints and the know-how from the community, and can even download the tailored software (see "Internet Links," below). The RobotCub consortium comprised EU partners from a wide range of disciplines, spanning developmental psy-

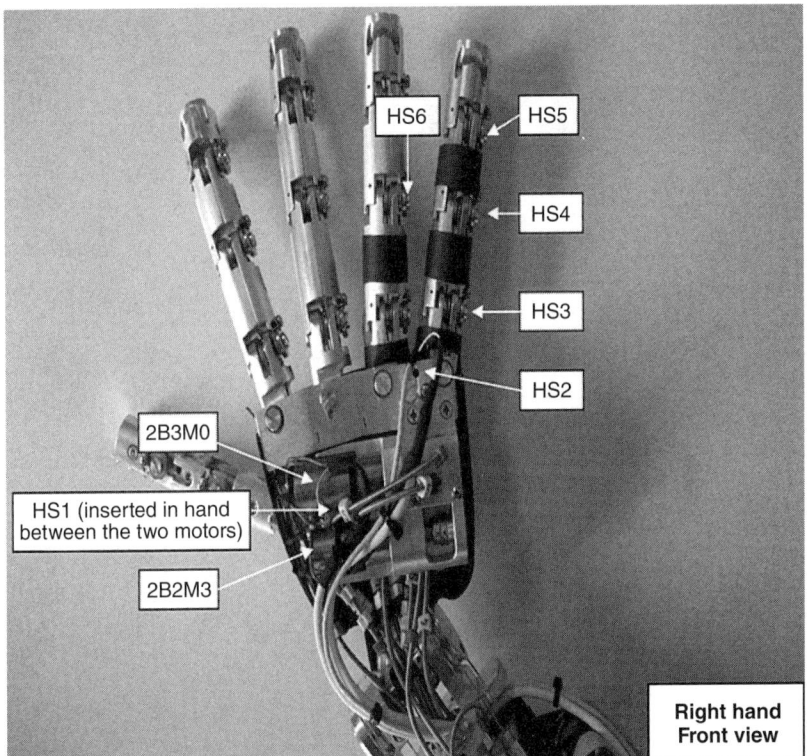

Figure 17.3
Right hand of the iCub. Labels indicate individual read-out sensors. The communication between the motors and the joints is performed through tendon-like connections. Each hand has nine actuated degrees of freedom, and seven in each arm. Even though it is made of metal, it is still very fragile; I don't think anyone dared to try measuring how much weight it could carry.

chologists to mechanical engineers, whose aim was not to build the "embodied cognition" of a three year old, but to construct the platform—the empty shell, if you will—that would allow researchers to investigate that cognition. The emphasis was therefore set on producing a device "that works," so that it reliably produces a range of behaviors. If programmed to do so, for instance, the iCub can move its hand to a particular location in space, through a predetermined sequence of movements, to grasp a nearby object. In this situation, the content of the environment is known beforehand or expected. The information available to the iCub, however, at best combines information from the iCub's visual input, the crude

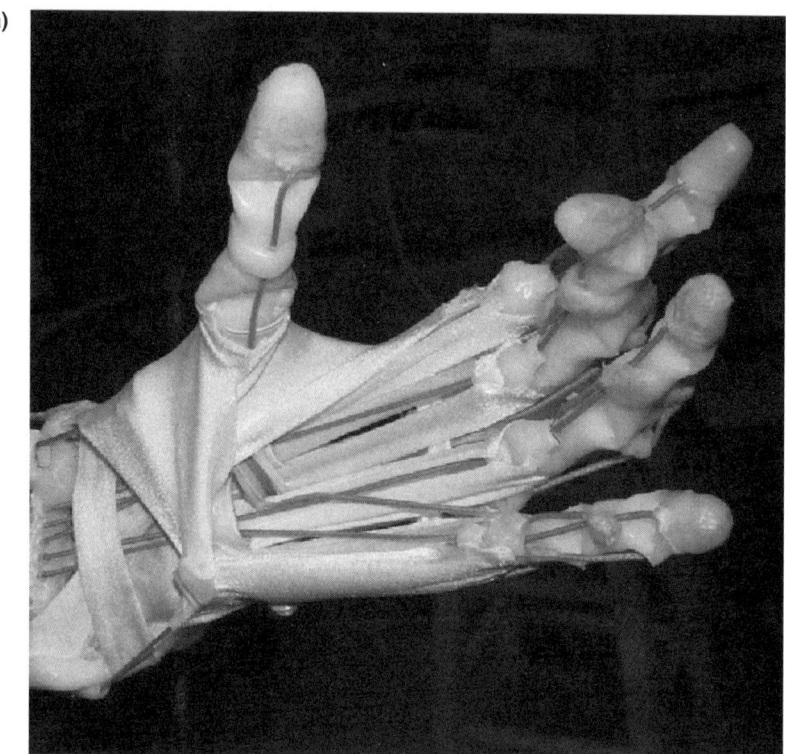

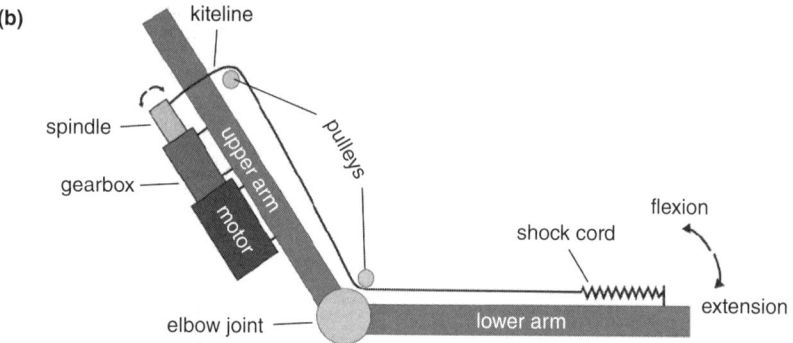

Figure 17.4

The aim of the CRONOS/ECCEROBOT project was to design a robot based on human anatomy. Prof. Holland and colleagues thus developed a robot with bones made of molded plastic, which gives a "tough and springy" skeleton. Limbs and fingers are actuated through flexible tendons. (a) The left hand of CRONOS, showing a tendon-actuated movement being performed on the middle finger. (b) Schematic demonstrating the workings of the tendon-based actuation for an arm: a rotating motor exert some tension on the tendon, which consequently leads to the contraction of the arm; the same principle is applied to the rest of the skeleton, including the arms and fingers.

tactile sensors in its palm, and the current state of activity of its motors. The ability of the iCub to respond to the tiniest changes in its environment is therefore limited by the programmer's ingenuity and his or her ability to predict unforeseen circumstances. If, for instance, the weight distribution of the object is placed in an odd position, there is nothing that will initiate an adjustment of the limb or the fingers to support the grasp. In this respect, the iCub's abilities are very similar to that of industrial robots.

Another influential project is the CRONOS project funded by the UK Engineering and Physical Sciences Research Council, now extended in the ECCEROBOT project funded by the European Commission. The aim of this project was rather different from that of RobotCub; whereas in RobotCub researchers tried to advance the field of mechanical engineering, ingenuously putting together the bits and pieces to build the most functional skeleton, researchers in the CRONOS project aimed at developing a humanoid robot that would be modeled after the human body, in the most biologically plausible way. Researchers used medical textbooks (Gray, Williams, and Bannister 1995) to build a reactive skeleton entirely driven by flexible tendons (fig. 17.4). For these reasons, CRONOS/ECCEROBOT is less rigid than, say, the iCub, and as a consequence it is much more responsive to changes in the environment. Yet, CRONOS/ECCEROBOT is not agile enough to perform complex tasks with its hands.

The list of humanoid robots is growing rapidly (see "Internet Links", below, for other successful examples). My goal thus far has been to introduce the reader to the wide spectrum of robotic devices available, and to point out how the aims of different projects can have qualitative consequences on the way robotic hands are conceptualized. These differences are further evident in the accompanying figures. Generally, researchers optimize the reliability–flexibility trade-off, in one way or another; depending on the balance of this trade-off, their decision may yield a very stiff but precise device, like WABOT-2 and to some extent the iCub, or a very flexible but clumsy device, like CRONOS/ECCEROBOT. In taking a broad look at the humanoid robots I've introduced in this section, which, at the time of writing, represent the state-of-the-art, one is particularly surprised by the variety of strategies that are put in place to build such robots in general, and robotic hands in particular. None of these efforts, otherwise great successes of engineering, comes close to being able to perform movements that are considered the "bread and butter" for humans.

So far, I've described a number of tasks that robot hands are programmed to perform, alluding to sensors that might be used to adapt a
behavior to a changing environment only once, without discussing the
mechanisms that give rise to these behaviors, which are at least as important as using a reliable and flexible robotic platform. In the following
section, I explore the mechanisms that engineers employ to get robots to
"do stuff," and ultimately use their hands. In the next sections, I discuss
the concept of embodiment, which takes on a particular meaning in cognitive robotics, and consider its importance not only for the way one builds
robots, but also for the way one would describe our cognitive system.

3   How Do You Make a Robot Do Stuff?

First of all, one has to define not only what "stuff" is, but also what to
"do" with it. It sounds like a trivial first step, but it is not. In fact, I will
argue later that defining the context of action in which one's robot is to
"perform" "something" not only determines the way one will design the
robot and go about making it perform a particular action, but it also
(mirrors and) influences the way one will perceive one's own abilities in
the environment. In the context of robotic hands, common tasks include
grasping, moving objects, and manipulating artifacts (Iberall 1987, 1997).
On deciding on a task to accomplish, and the types of objects to manipulate, roboticists will most likely attempt a description of the actions required
to perform the task. They will thus try to imagine the unfolding of the
hand–object interaction, isolating the moments in time—that's important—
during which a significant part of the interaction takes place, and characterizing the ins and outs of this dynamical system. In other words, they
will focus on identifying the behavioral regularities supporting the interaction with the environment. Engineers and roboticists interested in artificial intelligence will then seek the help of various tools, such as control
theory, planning algorithms, formal logic, probabilistic inference, or statistical learning, to derive a sequence of operations, which they will then
implement in their favorite programming language, or whichever language
happens to be trendy at the time.[1]

At this point, roboticists will have made a great number of (what seem
to be at the time) reasonable assumptions and compromises: with regard
to the microworld that the robot is meant to encounter, the way to algo-
rithmically *represent* both the object and the task at hand, the types of
information that will be required to perform the task and ensure its unfolding over time (which will be conceptualized at a particular resolution), and

eventually the way to handle situations when things don't go as planned. It is easy to appreciate how this initial "trivial" first step has been applied in the design of the humanoid robots I presented in the previous section, and the consequences for their respective abilities.

The engineering of the robot, and the design of its control architecture usually happen in parallel, one constraining the other: The synergy of these two systems constitutes what is called the perception-action cycle, whereby a robot perceives its environment through sensors (e.g., pressure sensors) and acts upon its world through actuators (e.g., motors). In other words, it is the task of the control architecture to figure out what to do with the information provided by this collection of individual devices. In recent days, progress has been faster in the engineering of the hardware than in the design of control architectures: Robotic components have become smaller and smaller, energy consumption has become much more efficient, and so forth. However, tackling a complex task such as grasping, in an environment that is not known in advance, is still an impossible challenge. Why is that?

This realization is really not new; it relates to more fundamental criticisms of computationalism in general, and of artificial intelligence in particular (Bishop 2009; Dreyfus 1992, 2007; Preston and Bishop 2002; Searle 1980). The gist of these criticisms emerged from the inability of the traditional paradigm to adequately account for the interaction with the (real-world) environment. According to Wheeler, traditional computing is "digital, deterministic, discrete, effective and temporally austere (in that time is reduced to mere sequence)" (Wheeler 2002, 345), to which I would add that it is based on environmental invariance, yielding static notions of representations (of the environment, objects, tasks, etc.) that are then manipulated. In practical terms, this dependence on invariance is particularly troublesome; the so-called frame problem, for instance, whereby intractable computations are required to take into account even the most evident elements contained in the world, de facto limits the complexity of what can be expected from a control system. The necessity to represent the state of the world as it is, keep it up to date, and search through this space yields combinatorial problems that quickly prove impossible to handle (Shanahan 1997).

The frame problem is all the more relevant when one takes timing seriously. Interaction in the real world relies on real-time events. In this context, time is a limited resource. This mere fact has dramatic consequences, and the sole use of faster CPUs will not solve the frame problem— as put in Dennett's famous example: "'Do something!' [the designers] yelled at it. 'I am,' [the robot] retorted. 'I'm busily ignoring some thousands of implications I have determined to be irrelevant. Just as soon as I find an irrelevant implication, I put it on the list of those I must ignore, and . . .' the bomb went off" (Dennett 1987).

For robotics in general, and for robotic hands in particular, the representation of time as mere sequence is particularly worrying. In fact, I posit that the process of defining the moments in time that are significant for hand–object interaction is intrinsically flawed. Formulating the assumption that an unfolding interaction can be decomposed in sequences of moments, and that in between these moments lies the opportunity for some computation that will support and give rise to the remain of the interaction in the next moment, is to misconstrue what the interaction is all about. An interaction is no more about a sequence of reactions than a dialogue is about alternating monologues. A dialogue is about exchanging, listening, and sharing at the same time, in a very multimodal yet orderly fashion, not just taking turns speaking; in much the same way, an interaction is about the continuous unfolding of interacting enactions (Port and van Gelder 1995; van Gelder and Port 1995).

Similar views have been expressed in the 1990s, in the radical antirepresentational shift in AI and robotics (Brooks 1991; Chapman and Agre 1987). To this date, this approach remains, however, largely controversial, mainly because of a lack of results. Brooks posited that human intelligence lies on a continuum, and that scaling up models of insects to humans will give rise to a new, representation-less AI (Brooks 2001). These models set a particular focus on, and take advantage of, the embodiment at play in the interaction with the environment. A classic example is that of the reflex exhibited by cockroaches when facing an obstacle (fig. 17.5). The insect's antennas initiate a bending reflex of the entire body, thus positioning the legs of the insect in such a way that permits climbing over the obstacle (Watson et al. 2002). This reflex constitutes a clever way to outsource the complex manipulations required for the reorientation of the individual parts of the body. One could imagine that such "outsourcing" is at play in the human body. Think of the palmar grasp reflex in newborns, for instance, which occurs when an object is placed in the palm of the baby's hand. The ensuing grip is so strong that it can support the child's own weight. Wouldn't such mechanisms be important for the unfolding of our interactions with the outside world? How much of these reflexes are at play when I grasp my morning coffee, without even looking at the mug? How could roboticists take advantage of this "outsourcing" to build more efficient, more reactive artificial hands? And, more importantly, what

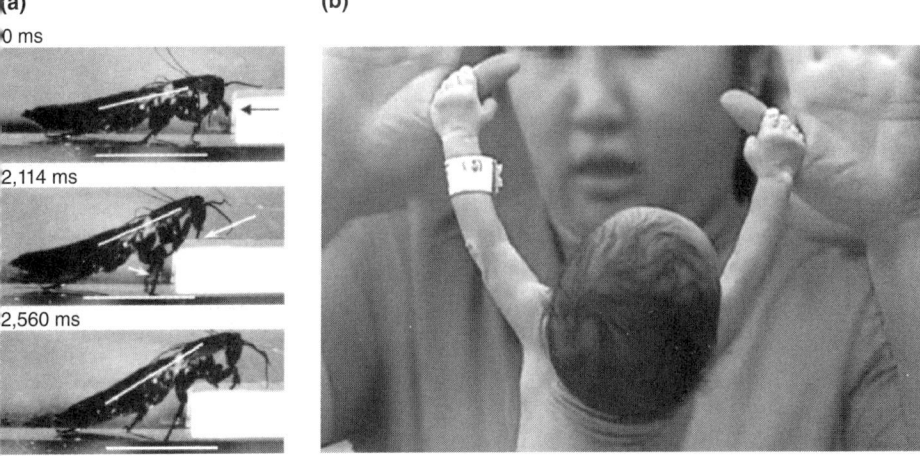

Figure 17.5
(a) Body adjustments occurring when a cockroach climbs an 11mm high obstacle. The front antennas initiate a bending reflex of the whole body (0ms), which positions the front legs over the obstacle (2,114 ms after collision). The legs of the cockroach do not have sufficient DOFs to allow this maneuver by themselves (Watson et al. 2002). (b) Palmar grasp reflex in a days old newborn.

does it say about ourselves, and the cognition that supports our interaction with the world?

However, even though it makes a lot of sense to restate the biological nature of human intelligence in such a way, it is not nearly enough to model a human body or a human mind (Dreyfus 2007), and much is left to do before we manage to include these concepts in our theories of (what should be embodied) cognition.

In this section, I've summarized some of the ways available to roboticists for controlling a robotic device such as a hand. By spelling out and programming the sequence of exact movements that must be performed, in a known and static environment, however, designers are forced to make decisions and compromises that obscure the synergetic nature of interaction with the real-world environment. This focus on planning, omitting the foundational role of the body in interaction with the environment, parallels the emphasis on the brain in cognitive science. Historically, these two disciplines have been reinforcing the same biases for computation as the best way to represent cognition (van Gelder 1995). In closing, I will discuss some of the consequences that this focus has had on the way we build theories and ultimately on the way we understand ourselves.

4   What Have We Learned, and Where Do We Go from Here?

Research in robotics goes beyond motor-related tasks, because it has long been recognized that a successful robot needs to be more than a mere golem—that mythical creature made of clay that would dutifully perform actions described on scrolls. Taking Nature as an example, researchers thus attempt to perfect the robots' visual perception abilities, to make sense of the world and plan trajectories (Davison 2003; Newcombe, Lovegrove, and Davison 2011), for instance. Other efforts are invested in making the robots look cute and teaching them to read social signals from us (Breazeal and Scassellati 2002), with a view to facilitate human–robot interaction (Castellano and Peters 2010). These threads of research leave the implementation of motor-related tasks to engineers and low-level roboticists. It generally assumes that motor abilities are or will be mastered in some way, and thus sets the focus on higher-level abilities. What we've come to realize, however, is that we are nowhere near producing a robot with humanlike motor abilities. Robotic hands are probably the best example of this failure: We build ever-sophisticated robotic devices, but we cannot figure out how to use them in the type of seamless interactions with our world that they are meant to enact.

I submit that the way we have been designing robots mirrors the way we perceive and understand our body and our mind, and that this picture is incomplete. Over the years, cognitive science has come to focus primarily on the workings of the brain as the prime location for information processing, supporting our abilities to plan and to react to the outside world. This sustained attention on the brain has somehow occluded the role of the body and the crucial embodiment of our interaction with the world. For matters of the hand, this has meant understanding what kinds of information is available to the brain for the planning, execution, and online correction of unfolding gestures, instead of investigating the levels of adjustments that occur in the organism as a whole that ensure coherent gestures. For robots modeled after our understanding of the brain, this has meant predicting and planning the sequence of movements that would yield a successful gesture, instead of grounding interactions in a body-world synergy. We have forgotten that our body is intrinsically enactive—it is embedded in the environment in tightly coupled feedback loops. Not only is it our brain's interface with the environment, it is also an integrative component of our embodiment in the world. This layered architecture provides flexibility and adaptability, in the form of an embodied, decoupled perception-action synergy with the environment.

In contrast, traditional ways of implementing control architectures for robotic devices create an "uncoupled" situation, whereby even though a control architecture is kept informed of what is going on in the outside world through sensors, computational (i.e., algorithmic) operations impose an overhead in time and space that creates a distance between the agent and the environment that impedes the interaction and prevents any sort of interactive synergy. Of course, given an infinite resource paradigm (i.e., faster CPUs), one could imagine a computational device that would mimic the behavior of an interactive organism down to the smallest granularity. Reality is such, however, that the possibility of devices of that sort is next to nil, and the opportunity to even conceive the smallest granularity seems improbable at best. In other words, the nature of our interaction with the environment does not allow for uncoupled systems but can accommodate decoupled ones, provided that operations are constantly flowing, dynamically informing the organism of its state in its lived world. This idea leads to the notion of phenomenal embodiment and has strong consequences for theories of cognition.

Phenomenal embodiment does not exist in robots, because a robot-body "is merely a collection of inanimate mechanisms and non-moving parts that form a loosely integrated physical entity" (Sharkey and Ziemke 2001, 256). In fact, even though the concept of embodiment is, for obvious reasons, closely related to robotics, individual robotic projects are only gradually taking up features of embodiment (Sharkey and Ziemke 2001). These projects can be placed on a continuum, spanning *situated*, *weakly*, *strongly*, and *enactive* forms of embodiment:

- By situated, I refer to devices that are programmed to perform a number of actions, in an environment that is known, expected, and static, but which are completely unaware of their surroundings. Most robots are situated: industrial robots, of course, that rely on the environment to be in a very particular state to be able to perform tasks, but I also include in this class robots like WABOT-2 and earlier versions of the iCub, which do not make use of the (often poor) sensors they are equipped with and cannot achieve much more than industrial robots. Notably, robots in that category would still attempt to perform the task they are programmed to perform, even if the environment is not adequate; for example, they would still attempt to play the keyboard in the absence of their favorite instrument, or try to grasp an object that is not there.
- Weakly embodied robots use sensors to adapt some of their behavior to the environment. These robots still need the environment to be known and static, and they rely heavily on programmed sequences of actions; but

they can make some adjustment if need be. They could, for instance, adjust their grip when sensors signal that an object is being held. Most if not all robotic hands fit in this category; the latest versions of the iCub's hands include arrays of tactile sensors that can be used to halt an ongoing grasping movement, if so programmed. I would also include CRONOS/ ECCEROBOT in this category, even though the attempt is clearly to produce a strongly embodied robot.

• Strongly embodied robots make use of natural dynamics that arise during the interaction with the environment as an integral component of gestures. These robots, for instance, may be made aware of their own weight, and may use this information to adjust and support a gesture in the environment, by employing a model of gravity, for instance. Over the last couple of years, Boston Dynamics released very impressive videos of quadruped ("Bigdog") and biped ("Petman") walking robots that recover from a fall in ways that look very similar to their natural counterparts. For each of these robots, information about their own weight is crucial: combined to gyro and linear accelerometers, which provide a sense of motion in space, this information contributes to the control and the balance of the robot during the unfolding of gestures. In earlier videos, one can see that previous versions of these robots purposefully included extra weight, which seemingly contributed to grounding the robot in this "balance space."

• In my opinion, enactive robots do not yet exist. Efforts by Boston Dynamics might be the closest attempt in that direction, but their robots still rely primarily on predetermined algorithms that drive the unfolding of the legs in what is meant to look like natural motion. The big difference I make between strongly embodied and enactive robotics lies in the importance given to the body-world synergy, implemented in the enactive paradigm as layered, decoupled perception-action processes grounded in the homeostasis of the robot/organism. For an enactive robot, not only do the architectural characteristics of the robot, say, its weight in space, play an active role in the unfolding of gestures, but the coherence of these gestures is also warranted by the homeostatic dynamics within and between each architectural level, thus approaching phenomenal embodiment. Ultimately, these dynamics have nothing to do with predetermined algorithms that describe the sequences of actions yielding particular gestures.

So, what would an enactive robotic hand be like? First, every bit of it would be used to gather information. This information would be multimodal and would refer to properties such as weight distributions, temperature, tension, and pressure, and so on, that together arise from the interaction with the environment, the object to manipulate, and also from the internal state of the hand, its "actuators" and its homeostasis. Second, this information would originate from and inform each of the levels that compose the device. The human body in general can be described in terms of concentric, interacting systems, including, for instance, the internal dynamics of the cell found at the tip of the finger that responds to pressure, the neuromuscular system that evokes the contraction of a small patch of muscle fibers, or the interacting immune systems that ensure the integrity of the system as a whole. An enactive robotic hand would be composed of similar layered systems, which would act in concert to adjust gestures over time. Finally, because the system would ground gestures on this massive amount of information, precision wouldn't be so much of an issue. In fact, there wouldn't be any need to compute the exact distance the hand needs to move to reach an object because, at any point in time, the system would adjust itself in response to the interaction with the environment. Much like when I grab my morning coffee: I do not make a decision about where my little finger should be placed on the mug—it just naturally lends itself to the most appropriate position for the ongoing interaction with the handle of my mug.

When it comes to everyday actions, it is a mistake to think that precision is important (Agre 1988). Of course, when one is asked to write a control algorithm to describe the unfolding of a given gesture for a particular robotic device with next to null sensor abilities, a very small number of actuators, and the added constraint to use a particular application programming interface (API) with whatever programming language happens to be trendy at the time, one is forced to be as precise as possible. Each line of code is meant to represent a very particular situation, and nothing else. That is not, however, what a natural interaction is about. In Nature, an interaction is about being good enough, not about being perfect.

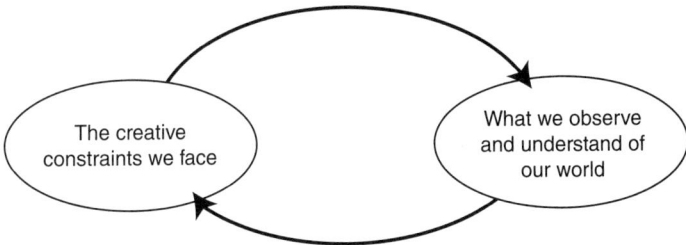

Figure 17.6
The creative process in cognitive science.

5 Conclusion

It is incredibly easy to be fooled by intelligence. We attribute perfection in Nature to situations that look precise and calculated, but often these situations are simply the "good enough" result of that particular interaction, as endlessly experienced and fine-tuned through evolution. There is nothing explicitly intelligent about the cockroach bending its spine to level up its legs. Yet, it achieves more than most mobile robots. Intelligence arises from interaction with the environment; it is not necessarily the result of clever planning. Going beyond "Good Old-Fashioned AI," modern approaches to AI explore intelligence along many dimensions, which, taken together, represent as many alternative routes. Associative intelligence (e.g., data mining, statistical and stochastic processes), swarm intelligence, emotional and social intelligence, and morphological intelligence are among these new perspectives.

At the time I am writing these lines, the fields of cognitive science(s) and artificial intelligence are dominated by a view that describes the brain, the mind, and ourselves as information-processing machines. Computationalism (Fodor 1975, 1980, 1981; Putnam 1965, 1975), the idea that thinking is a form of computation, is latent in most theories that are put forth to explain who we are and how we function. I believe that the way we build robots is symptomatic of this ubiquitous paradigm, and that the technical steps involved in the creative process of building such an artificial device only reinforce this attraction for computation in a vicious circle (fig. 17.6). In other words, part of the reason why we carry on using concepts like representations or modules in our theories is because we do not see how else we would create models. Modularity is a self-realizing principle that fits our abilities to describe ourselves and to create models.

In this chapter, I have provided a brief overview of some successes in robotics; I have described the process involved in providing these robots with the ability to interact with the environment; and I have underlined the limits of this creative process. I have situated these limitations within the paradigmatic view that equates cognition with information processing; and, to conclude, I have tried to imagine an alternative that could arise from biology. Similar attempts can be found in the rise of enactive cognitive science (Varela, Thompson, and Rosch 1991), which emphasizes the grounding role of the body in the interaction with the environment and extends cognitive science by proposing new methods, new concepts, and new theoretical perspectives. Only time will tell whether these efforts will be successful.

Websites

Android World is a website dedicated to humanoid robots. It gathers as much information as possible about attempts at constructing anthropomorphic robots. It has a specific section about robotic hands: http://www.androidworld.com/prod76.htm
The WABOT series, by the Humanoid Robotics Institute at Waseda University, Japan: http://www.humanoid.waseda.ac.jp/history.html.
An archive of the RobotCub project (2005–2010): http://robotcub.org.
The details of iCub robot can be found at http://icub.org, where the reader will find more information about this open source humanoid robot, the blueprints to build the parts, a community wiki and the software necessary to interact with the robot, including an iCub simulator.
The website for the CRONOS project (2004–2007) can still be found at http://cswww .essex.ac.uk/staff/owen/machine/mchome.html. It is now extended in ECCEROBOT (2009–), which can be found at http://eccerobot.org/.
The YouTube channel of the U.S.-based company Boston Dynamics, showing videos of their walking robots: http://www.youtube.com/user/BostonDynamics.
A nonexhaustive list of humanoid robots is available at http://en.wikipedia.org/ wiki/Humanoid_robot.

Note

1. I leave to the reader to see how the philosophy of the chosen programming language will influence the way one conceptualizes interaction; compare, say, an event-based programming language like Java, with a function-based language, like Lisp. Note: For forty years, Lisp was the language of choice for AI researchers and roboticists.

References

Agre, P. E. 1988. *The Dynamic Structure of Everyday Life*. Ed. M. J. Brady. Cambridge, MA: MIT Press.

Bishop, J. M. 2009. A cognitive computation fallacy? Cognition, computations, and panpsychism. *Cognitive Computation* 1 (3):221–233. doi:10.1007/s12559-009-9019-6.

Breazeal, C., and B. Scassellati. 2002. Robots that imitate humans. *Trends in Cognitive Sciences* 6 (11):481–487.

Brooks, R. A. 1991. Intelligence without representation. *Artificial Intelligence* 47 (1–3):139–159.

Brooks, R. A. 2001. The relationship between matter and life. *Nature* 409 (6818): 409–412.

Castellano, G., and C. Peters. 2010. Socially perceptive robots: Challenges and concerns. *Interaction Studies: Social Behaviour and Communication in Biological and Artificial Systems* 11 (2):201–207.

Chapman, D., and P. E. Agre. 1987. Pengi: An implementation of a theory of activity. In *Proceedings of the Sixth National Conference on Artificial Intelligence*, 268–272. Menlo Park, CA: American Association for Artificial Intelligence.

Davison, A. J. 2003. Real-time simultaneous localisation and mapping with a single camera. In *ICCV 2003: International Conference on Computer Vision*, vol. 2: *IEEE Computer Vision*, 1403–1410. doi:10.1109/ICCV.2003.1238654.

Dennett, D. C. 1987. Cognitive wheels: The frame problem in AI. In *The Robot's Dilemma: The Frame Problem in Artificial Intelligence*, ed. Z. Pylyshyn. Norwood, NJ: Ablex.

Dreyfus, H. L. 1992. *What Computers Still Can't Do: A Critique of Artificial Reason*. Cambridge, MA: MIT Press.

Dreyfus, H. L. 2007. Why Heideggerian AI failed and how fixing it would require making it more Heideggerian. *Artificial Intelligence* 171 (18):1–43.

Fodor, J. A. 1975. *The Language of Thought*. Cambridge, MA: Harvard University Press.

Fodor, J. 1980. Methodological solipsism considered as a research strategy in cognitive science. *Behavioral and Brain Sciences* 3:63–73.

Fodor, J. A. 1981. *Representations: Philosophical Essays on the Foundations of Cognitive Science*. Cambridge, MA: MIT Press.

Gray, H., P. L. Williams, and L. H. Bannister. 1995. *Gray's Anatomy. The Anatomical Basis of Medicine and Surgery*. New York, NY: Churchill-Livingston.

Iberall, T. 1987. Grasp planning from human prehension. In *International Joint Conference in Artificial Intelligence*, vol. 2, 1153–1156. Menlo Park, CA: American Association for Artificial Intelligence.

Iberall, T. 1997. Human prehension and dexterous robot hands. *International Journal of Robotics Research* 16 (3):285–299.

Newcombe, R. A., S. J. Lovegrove, and A. J. Davison. 2011. DTAM: Dense Tracking and Mapping in Real-Time. In *ICCV 2011: International Conference on Computer Vision*.

Pfeifer, R., and J. Bongard. 2007. *How the Body Shapes the Way We Think: A New View of Intelligence*. Cambridge, MA: MIT Press.

Port, R. F., and T. van Gelder. 1995. *Mind as Motion: Explorations in the Dynamics of Cognition*. Cambridge, MA: MIT Press.

Preston, J., and J. M. Bishop, eds. 2002. *Views into the Chinese Room: New Essays on Searle and Artificial Intelligence*. Oxford: Oxford University Press.

Putnam, H. 1965. Brains and behavior. In *Analytical Philosophy*, ed. R. J. Butler. New York: Barnes and Noble.

Putnam, H. 1975. The nature of mental states. In *Mind, Language, and Reality: Philosophical Papers*, vol. 2. Cambridge: Cambridge University Press.

Searle, J. R. 1980. Minds, brains, and programs. *Behavioral and Brain Sciences* 3:417–457. doi:10.1017/S0140525X00005756.

Shanahan, M. 1997. *Solving the Frame Problem: A Mathematical Investigation of the Common Sense Law of Inertia*. Cambridge, MA: The MIT Press.

Sharkey, N. E. and T. Ziemke. 2001. Mechanistic versus phenomenal embodiment: Can robot embodiment lead to strong AI? *Cognitive Systems Research* 2 (4):251–262. doi: 10.1016/S1389-0417(01)00036-5.

van Gelder, T. 1995. What might cognition be, if not computation? *Journal of Philosophy* 92 (7):345–381.

van Gelder, T., and R. Port. 1995. It's about time: An overview of the dynamical approach to cognition. In *Mind as Motion: Explorations in the Dynamics of Cognition*, ed. R. Port and T. Van Gelder. Cambridge, MA: MIT Press.

Varela, F. J., E. Thompson, and E. Rosch. 1991. *The Embodied Mind: Cognitive Science and Human Experience*. Cambridge, MA: MIT Press.

Watson, J. T., R. E. Ritzmann, S. N. Zill, and A. J. Pollack. 2002. Control of obstacle climbing in the cockroach, *Blaberus discoidalis*. I. Kinematics. *Journal of Comparative Physiology A: Neuroethology, Sensory, Neural, and Behavioral Physiology* 188 (1):39–53.

Wheeler, M. 2002. Change in the rules: Computers, dynamical systems, and Searle. In *Views into the Chinese Room: New Essays on Searle and Artificial Intelligence*, ed. J. Preston and M. J. Bishop. Oxford: Clarendon Press.

Wittgenstein, L. 1953. *Philosophical Investigations: The German Text, with a Revised English Translation*. Trans. G. E. M. Anscombe. Oxford: Blackwell.

Postscript: Rehabilitating the Hand: Reflections of a Haptic Artist

Rosalyn Driscoll

The hand heals the fissure the eye creates.
—Michael Brenson, art critic

For twenty years I have been making sculptures that provide a new role for the hand, a role that draws on our existing abilities but has not been fully explored: aesthetic touch. So much hand use is functional—manipulating things and wielding tools—but the hand can be receptive and inquiring when exploring an artwork through touch. This kind of touch creates a direct experience of forms and materials, unmediated by tools or separated by distance. Aesthetic touch generates meaning through the body, its actions and its memories. Intimate, contemplative, enactive, embodied, and creative, aesthetic touch provides a way of knowing that proves to be radically different from seeing, which tends to be faster, more distant, and less embodied.

When I began making haptic sculpture, I observed my own hands as well as the hands of others (both sighted and blind) who encountered my sculptures. I wanted to understand how hands explore an artwork so that what I made would both harmonize with and challenge those abilities. I considered the scale of hands and fingers, the ways hands move, how they conform to shapes, what feels good to hold or touch, how the hands work together or independently, how they manage transitions between materials. I asked people about their experience through interviews and comment books. I read whatever I could find on the sense of touch and on hands themselves.

I learned that hands' sensitivity, structure, and movement range make them extraordinarily versatile and articulate instruments for manipulation, sensing, expression, communication, and creation. With their remarkable mobility and subtlety of articulation, hands contain infinite possibilities for haptic exploration. So central is the hand to our notions of touch that perceptual psychologist David Katz designates the hand as the organ of touch, rather than the skin, for several reasons: the hand's ability to make many kinds of grasping movements and to take many shapes; the hand's equivalence to the other discrete sense organs (eyes, ears, tongue and nose); the way the hand, in spite of the gaps between its fingers, integrates stimuli so that a surface is read as continuous and unified; the way we form memory images rich and specific enough to recognize the object with less sensitive parts of the body; the close connection between tactile activity and mental activity; and finally, that a tactile memory image can be more potent than a visual one.

The structure of the hand is incredibly complex. Fingers themselves contain no muscles, which would render them too bulky and unwieldy. Their astonishing strength, speed, and agility are directed by muscles in the palm of the hand and as far away as the forearm. Complex lacings of tendons connect fingers and their distant muscles like marionettes on strings. The hands are global in structure and function, curving as if to hold a sphere; to flatten, the hand requires effort by muscles on the back of the hand. This natural sphericity makes the hands spatially keen. The many small joints provide a range of mobility, allowing the hand to conform itself to an object to a remarkable degree. The hand can swivel in a full hemisphere at the wrist as well as rotate axially. Add hinged movement at the elbow, three-dimensional rotation at the shoulder, a floating shoulder blade, and a flexible spine, and we have an extraordinarily versatile organism for gathering information and engaging with the world.

Neurologist Frank Wilson proposes that the human hand coevolved with the brain, and that hand use shapes the cognitive, emotional development within each individual. The relationship between hand and brain, Wilson says, is a two-way exchange:

The hand is so widely represented in the brain, the hand's neurological and biomechanical elements are so prone to spontaneous interaction and reorganization, and the motivations and efforts which give rise to individual use of the hand are so deeply and widely rooted, that we must admit we are trying to explain a basic imperative of human life. (Wilson 1998, 10)

Wilson aims to inject into theories of brain, mind, language, and human behavior the grounding, driving reality of action by the hand. He finds associations between the use of one's hands and an emotional connection with one's inner life. He finds that developing motor skills stimulates the development of cognitive skills and that we have hardwired instinctual strategies for tactile skills, though it takes years to finely tune these abilities.

After years of exploring the sense of touch and the hand through my artwork and research, a first-hand understanding of the structure of the hand and its powers came to me by accident. While rafting through the Grand Canyon on the Colorado River, my boat plunged into a huge wave in one of the rapids. I grabbed a rope lest I be swept out of the boat, but my hand was wrenched from the rope, severely dislocating the third finger of my left hand. Back home, surgery restored the right relationship of tendons, ligaments, and joint. When cast and pin were removed, my whole hand needed rehabilitation.

The scarring and swelling that immobilized the injured finger allowed tissue repair but also blocked sensation. My finger hungered for feeling, even pain. With so little feedback, I could neither connect with nor direct my finger. Proprioceptive sensation, I learned, is critical for a sense of aliveness and ownership, let alone action.

I learned that an injury to a part is an injury to the whole. The stiff, unyielding finger compromised the sensory abilities of the whole hand and curtailed my ability to do simple things like wash dishes, open a jar, or hammer a nail. I discovered by its disruption the unity of the hand and the seamlessness of sensing and action.

I was amazed that an injury to such a seemingly minor figure in the pantheon of the hand could engender such profound effects. The injury altered my physical agility but also my mental abilities. My body image was disrupted. I ran my hand into doorframes, my feet misstepped. My sense of self was wounded. I felt disoriented and depressed. When I regained even the tiny ability to bend the outermost joint of the injured finger, my whole self-image became more articulated. Remembering the proportionately large space in the brain allocated to the sensory and motor functions of the hand helped explain why my self-image was so disturbed. As my hand recovered its lost abilities, I also recovered my sense of self.

Because I am left-handed, my nervous system had to forge new connections to direct the right hand in unfamiliar tasks. The left taught the right new actions and the right reeducated the left in old ones. Normally my left hand is the diamond-cutter, the one with finer, more precise manipulative abilities such as drawing, writing, and sewing. My right hand is the workhorse, stronger and less refined, good for carrying buckets, sanding wood and wielding power tools. The sculptor Barbara Hepworth, who worked largely in stone, spoke of her hands in a similar division of labor:

My left hand is my thinking hand. The right is only a motor hand. This holds the hammer. The left hand, the thinking hand, must be relaxed, sensitive. The rhythms of thought pass through the fingers and grip of this hand into the stone. It is also a listening hand. It listens for the basic weaknesses or flaws in the stone, for the possibility or immanence of fractures. (Hepworth 1970, 79)

Working with a hand therapist to restore motion and flexibility, I became fascinated by the interaction of my therapists' hands and my hand. I knew about the reciprocity of touch—that as I touch I am also touched—but this reciprocity was active, circular, and mutual. Like Escher's image of two hands drawing each other, my therapist's hands and mine worked together to create a new hand.

One day the therapist gently moved my whole arm in an exploratory way. As my arm rose and fell, I was suddenly back in the boat on the river, reexperiencing the moment of injury. This changed the entire project for me. I realized there was more to rehabilitation than physical repair and more to it than the hand. I began moving with my whole body in this receptive, inquiring way, with the therapist and by myself, moving into sensations, images, and blank places in my body. The hand, I learned, does not end at the wrist but is continuous with the forearm, shoulder, scapula, spine, ribs, heart, throat, and lungs. I explored each of these and their relationship to my hand.

These insights were bound to express themselves in my art. The first piece I made after the injury was a carving of a wooden hand resting on an alabaster head. It represents the sensory as well as the expressive abilities of the hand. As the piece is turned on its various sides, the gesture carries different meanings: hiding, cradling, caressing, protection, dismay. When someone touches the sculpture, his hand rests on the carved hand, hand upon hand. This sculpture evokes many uses of the hand: touching a person, touching an object, touching oneself, and communicating a mental state.

For months I had watched my therapist's hands on my hand, which made me think about all the hands that touch each of us in a lifetime, for good or ill, and how we are literally shaped by these countless touches. I constructed a sculpture of dozens of white, life-size hands, made from many diverse materials, that hang from a light steel framework and form an invisible body within their gestures of touch. Then I videoed the working hands of my therapist, a potter, baker, cellist and sign-language interpreter so their hands seem to be performing their own particular dance. The moving hands are then projected onto the white sculpted hands from two sides and above so they flicker onto the hands, the walls, and down through the core of the body formed by the sculpted hands. I edited the video so the images move slowly, allowing the viewer to *feel* the motion, not just see it. The moving hands and light seems to touch the many hands of the sculpture.

For the first time I had made something too fragile to touch, yet it was all about touch. People seeing it said they could sense the movement of the video hands, imagine the touch of the sculpted hands on their bodies, and feel the light as a kind of touch. A woman said she felt a shiver as images passed through the body of the sculpture: "I felt as if I were being touched on the inside."

For the next piece, still thinking about the therapist's hands on mine, I made two six-foot hands in rawhide and hung them from a wooden frame like a marionette's cradle twelve feet above the ground. From these big cupped hands hang more life-size white hands that support and delineate an invisible body within their touch (like the previous sculpture), this time a figure falling or floating. Hands are so complex in their meanings that these various hands and their gestures can be interpreted in many ways: as the hands of a god, as the hands of a puppeteer, as qualities of being, as actions, intentions, emotions, or states of mind.

Rainer Maria Rilke describes the complexity of the hand in writing about the sculptured hands of Rodin:

Hands are a complicated organism, a delta in which much life from distant sources flows together and is poured into the great stream of action. Hands have a history of their own, they have indeed, their own civilization, their special beauty; we grant them the right to have their own development, their own wishes, feelings, moods and occupations. (Rilke 2004, 45)

French philosopher Michel Serres describes the poetic, empathetic capacity of the hand:

The hand is no longer a hand when it has taken hold of the hammer, it is the hammer itself, it is no longer a hammer, it flies transparent between the hammer and the nail, it disappears and dissolves, my own hand has long since taken flight in writing . . . Our hand . . . can make itself into a pincer, it can be a fist and hammer, cupped palm and goblet, tentacle and suction cup, claw and soft touch. So what is a hand? It is not an organ, it is . . . a capacity for doing. (Serres 1995, 30–35)

The hand is a capacity for shaping the world, but also for shaping one's own body, mind, and evolution, as my injured hand reshaped my body, my mind, and my work. The irony of a tactile sculptor suffering a hand injury was not lost on me. I understood my impulse to make tactile art as an effort to forge a stronger connection between my inner life and the world. An inveterate introvert, I sensed this injury provided an opportunity to create a more open passageway between me and the world. To my surprise, I discovered that my task was not to move outward but rather to move *deeper* into my body, to inhabit it more fully. I first grasped this possibility while exploring my visual perception. By letting my gaze sink back into my eyes rather than reaching out toward what I see, I realized my perceptual habit had been to operate as a disembodied eye, extending myself into the world through sight, whether in empathy, identification, appreciation, or curiosity. This excursion can leave me feeling stranded, out on a limb. Shifting to a more *embodied* way of seeing, I become more stable and the world more palpable and three-dimensional. Perception is specific and detailed rather than generalized. Space is tangible, a substance rather than a void. Patterns and connections emerge spontaneously. By more fully inhabiting the interior of my body, I become more proprioceptively aware of my body and all its parts. Internal feelings of depth and spatial acuity are projected onto the world around me, generating a richer sensing of the environment.

I discovered that I use my hands the same way I use my eyes, as if disembodied. The sculptures I had made all had disembodied hands. By operating in my periphery, rather than from my core, I distance myself from my emotions and intuitions. By inhabiting my whole body, I learned that hands form a joint with the world. When I touch something, I temporarily become joined to that object in a relationship of mutual exchange. Hands are sites of interaction and creativity as well as vulnerability and disconnection. Through the disruption of injury, I grasp the hand's affordance for connection. The sense of myself as a discrete, separate organism yields to a feeling of being embedded in a web of connections and reciprocity. Inside and outside fuse into a larger whole. From the new perspective of an inhabited, grounded core, my hand turns out to be not peripheral at all. My hands are not the outer reaches of my body, but rather the channels through which my body sees and acts—a medium between self and world that participates in both.

References

Hepworth, B. 1970. *A Pictorial Autobiography*. Bath: Adam & Darts.

Rilke, R. M. 2004. *Auguste Rodin*. New York: Archipelago Books.

Serres, M. 1995. *Genesis*. Trans. G. James and J. Nielson. Ann Arbor: University of Michigan Press.

Wilson, F. R. 1998. *The Hand: How Its Use Shapes the Brain, Language, and Human Culture*. New York: Pantheon.

Index